WORKBOOK AND ANSWERBOOK

for

Geometric Dimensioning and Tolerancing

[per ASME Y14.5-2009]

James D. Meadows

GD&T Specialists
James D. Meadows & Associates, Inc.

WORKBOOK AND ANSWERBOOK

for

GEOMETRIC DIMENSIONING AND TOLERANCING

Table of Contents

WORKBOOK AND ANSWERBOOK for Geometric Dimensioning and Tolerancing [per ASME Y14.5-2009]

Published and distributed by:
James D. Meadows & Associates, Inc.
170 E. Main, D-137
Hendersonville, TN 37075
Phone: (615) 824-8644
FAX: (615) 824-5262
www.geotolmeadows.com

ISBN: 0-9714401-7-4
Printed in the United States of America

3 4 5 6 7 8 9 10 Printing

I. INTRODUCTORY SECTION
to the
Geometric Dimensioning and Tolerancing
(per ASME Y14.5-2009)

Workbook and Answerbook

This workbook-answerbook includes an Introductory Section on Geometric Dimensioning and Tolerancing Principles, Symbols and Rules per ASME Y14.5-2009. Topics covered include:

- Symbols
- Rules
- Symbol Meanings
- Formulas
- Datum Feature Symbols
- Boundaries
- Virtual Conditions
- Resultant Conditions
- Material Condition Symbols
- Bonus Tolerance

NOTE: This short introductory information section summarizes material necessary to complete some of the subsequent worksheets. For a more complete explanation of these concepts, see the 574-page textbook entitled ***Geometric Dimensioning and Tolerancing: Applications, Analysis & Measurement per ASME Y14.5 2009*** © James D. Meadows.

GEOMETRIC CHARACTERISTICS

Category and Geometric Characteristic Symbol		Material Condition Symbols Allowed (Modifiers)	Datum Reference Allowed
FORM			
Flatness	⏥	Ⓜ or Ⓛ for derived median plane; No for surface	No
Straightness	—	Ⓜ or Ⓛ for derived median line; No for surface	No
Circularity (Roundness)	○	No	No
Cylindricity	⌭	No	No
PROFILE			
Profile of a surface	⌓	No for feature; Ⓜ or Ⓛ allowed for datum features of size	Allowed but not required
Profile of a line	⌒	No for feature; Ⓜ or Ⓛ allowed for datum features of size	Allowed but not required
ORIENTATION			
Parallelism	//	Ⓜ or Ⓛ allowed for feature of size Ⓜ or Ⓛ allowed for datum features of size	Required
Perpendicularity	⊥	Ⓜ or Ⓛ allowed for feature of size Ⓜ or Ⓛ allowed for datum features of size	Required
Angularity	∠	Ⓜ or Ⓛ allowed for feature of size Ⓜ or Ⓛ allowed for datum features of size	Required
RUNOUT			
Total Runout	⌰	No	Required
Circular Runout	↗	No	Required
LOCATION			
Position	⌖	Ⓜ or Ⓛ allowed for all features Ⓢ implied if none specified Ⓜ or Ⓛ allowed for all datum features of size. Ⓢ implied if none specified	Required except for feature-to-feature only situations
Concentricity	◎	No	Required
Symmetry	⌯	No	Required

[RFS concept is implied for all geometric tolerances and RMB is implied for all datum features of size where no modifier is specified.]

ADDITIONAL SYMBOLOGY
IN Y14.5-2009

SYMBOL FOR	ASME Y14.5
Arc Length	$\overset{\frown}{100}$
Maximum Material Condition	Ⓜ
Least Material Condition	Ⓛ
Regardless of Feature Size	Ⓢ implied
Projected Tolerance Zone	Ⓟ
Tangent Plane	Ⓣ
Free State	Ⓕ
Reference Dimension (Parenthesis on each side of any dimension)	(1.25)
Statistical Tolerance	⟨ST⟩
Dimension is Not to Scale	$\underline{20}$
Datum Target	A1 OR Ø5 over A1 OR Ø5 with A1
Basic Dimension (theoretically exact dimension to show size, orientation or location with its tolerance given in feature control frames, established tooling or gaging tolerances or notes.)	$\boxed{100}$

CHART OF SYMBOLS COMMON IN ASME Y14.5

SYMBOL FOR	ASME Y14.5
Counterbore	⌴
Countersink	⌵
Depth or Deep	↧
Times, Places or By	X
Square Shape	□
All Around (Profile)	
Radius	R
Spherical Radius	SR
Controlled Radius	CR
Diameter	Ø
Spherical Diameter	SØ
Between, such as C ⟷ D, commonly used to show extent of control as in profiles.	⟷
Dimension Origin	⌱
Slope (Flat Taper)	⌳
Conical Taper	⌲

New Symbols

▷	Translation Symbol
Ⓤ	Unequally Disposed or Unilateral Tolerance
Ⓘ	Independency (Size Tolerance Does Not Control Form)
⟨CF⟩	Continuous Feature (Treat Multiple Features as One)
	All Over (Control Applies in all Views)

Old Symbols, New Meanings

Ⓜ	Means: Applies at Maximum Material Condition when it is used after a geometric tolerance. Means: Applies at Maximum Material Boundary when it is used after a datum reference.
Ⓛ	Means: Applies at Least Material Condition when it is used after a geometric tolerance. Means: Applies at Least Material Boundary when it is used after a datum reference.
Ⓢ Implied	When no material condition symbol is specified, it means: a) Applies at Regardless of Feature Size when it is implied after a geometric tolerance. b) Applies at Regardless of Material Boundary when it is implied after a datum reference.

NEW SYMBOL

SPOTFACE	⌊SF⌋Ø20

When the symbol for spotface is used, either the remaining thickness of material may be specified or the depth of the spotface may be given. The spotface depth is the minimum needed to clean up the material of the surface to the diameter of the spotface. The spotface may be noted using the symbol and its diameter without showing it on the design drawing or in the CAD model. The surface to receive the spotface must be clearly indicated. If desired, a filet radius may be given on the drawing to relieve sharp corners for the spotface.

For example:

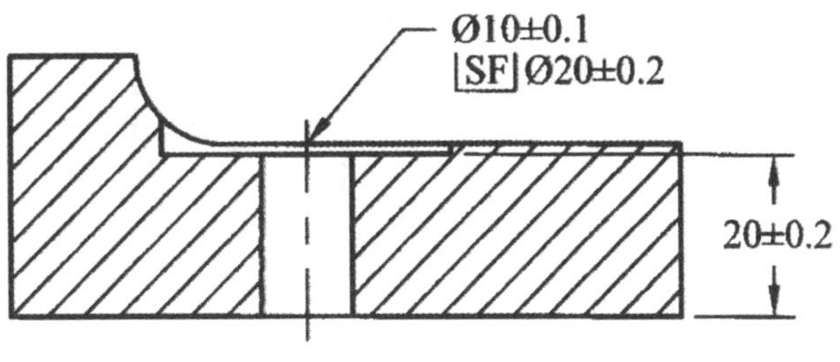

Datums and Datum Features

Datum Feature Symbols

[Datum Feature Symbols - Some Typical Ways to Specify a Datum Feature]

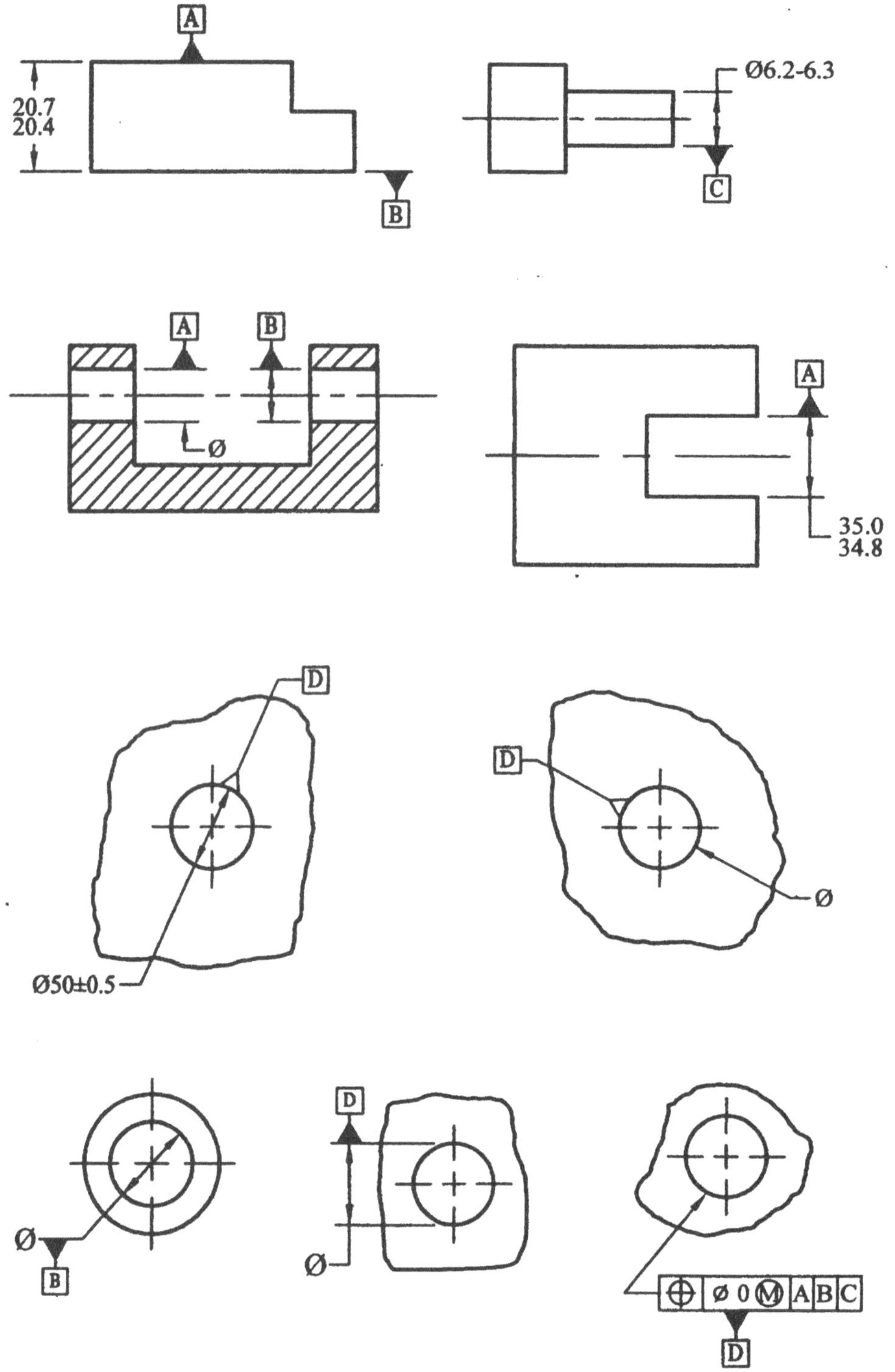

Food Chain of Symbology

if Used on Planar Surfaces as a Surface Control

⌓ = **Profile of the surface.** Tolerance zone consists of two uniform boundaries that follow the basically dimensioned or implied profile. All elements of the controlled surface(s) must reside within the tolerance zone. Datums may be referenced but are not required. It always controls 3D form but can be used to control size, angles and location.

↑

∠ = **Angularity of the surface.** Tolerance zone consists of two parallel planes separated by the angularity tolerance between which all elements of the surface must reside. These parallel planes are at the specified basic angle(s) to the datum(s) referenced. It controls flatness and the specified basic angle(s) to the datum(s) referenced in the feature control frame.

↑

⊥ = **Perpendicularity of the surface.** Tolerance zone consists of two parallel planes separated by the perpendicularity tolerance between which all elements of the surface must reside. These parallel planes are perpendicular (90°) to the primary and perhaps a secondary datum plane. It controls flatness and 90° angles.

↑

// = **Parallelism of the surface.** Tolerance zone consists of two parallel planes separated by the parallelism tolerance between which all elements of the surface must reside. These parallel planes are parallel (0°) to the primary datum plane. It controls flatness and 0°/180° angles.

↑

▱ = **Flatness of the surface.** Tolerance zone consists of two parallel planes separated by the flatness tolerance between which all elements of the surface must reside. It controls pits, bumps and curves, which is straightness in all directions.

↑

— = **Straightness of linear line elements.** Tolerance zone consists of two parallel straight lines separated by the straightness tolerance. Each line element in the plane of the view has its own separately verifiable tolerance zone. It controls pits, bumps and curves on each line under control.

Food Chain of Symbology

Commonly Used on Round Surfaces as a Surface Control

⌰ = **Total Runout of the surface.** If used on a cylindrical surface, the tolerance zone consists of two concentric cylinders separated by the runout tolerance between which all elements of the surface must reside. The tolerance zone is centered on the datum axis. As the part continuously revolves 360° while moving the indicator longitudinally down the entire surface, the full indicator movement may not exceed the runout tolerance. It controls cylindricity and concentricity.

↗ = **Circular Runout of circular cross sections of the surface.** Like controlling a stack of coins, each circular cross section, while centered on a datum axis, has a tolerance zone consisting of two concentric circles between which all points on the circle under test must reside. The full indicator movement for each cross section independently verified may not exceed the runout tolerance. It controls pits, bumps, flats, ovals, off-center of the datum axis and feature axis straightness.

These two geometric characteristics are not surface controls.

◎ = **Concentricity of the median points.** This coaxiality-type control has a cylindrical tolerance zone which confines all median points created by taking differential measurements on the part surface 180° apart (opposed points). The tolerance zone is centered on the datum axis. Unlike the other symbols on this page, concentricity is not a surface control, but rather tries to balance material on each side of the datum axis. See the note at the bottom of the page.

| — | Ø | = **Straightness of a derived median line.** This axis control has a cylindrical tolerance zone which confines the central points taken normal to the minimum circumscribed cylinder (if a shaft) or to the maximum inscribed cylinder (if a hole). These central points reflect the cross-sectional axes and together constitute a derived median line. Confines axial bowing. Like concentricity, this is not a surface control. See the note at the bottom of the page.

⌭ = **Cylindricity of a surface.** Tolerance zone consists of two concentric cylinders separated by the cylindricity tolerance between which all elements of the surface must reside. The full indicator movement in continuous 360° revolutions, while moving longitudinally down the surface, may not exceed the cylindricity tolerance. It controls circularity, straightness and taper.

○ = **Circularity of round line elements.** Tolerance zone consists of two concentric circles separated by the circularity tolerance. All points on the circle under consideration must reside between the two concentric circles. Each circular cross-section (like a stack of coins) of the surface has its own separately verifiable tolerance zone. The full indicator movement, in a 360° revolution, may not exceed the circularity tolerance. It controls pits, bumps, flats and ovals.

Note: As mentioned in the definitions above, unlike the other symbols on this page, concentricity and straightness of the derived median line are not surface controls.

A Few Basic Definitions, Formulas and Guidelines

Maximum Material Condition (MMC) Definition per ASME Y14.5-2009: It is the condition in which a feature of size contains the maximum amount of material within the stated limits of size. For example, the minimum hole diameter or the maximum shaft diameter is its maximum material condition.
Ⓜ = MMC = Smallest hole or largest shaft.

Least Material Condition (LMC) Definition per ASME Y14.5-2009: It is the condition in which a feature of size contains the least contains the least amount of material within the stated limits of size. For example, the maximum hole diameter or the minimum shaft diameter is its least material condition.
Ⓛ = LMC = Largest hole or smallest shaft

Regardless of Feature Size (RFS) Definition per ASME Y14.5-2009: Indicates a geometric tolerance applies at any increment of size of the actual mating envelope of the feature of size. They symbol for regardless of feature size (formerly a circled S) is implied for all geometric tolerances unless specified otherwise.

Regular Feature of Size One cylindrical or spherical surface, a circular element, a set of two opposed elements or opposes parallel surfaces, each of which is associated with a size dimension.

Irregular Feature of Size The two types of irregular features of size are: a) a feature or collection of features that may contain or be contained by an actual mating envelope which is a sphere, cylinder or pair of parallel planes, and b) a feature or collection of features that may contain or be contained by an actual mating envelope other than a sphere, cylinder or pair of parallel planes.

Rule #1
Size limits control surface form. Unless otherwise specified, for rigid features the LMC is measured for violations at cross-sections and the MMC is measured to verify compliance with an envelope of perfect form at MMC. The new Independency symbol placed near the size dimension of a feature negates this rule.

Rule #2 (RFS for geometric tolerances and RMB for datum features is implied)
For all geometric characteristic symbols used, where no Ⓜ (maximum material condition symbol) or Ⓛ (least material condition symbol) is specified in the feature control frame after the geometric tolerance, the regardless of feature size (RFS) concept is implied for the geometric tolerance.

For all geometric characteristic symbols used, where no Ⓜ (maximum material boundary symbol-MMB) or Ⓛ (least material boundary symbol-LMB) is specified in the feature control frame **after datum features**, the regardless of material boundary concept (RMB) is implied.

Definition of and Guidance for the Selection of Datum Features
Datum features are real. They are physical features (surfaces) that generate datums (theoretical axes or planes) from which we measure either angles, location or both.

Datum features should most often be:
1) Functional (serving some purpose in the way the part works).
2) Representative of seating features, mating features and alignment features.

3) Accessible (which is influenced by it having sufficient surface area to stabilize the part in the way it fits into the assembly).
4) Repeatable (which is greatly influenced by the feature's precision of form).

Virtual Condition MMC Concept

The virtual condition (MMC Concept) is the worst mating boundary.

The virtual condition of mating features should be compatible. The virtual condition of the mating hole and shaft should be the same if one is using 100% of the available arithmetically calculated tolerance.

Functional gages are sized at the maximum material boundary (MMB) of the features they are to gage. The MMB is calculated using the virtual condition formulas as follows.
To calculate the virtual condition (MMC Concept):

- for holes
  MMC of the hole minus the geometric tolerance at MMC = virtual condition

- for shafts
  MMC of the shaft plus the geometric tolerance at MMC = virtual condition

Floating Fastener Formula:

MMC hole
- <u>MMC shaft (or screw)</u>

Geometric Tolerance for all holes

Fixed Fastener Formulas:

MMC hole
- <u>MMC shaft (or screw)</u>

Geo. Tol. to be divided between the two mating features (parts)

Virtual Condition hole (MMC Concept)
- <u>MMC shaft (or screw)</u>

Geo. Tol. for shaft

MMC hole
- <u>Virtual Condition shaft (or screw) (MMC concept)</u>

Geo. Tol. for hole

Virtual Condition-ASME Y14.5-2009

A constant boundary generated by the collective effects of a considered feature of size's specified MMC or LMC material condition and the geometric tolerance for that material condition.

Resultant Condition-ASME Y14.5-2009

The single worst-case boundary generated by the collective effects of a feature of size's specified MMC or LMC, the geometric tolerance for that material condition, the size tolerance, and the additional geometric tolerance derived from the feature's departure from its specified material condition.

In ASME Y14.5M-1994, resultant condition was defined as "The variable boundary generated by the collective effects of a size feature's specified MMC or LMC material condition, the geometric tolerance for that material condition, the size tolerance, and the additional geometric tolerance derived from the feature's departure from its specified material condition.

New Rule Regarding the Use of the Ⓢ Symbol

The old Rule 2a which stated that a position tolerance was allowed to use a redundant Ⓢ symbol as a clarification of the applicability of the regardless of feature size concept is no longer true. Unless otherwise specified at MMC or LMC, all geometric tolerances are implied at RFS (regardless of feature size). Unless otherwise specified at MMB (maximum material boundary) or LMB (least material boundary) all datum features are implied to apply at regardless of material boundary (RMB).

New Terms for Existing Concepts: MMB, LMB and RMB

Maximum Material Boundary (MMB) is the limit that is defined by a tolerance or combination of tolerances, existing outside of the material of a feature or features. It is calculated by the following formulas:

For holes, slots (or other internal features of size), the maximum material boundary is the MMC of the feature minus the applicable geometric tolerance at MMC. The applicable geometric tolerance is (unless otherwise specified) the smallest geometric tolerance that exists between the datum feature being considered and any datums that precede it in the feature control frame. This boundary is also referred to as the inner boundary.

For shafts, tabs (or other external features of size), the maximum material boundary is the MMC of the feature plus the applicable geometric tolerance at MMC. The applicable geometric tolerance is (unless otherwise specified) the smallest geometric tolerance that exists between the datum feature being considered and any datums that precede it in the feature control frame. This boundary is also referred to as the outer boundary.

Least Material Boundary (LMB) is the limit defined by a tolerance or combination of tolerances, existing inside the material of a feature or features. It is calculated by the following formulas:

For holes, slots (or other internal features of size), the least material boundary is the LMC of the feature plus the applicable geometric tolerance at LMC. The applicable geometric tolerance is (unless otherwise specified) the smallest geometric tolerance that exists between the datum feature being considered and any datums that precede it in the feature control frame. This boundary is also referred to as the outer boundary.

For shafts, tabs (or other external features of size), the least material boundary is the LMC of the feature minus the applicable geometric tolerance at LMC. The applicable geometric tolerance is (unless otherwise specified) the smallest geometric tolerance that exists between the datum feature being considered and any datums that precede it in the feature control frame. This boundary is also referred to as the inner boundary.

Regardless of Material Boundary (RMB) is a similar perfect counterpart of a tolerance limit which grows or shrinks through a tolerance zone from MMB towards LMB until it makes maximum contact with the extremities of a feature. It is calculated by the following formulas:

For holes, slots (or other internal features of size), the regardless of material boundary for datum features controlled with position, perpendicularity, angularity or parallelism is the actual mating size of the datum feature minus the applicable geometric tolerance used by the produced datum feature at that size.

For shafts, tabs (or other external features of size), the regardless of material boundary for datum features controlled with position, perpendicularity, angularity or parallelism is the actual mating size of the feature plus the applicable geometric tolerance used by the produced datum feature at that size.

Actual Minimum Material Envelope is an envelope that is within the material as shown in the illustrations that follow this definition.

a) It is a similar perfect feature(s) counterpart of largest size that can be inscribed within an external feature(s) so that it just contacts the surface at the lowest points.
b) It is a similar perfect feature(s) counterpart of smallest size that can be circumscribed about the feature so that it just contacts the surface at its lowest points.

There are two types of actual minimum material envelopes.

1) One type is not related to datums. This one is called the unrelated actual minimum material envelope.
2) The other type is related to datums. This one is called the related actual minimum material envelope.

The following illustration compares the "actual mating envelope" to the "actual minimum material envelope" for internal and external features of size.

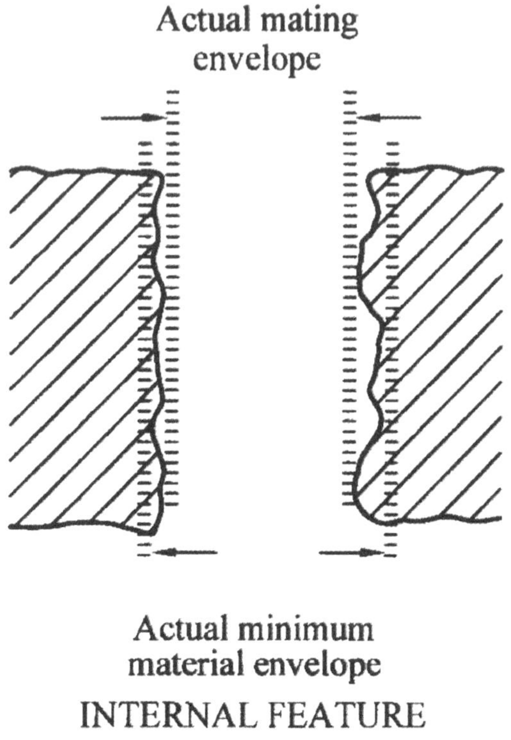

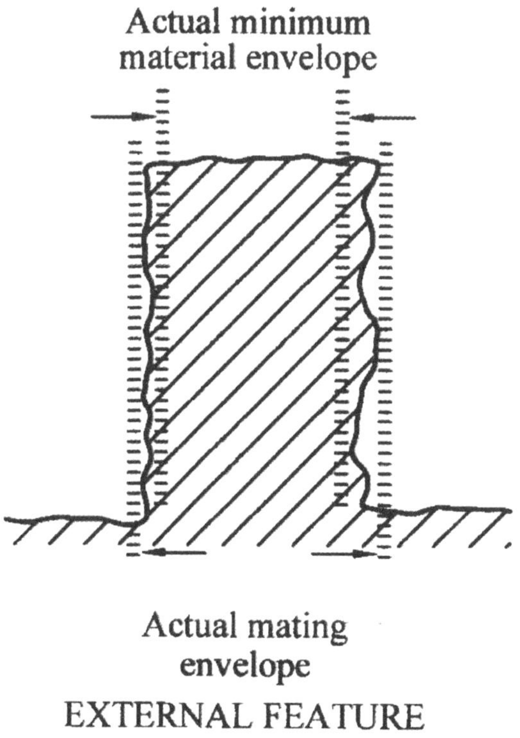

Boundaries

The concepts of boundaries generated by the collective effects of size and geometric tolerances are often referred to as simply inner boundaries and outer boundaries But they are also called virtual conditions and resultant conditions. Any of these boundaries that are constant are defined as virtual conditions and non-constant boundaries are resultant conditions.

Feature control frames that use maximum material condition symbols generate constant (virtual condition) boundaries on the inner side for holes and on the outer side for shafts. Any feature control frames that use least material condition symbols generate constant (virtual condition) boundaries on the outer side for holes and on the inner side for shafts.

In all of these conditions, the resultant, or non-constant, boundaries are generated on the opposite side of the virtual conditions. The regardless of feature size concepts, when used on holes or shafts in the feature control frames, generate only non-constant (resultant condition) boundaries.

FIGURE 1-1 [Holes-MMC Concept]

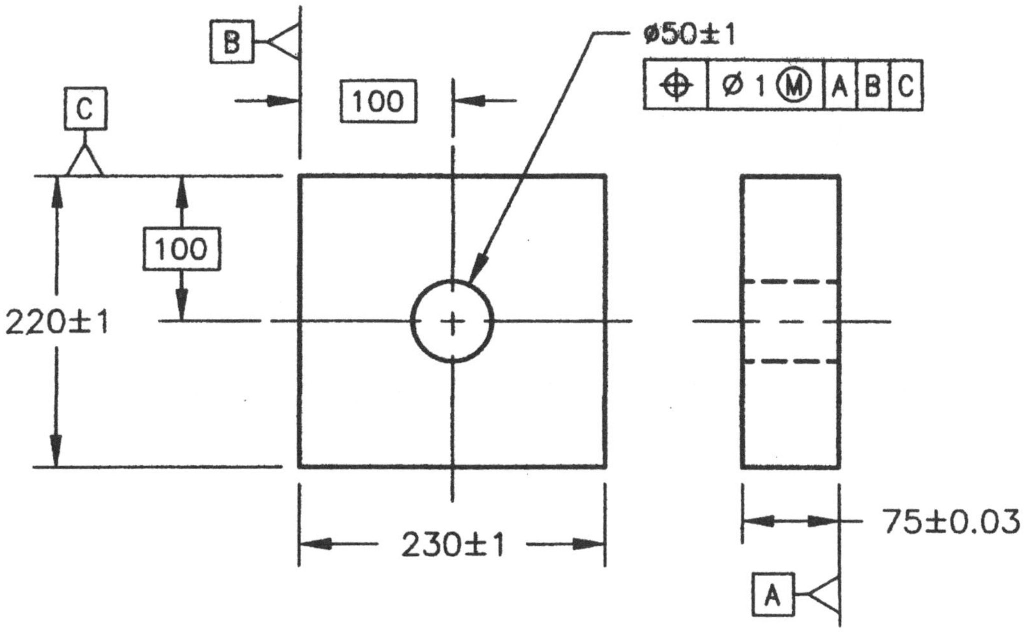

Constant vs. Non-Constant Boundaries

[FIGURE 1-1]

HOLE

Size		Geo. Tol.		Virtual Condition (constant boundary)
Ø49	-	1	=	Ø48) → worst case inner boundary
Ø50	-	2	=	Ø48)
Ø51	-	3	=	Ø48)

HOLE

Size		Geo. Tol.		Resultant Condition (non-constant boundary)
Ø49	+	1	=	Ø50
Ø50	+	2	=	Ø52
Ø51	+	3	=	Ø54 → worst case outer boundary

Creating an Equal Bilateral Toleranced Dimension from Virtual and Resultant Conditions

(MMC Concept Hole • FIGURE 1-1)

Resultant Condition Hole	=	Ø54
+ Virtual Condition Hole	=	Ø48
Sum	=	Ø102

Resultant Condition Hole	=	Ø54
- Virtual Condition Hole	=	Ø48
Difference	=	Ø6

Then:

$\frac{102}{2} = 51$ & $\frac{6}{2} = 3$

So,
Ø51±3 is an equal bilateral expression of the dimension and its tolerance.

FIGURE 1-2 [LMC Concepts]

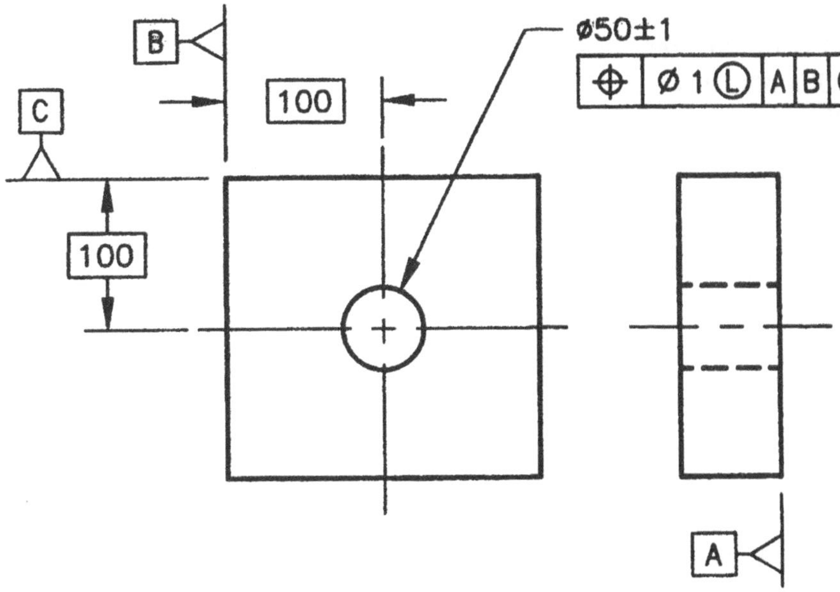

Constant vs. Non-Constant Boundaries

[FIGURE 1-2]

HOLE

Size	Geo. Tol.		Virtual Condition (constant boundary)
Ø51	+ 1	=	Ø52 → worst case outer boundary
Ø50	+ 2	=	Ø52
Ø49	+ 3	=	Ø52

HOLE

Size	Geo. Tol.		Resultant Condition (non-constant boundary)
Ø51	- 1	=	Ø50
Ø50	- 2	=	Ø48
Ø49	- 3	=	Ø46 → worst case inner boundary

Creating an Equal Bilateral Toleranced Dimension from Virtual and Resultant Conditions

[FIGURE 1-2]

Resultant Condition Hole	=	46
+ Virtual Condition Hole	=	52
Sum	=	Ø98

Resultant Condition Hole	=	52
- Virtual Condition Hole	=	46
Difference	=	6

Then:

$\frac{98}{2} = 49$ & $\frac{6}{2} = 3$

So,

Ø49±3 is an equal bilateral expression of the dimension and its tolerance.

FIGURE 1-3 [RFS Concept]

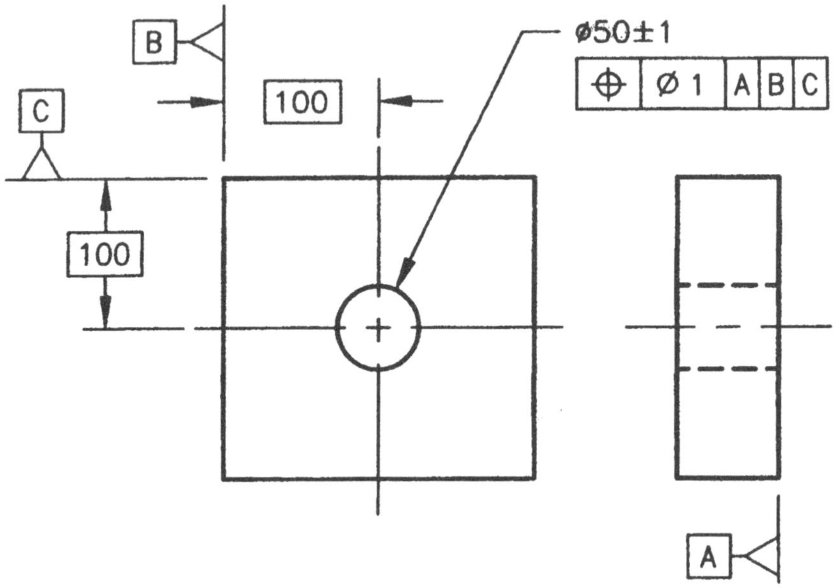

All Boundaries are Non-Constant in an RFS Call Out
[FIGURE 1-3]

HOLE

Size	Geo. Tol.		Inner Boundary
Ø49	- 1	=	Ø48 → worst case
Ø50	- 1	=	Ø49
Ø51	- 1	=	Ø50

HOLE

Size	Geo. Tol.		Outer Boundary
Ø49	+ 1	=	Ø50
Ø50	+ 1	=	Ø51
Ø51	+ 1	=	Ø52 → may be worst case

FIGURE 1-4

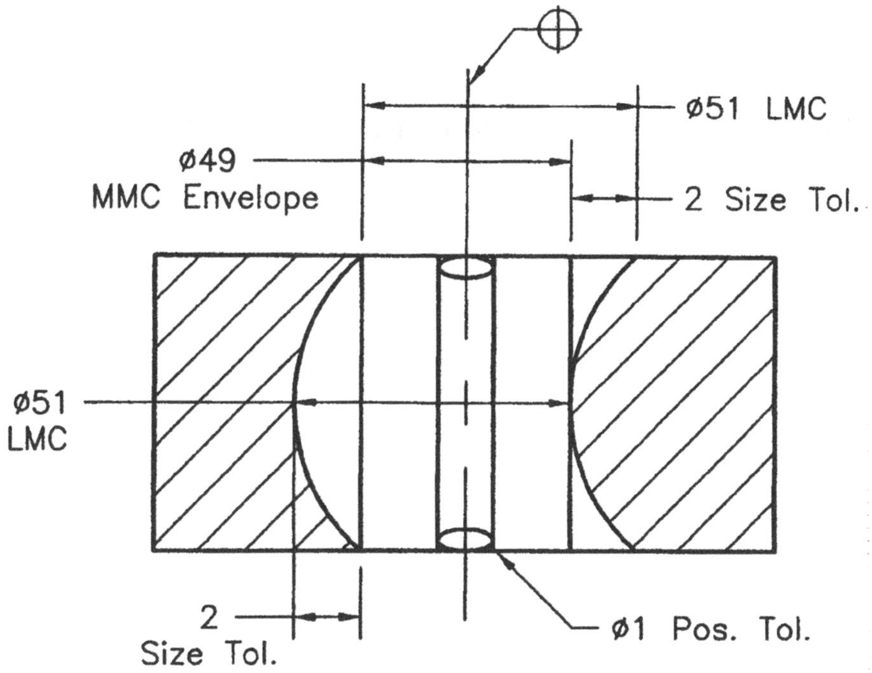

Ø49 + 2 + 2 + 1 = Ø54 worst case outer boundary

Only if the hole has a significant depth might this median line curvature (out-of-straightness) be a major factor. For very thin parts, such as sheet metal, it is probably not of concern in these analyses. In fact, many would contend that a banana-shaped hole is not likely to occur on most products. Therefore, for the analysis in this book, axially out-of-straightness holes will not be mentioned.

Creating an Equal Bilateral Toleranced Dimension from Inner and Outer Boundaries

[No Bananas - without the Out-of Axial Straightness Consideration - FIGURE 1-4]

Outer Boundary Hole	=	Ø52
+ Inner Boundary Hole	=	Ø48
Sum		Ø100

Outer Boundary Hole	=	Ø52
- Inner Boundary Hole	=	Ø48
Difference	=	Ø4

So,

$$\frac{Ø100}{2} = Ø50 \quad \& \quad \frac{4}{2} = 2$$

Therefore, expressed as an equal bilateral toleranced dimension = Ø50±2.

FIGURE 1-5 [Shafts-MMC Concept]

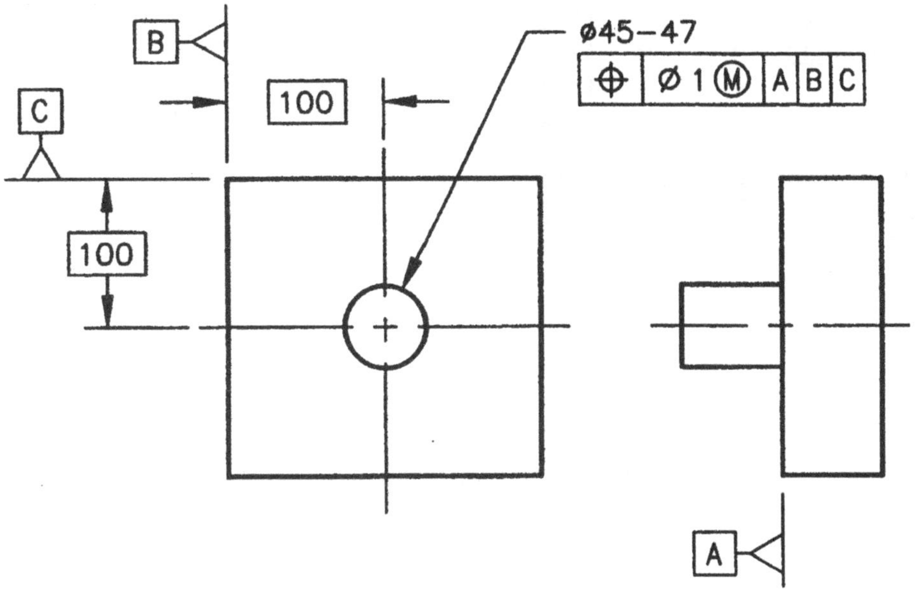

Constant vs. Non-Constant Boundaries

[FIGURE 1-5]

SHAFT

Size	Geo. Tol.		Virtual Condition
Ø47	+ 1	=	Ø48) → worst case outer boundary
Ø46	+ 2	=	Ø48)
Ø45	+ 3	=	Ø48)

SHAFT

Size	Geo. Tol.		Resultant Condition
Ø47	- 1	=	Ø46
Ø46	- 2	=	Ø44
Ø45	- 3	=	Ø42 → worst case inner boundary

Creating an Equal Bilateral Toleranced Dimension from Virtual and Resultant Conditions

[MMC Concept Shafts - FIGURE 1-5]

	Resultant Condition Shaft	=	Ø42
+	<u>Virtual Condition Shaft</u>	=	<u>Ø48</u>
	Sum	=	Ø90

	Resultant Condition Shaft	=	Ø48
-	<u>Virtual Condition Shaft</u>	=	<u>Ø42</u>
	Difference	=	Ø6

So;

$\frac{\text{Ø}90}{2} = \text{Ø}45$ & $\frac{6}{2} = 3$

Therefore,
Ø45±3 is an equal bilateral expression of the dimension and its tolerance.

FIGURE 1-6

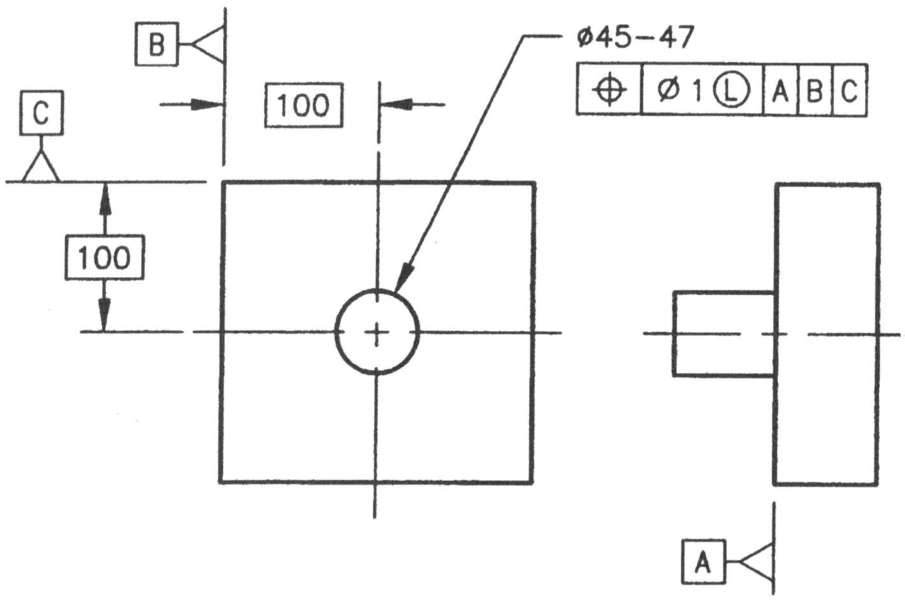

Constant vs. Non-Constant Boundaries

[FIGURE 1-6]

SHAFT

Size		Geo. Tol.		Resultant Condition
Ø45	+	1	=	46
Ø46	+	2	=	48
Ø47	+	3	=	50 → worst case outer boundary

SHAFT

Size		Geo. Tol.		Virtual Condition
Ø45	-	1	=	44) → worst case inner boundary
Ø46	-	2	=	44)
Ø47	-	3	=	44)

Creating an Equal Bilateral Toleranced Dimension from Virtual and Resultant Conditions

[FIGURE 1-6]

	Resultant Condition Shaft	=	Ø50
+	Virtual Condition Shaft	=	Ø44
	Sum		Ø94

	Resultant Condition Shaft	=	Ø50
-	Virtual Condition Shaft	=	Ø44
	Difference	=	Ø6

So,

$\varnothing\frac{94}{2} = \varnothing 47$ & $\frac{6}{2} = 3$

Therefore, expressed as an equal bilateral toleranced dimension = Ø47±3.

FIGURE 1-7

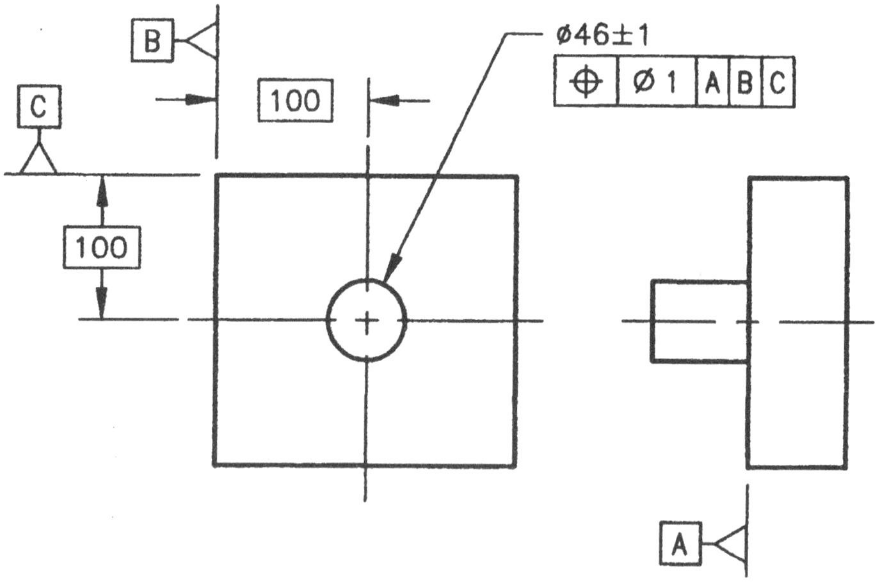

SHAFT

Size	Geo. Tol.			Outer Boundary
Ø45	+	1	=	Ø46
Ø46	+	1	=	Ø47
Ø47	+	1	=	Ø48 → worst case

SHAFT

Size	Geo. Tol.			Inner Boundary
Ø45	-	1	=	Ø44 → worst case (maybe)
Ø46	-	1	=	Ø45
Ø47	-	1	=	Ø46

The Ø44mm is identified only as *maybe* the worst case inner boundary. Regardless of feature size conditions have an oddity that must be considered in order to determine the absolute worst case inner boundary.

FIGURE 1-8

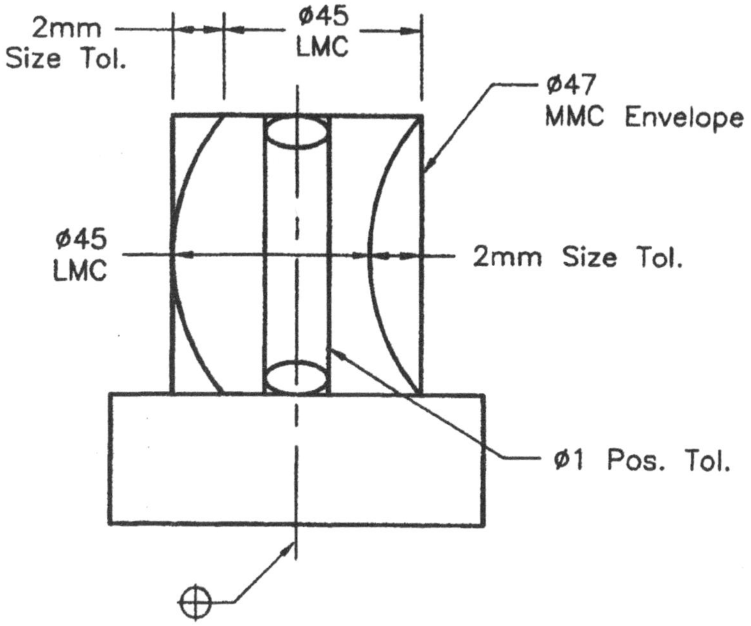

Ø47-2-2-1 = Ø42 worst case inner boundary
[with the median line out-of-straightness considered]

Only if the shaft has a significant length might this median line out-of-straightness be a major factor. For very short shafts, it is probably not of concern in the analysis. The analyses in this book will not consider this problem, but if your products run the risk of banana-shaped shafts, you may wish to use the above illustration to calculate the worst case inner boundary of your shafts.

Creating an Equal Bilateral Toleranced Dimension from Inner and Outer Boundaries

Outer Boundary Shaft	=	Ø48
+ Inner Boundary Shaft	=	Ø44
Sum		Ø92

Outer Boundary Shaft	=	Ø48
- Inner Boundary Shaft	=	Ø44
Difference	=	Ø4

So,

$$\frac{Ø92}{2} = Ø46 \quad \& \quad \frac{4}{2} = 2$$

Therefore, expressed as an equal bilateral toleranced dimension = Ø46±2.

I. WORKBOOK SECTION
to the
Geometric Dimensioning and Tolerancing
(per ASME Y14.5-2009)

Applications Problems
Word Problems
'NEW RULES' Problems

1. Please convert the following specifications to equal bilateral toleranced dimensions generated by their inner and outer boundaries. RFS boundaries have been calculated to reflect holes and shafts that are axially straight, and tabs and slots have straight centerplanes.

	Specification	Feature Control Frame	Outer Boundary	Inner Boundary
1)	Ø20±2 Hole	⌖ | Ø2 Ⓜ | A | B | C		
2)	Ø20±2 Hole	⌖ | Ø2 Ⓛ | A | B | C		
3)	Ø20±2 Hole	⌖ | Ø2 | A | B | C		
4)	Ø20 $^{+3}_{-2}$ Shaft	⌖ | Ø0.5 Ⓜ | A | B | C		
5)	Ø15-16 Shaft	⌖ | Ø0.4 Ⓛ | A | B | C		
6)	Ø30 $^{+0.4}_{-0.2}$ Shaft	⌖ | Ø0.3 | A | B | C		
7)	28.5 $^{+0.2}_{-0.1}$ Slot	⌖ | 0.8 Ⓜ | X | Y | Z		
8)	23.08-24.04 Tab	⌖ | 0.5 Ⓜ | T | E | C		
9)	75±2 Slot	⌖ | 1 Ⓛ | C | A | B		
10)	70±0.5 Tab	⌖ | 0.5 Ⓛ | D | E | F		
11)	12-13 Slot	⌖ | 0.3 | G | A | F		
12)	9.5-10.6 Tab	⌖ | 1.2 | H | J | K		

2. Please convert the following mating parts from plus and minus location tolerancing to position tolerancing. For Parts #1 and #2:
   Step 1) Assign a primary datum feature and give it a flatness tolerance of 0.01.
   Step 2) Assign a secondary datum feature and a perpendicularity tolerance of 0.03.
   Step 3) Assign a tertiary datum feature and a perpendicularity tolerance of 0.09.
   Step 4) Calculate and assign positional tolerances to both parts.

Part #1

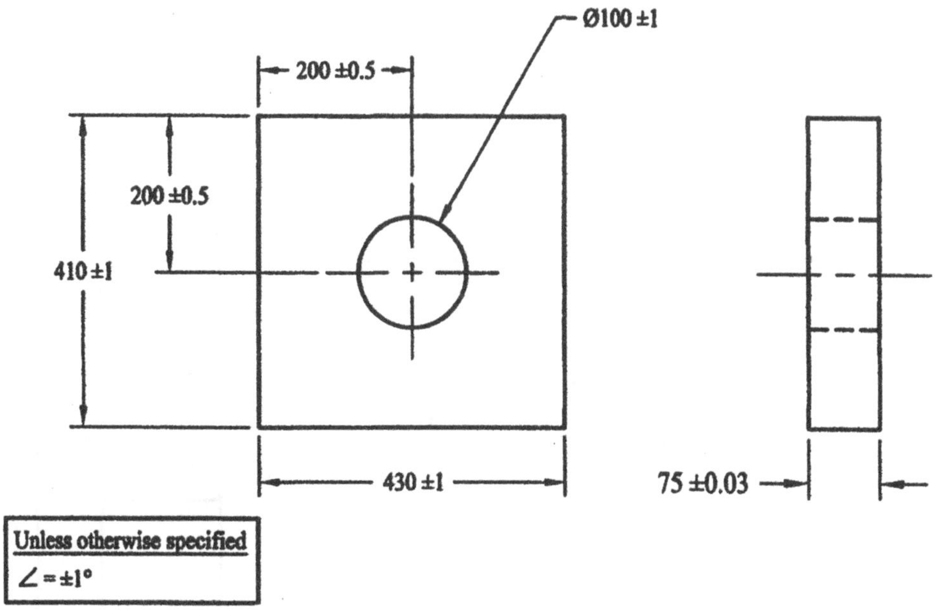

Part #2

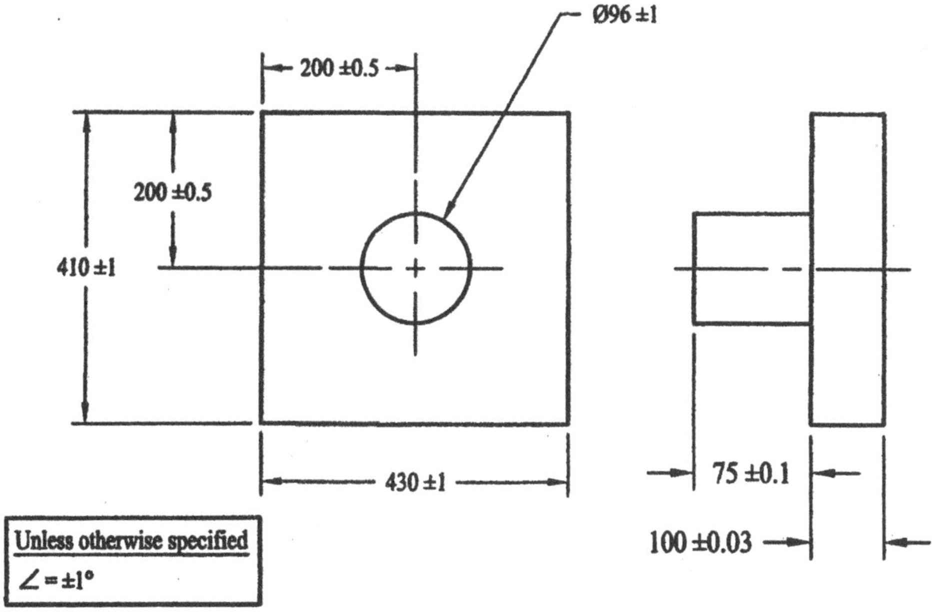

3. Please dimension and tolerance the following gage to inspect the position tolerance for your finished answer for Part #1 in Exercise 2.

Practical Absolute Functional Gage for Part #1

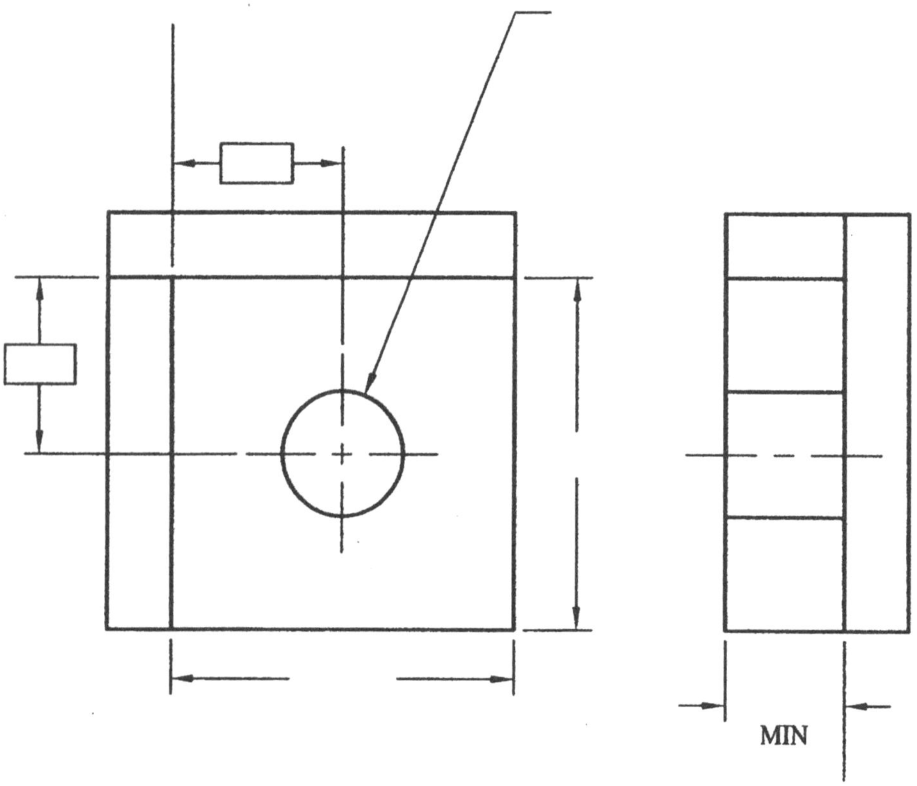

4. Using the illustration below, fill in the rest of the shaft's feature control frame to assure that it will mate with the hole.

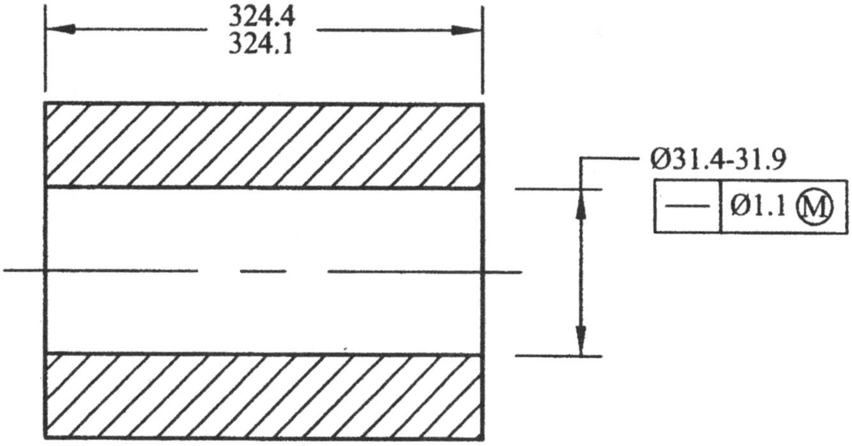

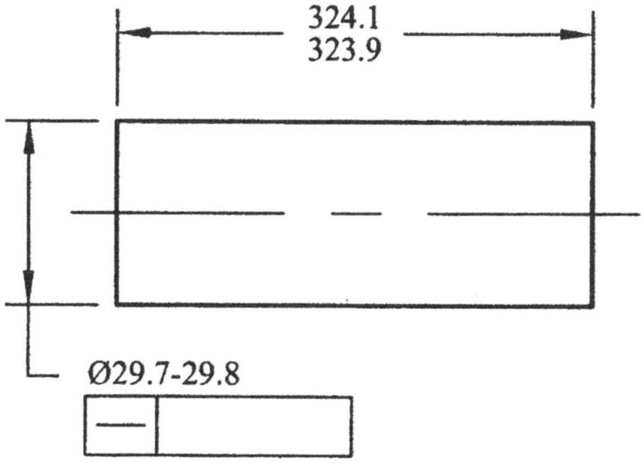

5. Using the illustration below, fill in the rest of the hole's feature control frame to assure that it will mate with the shaft.

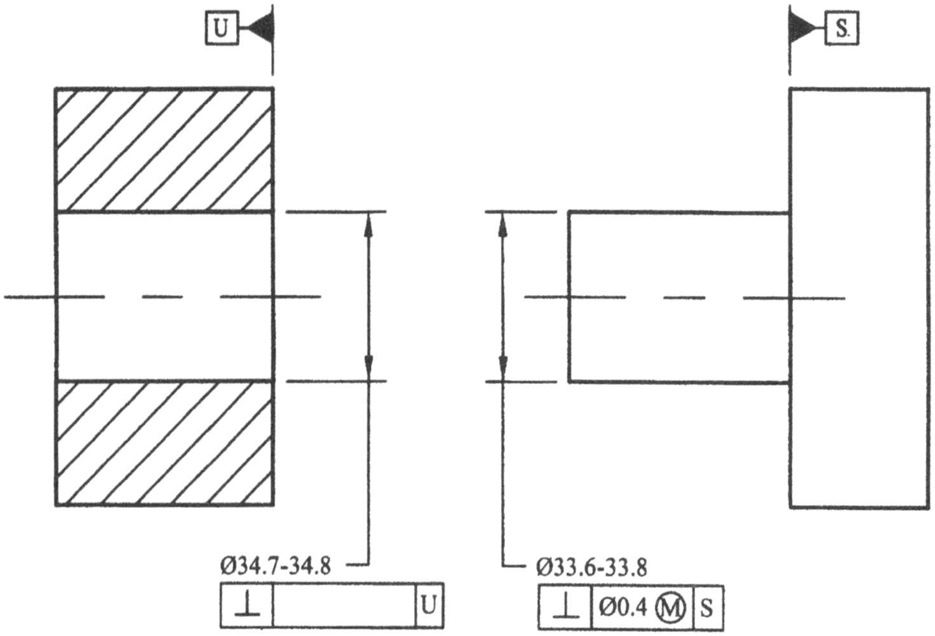

6. Fill in the rest of the shaft's feature control frame to assure that it will mate with the hole.

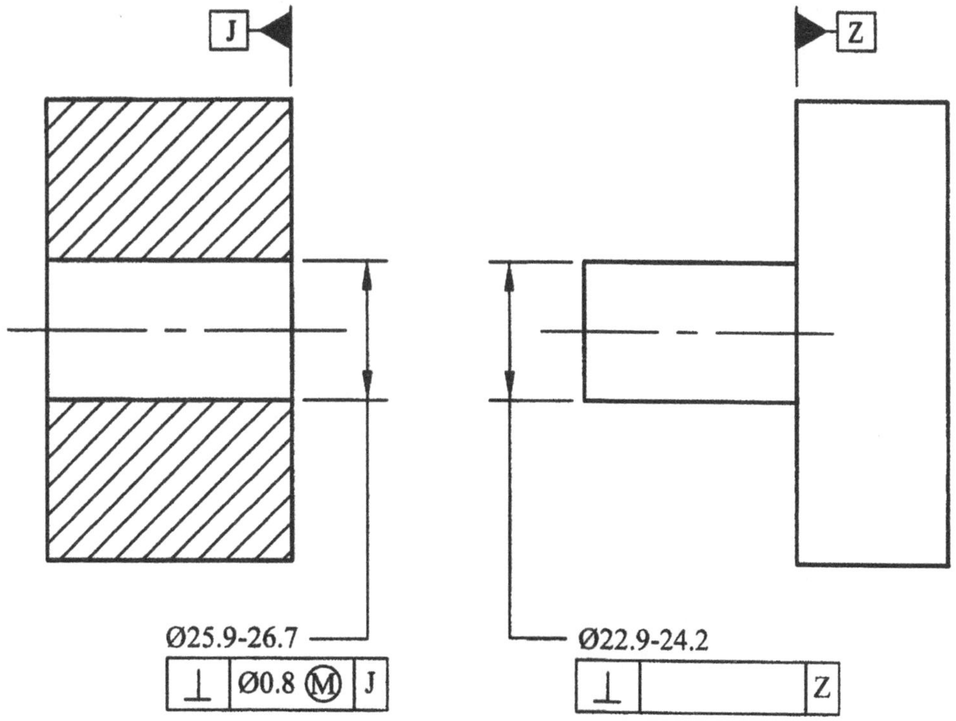

7. Using the illustration below, fill in the feature control frame so:
   a) the hockey puck (below the thick horizontal line) fits down inside of the cylindrical cavity (above the thick horizontal line).
   b) the holes in the hockey puck fit over the pins in the cavity.
   c) the load is not born by the pins but rather by L and T.

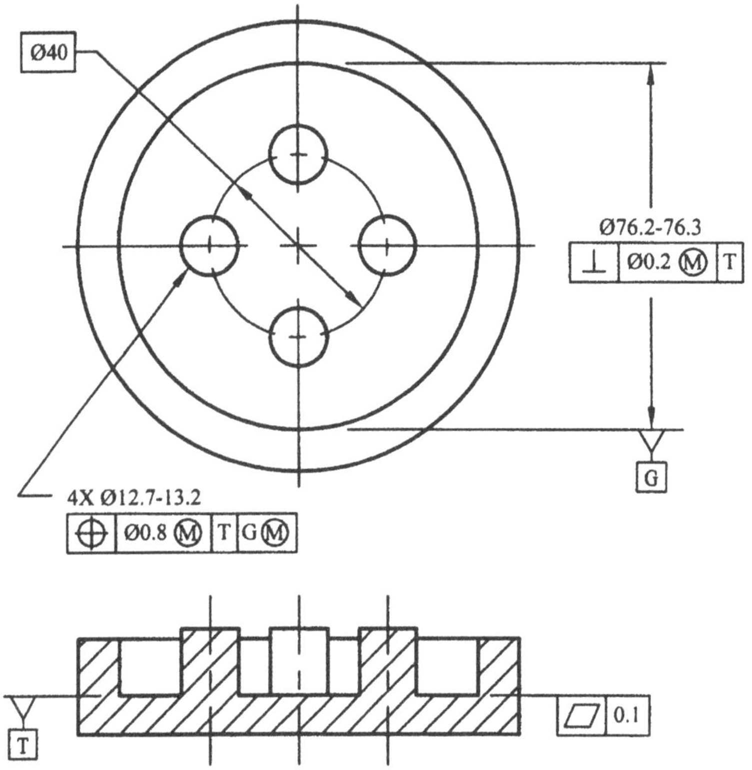

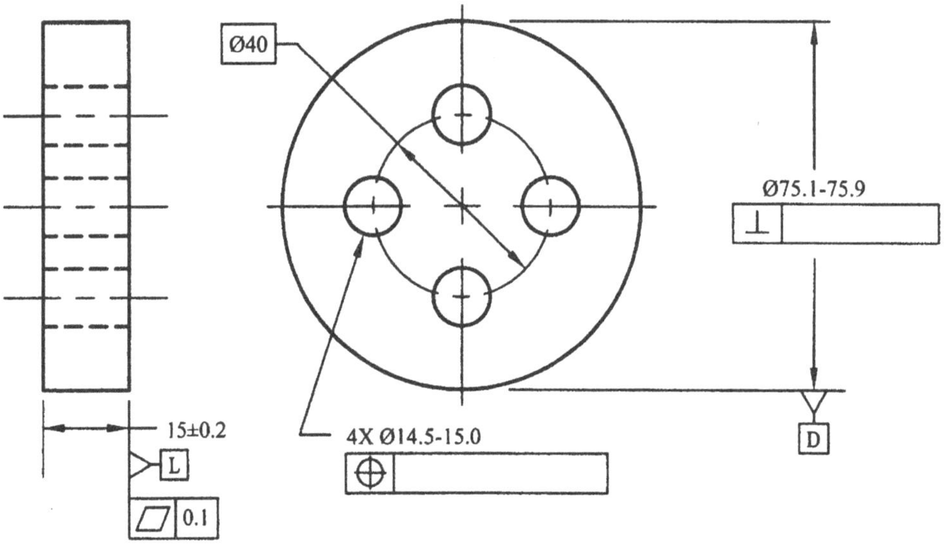

8.

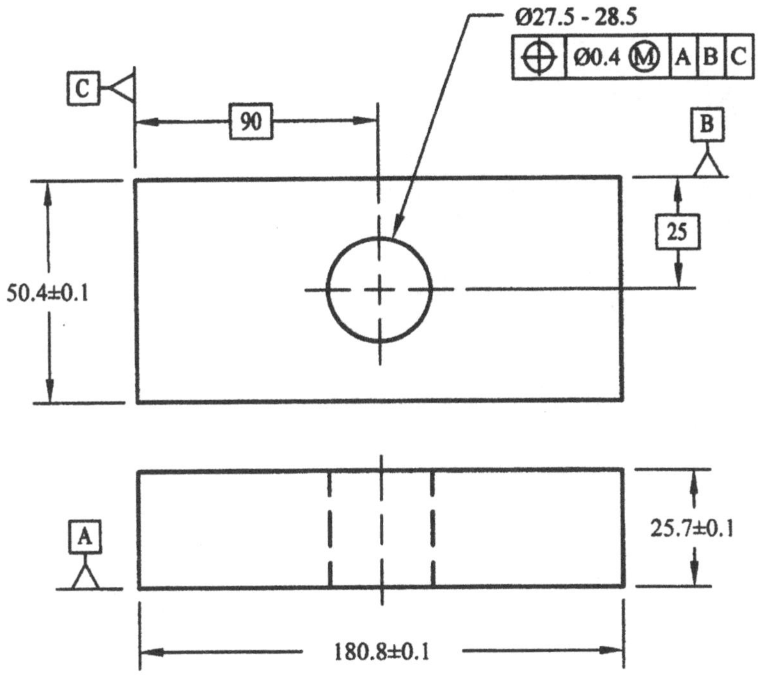

	MMC
-	Geo. Tol.
① ⬭	Virt. Cond. to protect
② ⬭	protected
③	

AND

①	MIC
-	MMC
	Bonus Geo. Tol.
+	Orig. Geo. Tol.
Ø ⬭	Total Geo. Tol.
②	MIC
-	MIC at ⌖
Ø ⬭	out of ⌖
③	

Part 1 as produced being inspected:

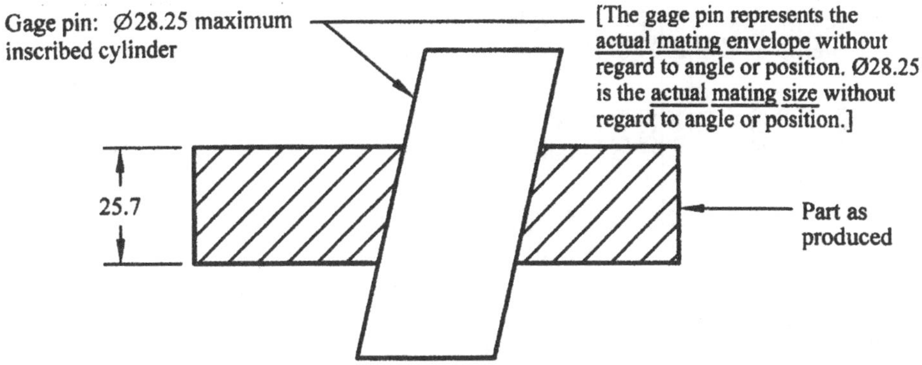

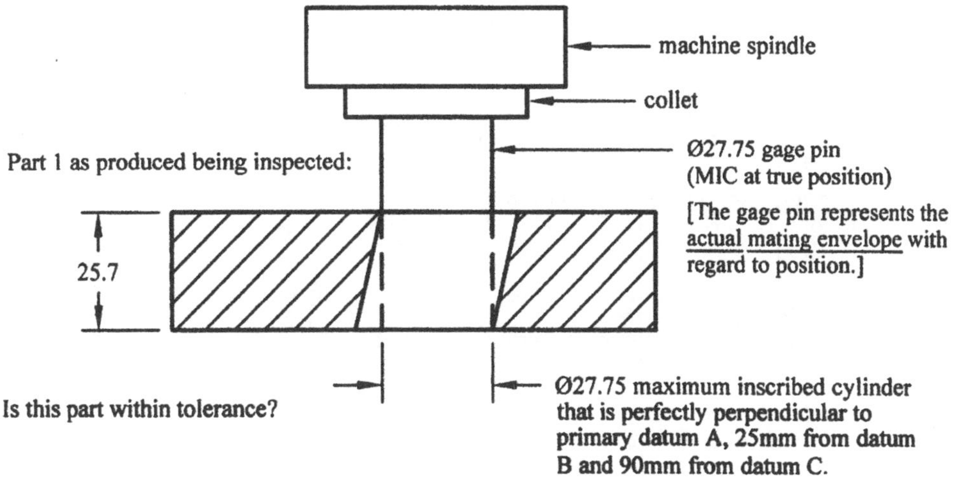

1. Is this part within tolerance?

Inch Conversion Chart

INCH CONVERSION CHART

Coordinate Measurement to Calculate the Diameter of the Actual Deviation from True Position (Z)

.020	.0400	.0402	.0404	.0408	.0412	.0418	.0424	.0431	.0439	.0447	.0456	.0466	.0477	.0488	.0500
.019	.0380	.0382	.0385	.0388	.0393	.0398	.0405	.0412	.0420	.0429	.0439	.0449	.0460	.0472	.0484
.018	.0360	.0362	.0365	.0369	.0374	.0379	.0386	.0394	.0403	.0412	.0422	.0433	.0444	.0456	.0469
.017	.0340	.0342	.0345	.0349	.0354	.0360	.0368	.0376	.0385	.0394	.0405	.0416	.0428	.0440	.0453
.016	.0321	.0322	.0325	.0330	.0335	.0342	.0349	.0358	.0367	.0377	.0388	.0400	.0412	.0425	.0439
.015	.0301	.0303	.0306	.0310	.0316	.0323	.0331	.0340	.0350	.0360	.0372	.0384	.0397	.0410	.0424
.014	.0281	.0283	.0286	.0291	.0297	.0305	.0313	.0322	.0333	.0344	.0356	.0369	.0382	.0396	.0410
.013	.0261	.0263	.0267	.0272	.0278	.0286	.0295	.0305	.0316	.0328	.0340	.0354	.0368	.0382	.0397
.012	.0241	.0243	.0247	.0253	.0260	.0268	.0278	.0288	.0300	.0312	.0325	.0339	.0354	.0369	.0384
.011	.0221	.0224	.0228	.0234	.0242	.0250	.0261	.0272	.0284	.0297	.0311	.0325	.0340	.0356	.0372
.010	.0201	.0204	.0209	.0215	.0224	.0233	.0244	.0256	.0269	.0283	.0297	.0312	.0328	.0344	.0360
.009	.0181	.0184	.0190	.0197	.0206	.0216	.0228	.0241	.0254	.0269	.0284	.0300	.0316	.0333	.0350
.008	.0161	.0165	.0171	.0179	.0189	.0200	.0213	.0226	.0241	.0256	.0272	.0288	.0305	.0322	.0340
.007	.0141	.0146	.0152	.0161	.0172	.0184	.0198	.0213	.0228	.0244	.0261	.0278	.0295	.0313	.0331
.006	.0122	.0126	.0134	.0144	.0156	.0170	.0184	.0200	.0216	.0233	.0250	.0268	.0286	.0305	.0323
.005	.0102	.0108	.0117	.0128	.0141	.0156	.0172	.0189	.0206	.0224	.0242	.0260	.0278	.0297	.0316
.004	.0082	.0089	.0100	.0113	.0128	.0144	.0161	.0179	.0197	.0215	.0234	.0253	.0272	.0291	.0310
.003	.0063	.0072	.0085	.0100	.0117	.0134	.0152	.0171	.0190	.0209	.0228	.0247	.0267	.0286	.0306
.002	.0045	.0056	.0072	.0089	.0108	.0126	.0146	.0165	.0184	.0204	.0224	.0243	.0263	.0283	.0303
.001	.0028	.0045	.0063	.0082	.0102	.0122	.0141	.0161	.0181	.0201	.0221	.0241	.0261	.0281	.0301
	.001	.002	.003	.004	.005	.006	.007	.008	.009	.010	.011	.012	.013	.014	.015

Deviation in X or Y Axis

Equation: $Z = 2\sqrt{X^2 + Y^2}$

z=the diameter of actual deviation from true position of the feature axis.

x=radial deviation from true position along the axis.

y=radial deviation from true position along y axis.

Metric Conversion Chart

METRIC CONVERSION CHART

Coordinate Measurement to Calculate the Diameter of the Actual Deviation from True Position (Z)

Y	Ø POSITION ZONE (Z)																								
0.50	1.001	1.003	1.007	1.013	1.020	1.028	1.038	1.050	1.063	1.077	1.093	1.109	1.127	1.146	1.166	1.187	1.209	1.232	1.256	1.281	1.306	1.332	1.359	1.386	1.414
0.48	0.961	0.963	0.967	0.973	0.981	0.990	1.000	1.012	1.025	1.040	1.056	1.073	1.092	1.111	1.132	1.154	1.176	1.200	1.224	1.250	1.276	1.302	1.330	1.358	1.386
0.46	0.921	0.923	0.928	0.934	0.941	0.951	0.962	0.974	0.988	1.003	1.020	1.038	1.057	1.077	1.098	1.121	1.144	1.168	1.193	1.219	1.246	1.273	1.301	1.330	1.359
0.44	0.881	0.884	0.888	0.894	0.902	0.912	0.923	0.936	0.951	0.967	0.984	1.002	1.022	1.043	1.065	1.088	1.112	1.137	1.163	1.189	1.217	1.245	1.273	1.302	1.332
0.42	0.841	0.844	0.849	0.855	0.863	0.874	0.885	0.899	0.914	0.930	0.948	0.967	0.988	1.010	1.032	1.056	1.081	1.106	1.133	1.160	1.188	1.217	1.246	1.276	1.306
0.40	0.801	0.804	0.809	0.816	0.825	0.835	0.848	0.862	0.877	0.894	0.913	0.933	0.954	0.977	1.000	1.024	1.050	1.076	1.103	1.131	1.160	1.189	1.219	1.250	1.281
0.38	0.761	0.764	0.769	0.777	0.786	0.797	0.810	0.825	0.841	0.859	0.878	0.899	0.921	0.944	0.968	0.994	1.020	1.047	1.075	1.103	1.133	1.163	1.193	1.224	1.256
0.36	0.721	0.724	0.730	0.738	0.747	0.759	0.773	0.788	0.805	0.824	0.844	0.865	0.888	0.912	0.937	0.963	0.990	1.018	1.047	1.076	1.106	1.137	1.168	1.200	1.232
0.34	0.681	0.685	0.691	0.699	0.709	0.721	0.735	0.752	0.769	0.789	0.810	0.832	0.856	0.881	0.907	0.934	0.962	0.990	1.020	1.050	1.081	1.112	1.144	1.176	1.209
0.32	0.641	0.645	0.651	0.660	0.671	0.684	0.699	0.716	0.734	0.755	0.777	0.800	0.825	0.850	0.877	0.905	0.934	0.963	0.994	1.024	1.056	1.088	1.121	1.154	1.187
0.30	0.601	0.605	0.612	0.621	0.632	0.646	0.662	0.680	.0700	0.721	0.744	0.768	0.794	0.821	0.849	0.877	0.907	0.937	0.968	1.000	1.032	1.065	1.098	1.132	1.166
0.28	0.561	0.566	0.573	0.582	0.595	0.609	0.626	0.645	0.666	0.688	0.712	0.738	0.764	0.792	0.821	0.850	0.881	0.912	0.944	0.977	1.010	1.043	1.077	1.111	1.146
0.26	0.522	0.526	0.534	0.544	0.557	0.573	0.591	0.611	0.632	0.656	0.681	0.708	0.735	0.764	0.794	0.825	0.856	0.888	0.921	0.954	0.988	1.022	1.057	1.092	1.127
0.24	0.482	0.487	0.495	0.506	0.520	0.537	0.556	0.577	0.600	0.625	0.651	0.679	0.708	0.738	0.768	0.800	0.832	0.865	0.899	0.933	0.967	1.002	1.038	1.073	1.109
0.22	0.442	0.447	0.456	0.468	0.483	0.501	0.622	0.544	0.569	0.595	0.622	0.651	0.681	0.712	0.744	0.777	0.810	0.844	0.878	0.913	0.948	0.984	1.020	1.056	1.093
0.20	0.402	0.408	0.418	0.431	0.447	0.466	0.488	0.512	0.538	0.566	0.595	0.625	0.656	0.688	0.721	0.755	0.789	0.824	0.859	0.894	0.930	0.967	1.003	1.040	1.077
0.18	0.362	0.369	0.379	0.394	0.412	0.433	0.456	0.482	0.509	0.538	0.569	0.600	0.632	0.666	0.700	0.734	0.769	0.805	0.841	0.877	0.914	0.951	0.988	1.025	1.063
0.16	0.322	0.330	0.342	0.358	0.377	0.400	0.425	0.453	0.482	0.512	0.544	0.577	0.611	0.645	0.680	0.716	0.752	0.788	0.825	0.862	0.899	0.936	0.974	1.012	1.050
0.14	0.283	0.291	0.305	0.322	0.344	0.369	0.396	0.425	0.456	0.488	0.522	0.556	0.591	0.626	0.662	0.699	0.735	0.773	0.810	0.848	0.885	0.923	0.962	1.000	1.038
0.12	0.243	0.253	0.268	0.288	0.312	0.339	0.369	0.400	0.433	0.466	0.501	0.537	0.573	0.609	0.646	0.684	0.721	0.759	0.797	0.835	0.874	0.912	0.951	0.990	1.028
0.10	0.204	0.215	0.233	0.256	0.283	0.312	0.344	0.377	0.412	0.447	0.483	0.520	0.557	0.595	0.632	0.671	0.709	0.747	0.786	0.825	0.863	0.902	0.941	0.981	1.020
0.08	0.165	0.179	0.200	0.226	0.256	0.288	0.322	0.358	0.394	0.431	0.468	0.506	0.544	0.582	0.621	0.660	0.699	0.738	0.777	0.816	0.855	0.894	0.934	0.973	1.013
0.06	0.126	0.144	0.170	0.200	0.233	0.268	0.305	0.342	0.379	0.418	0.456	0.495	0.534	0.573	0.612	0.651	0.691	0.730	0.769	0.809	0.849	0.888	0.928	0.967	1.007
0.04	0.089	0.113	0.144	0.179	0.215	0.253	0.291	0.330	0.369	0.408	0.447	0.487	0.526	0.566	0.605	0.645	0.685	0.724	0.764	0.804	0.844	0.884	0.923	0.963	1.003
0.02	0.057	0.089	0.126	0.165	0.204	0.243	0.283	0.322	0.362	0.402	0.442	0.482	0.522	0.561	0.601	0.641	0.681	0.721	0.761	0.801	0.841	0.881	0.921	0.961	1.001
X	0.02	0.04	0.06	0.08	0.10	0.12	0.14	0.16	0.18	0.20	0.22	0.24	0.26	0.28	0.30	0.32	0.34	0.36	0.38	0.40	0.42	0.44	0.46	0.48	0.50

FORMULA: $Z = 2\sqrt{X^2 + Y^2}$

9. Using the drawing below, complete the following charts.

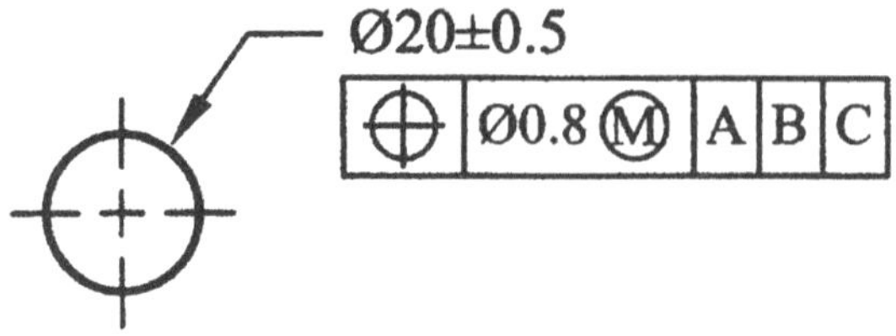

IF A HOLE

	Actual	Less MMC	Bonus Equals	Original Pos. Tol.	Total
A.	20.5				
B.	20.3				
C.	20.1				
D.	20				
E	19.8				
F.	19.6				
G.	19.5				

IF A SHAFT

	MMC	Less Actual	Bonus Equals	Original Pos. Tol.	Total
H.		19.5			
I.		19.6			
J.		19.8			
K.		20			
L.		20.1			
M.		20.3			
N.		20.5			

10. For a ∅25-26mm shaft, complete the table below to show the allowable total tolerance using the following formula:

    Bonus + Original Geometric Tolerance = Total Geometric Tolerance

Actual Feature Size	26	25.8	25.6	25.4	25.1
⌖ | Ø0.5 Ⓜ | A | B | C	A	B	C	D	E
// | Ø0.4 Ⓜ | A	F	G	H	I	J
⌖ | Ø0.5 | A | B | C	K	L	M	N	O
⊥ | Ø0 Ⓜ | A	P	Q	R	S	T
⌖ | Ø0.2 Ⓜ | L | D | C	U	V	W	X	Y
— | Ø0.3 Ⓜ	Z	AA	BB	CC	DD
⌭ | 0.05	EE	FF	GG	HH	II
⌖ | Ø0.08 Ⓜ | J | C | E	JJ	KK	LL	MM	NN

11. Complete chart sections A, B and C using the illustration below.
    a) What is the diameter of the allowed deviation from true position?
    b) What is the diameter of the deviation zone formed by the actual deviation from true position?
    c) Is the part good?

Dimension from Datum C	Dimension from Datum B	Actual Hole Size	Answers		
			A	B	C
(1) 1.901	.924	.500			
(2) 1.905	.929	.505			
(3) 1.902	.926	.510			
(4) 1.925	.932	.502			
(5) 1.895	.922	.500			
(6) 1.890	.925	.507			
(7) 1.900	.910	.509			
(8) 1.901	.926	.511			
(9) 1.870	.909	.509			

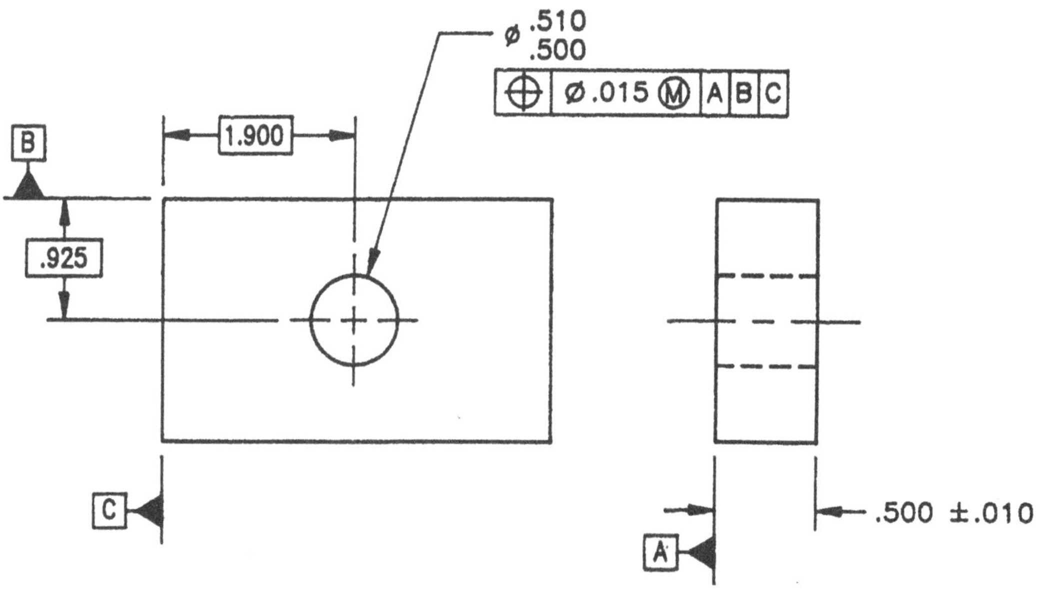

12. Size Tolerance: Using the drawing below, answer the following questions.

as drawn:

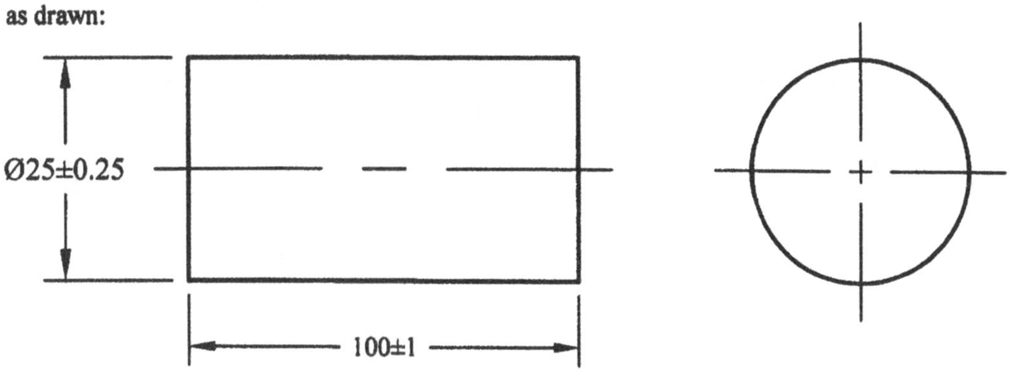

as produced:

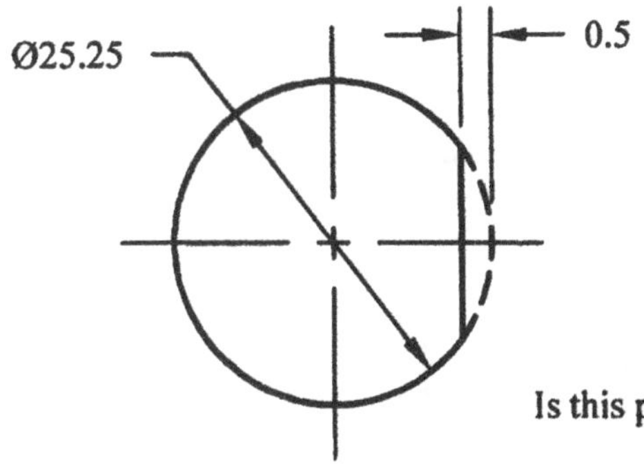

Is this part within tolerance ? ______

as produced:

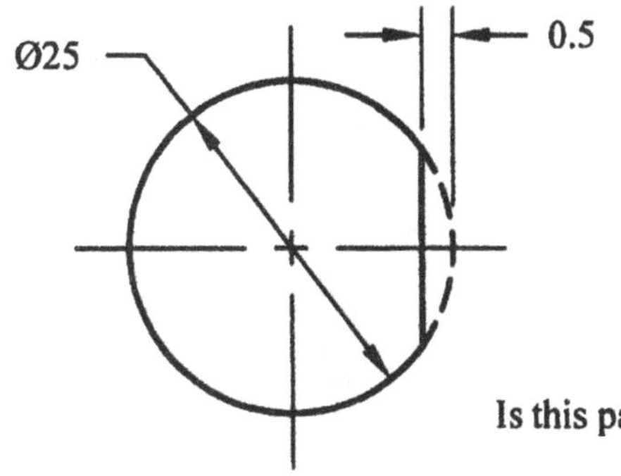

Is this part within tolerance ? ______

as produced:

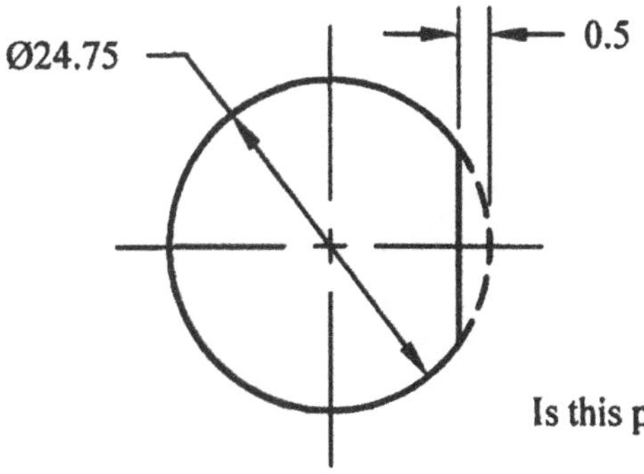

Is this part within tolerance ? ______

13. Size & Circularity Tolerance: Using the drawing below, answer the following questions.

as drawn:

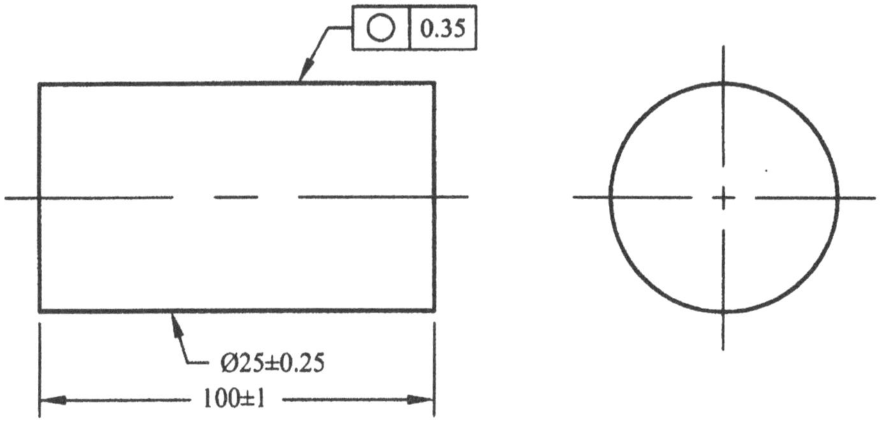

as produced:

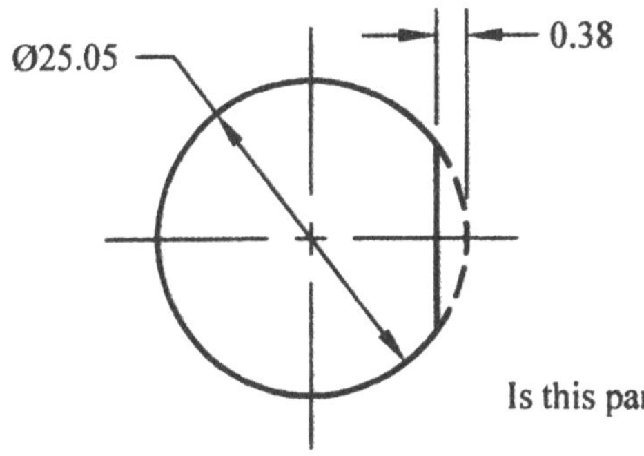

Is this part within tolerance ? _____

as produced:

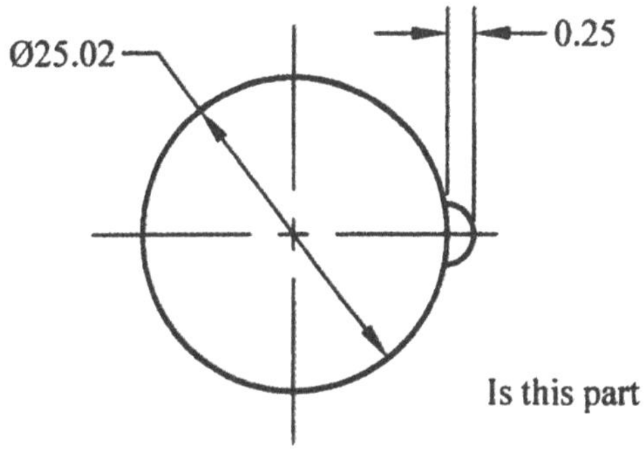

Is this part within tolerance ? _____

as produced:

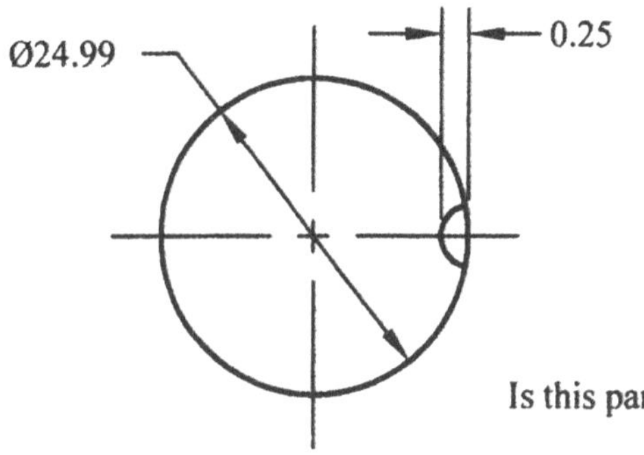

Is this part within tolerance ? _____

14. Cylindricity: Do any of the 'as produced' measurements shown indicate the part should be rejected? Why or why not?

Design Requirements:

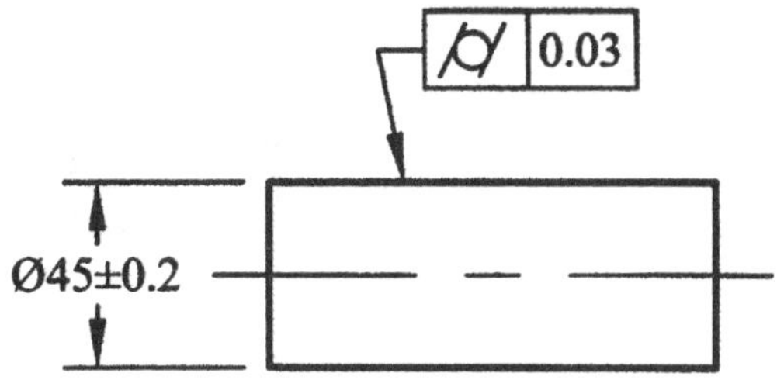

as produced:

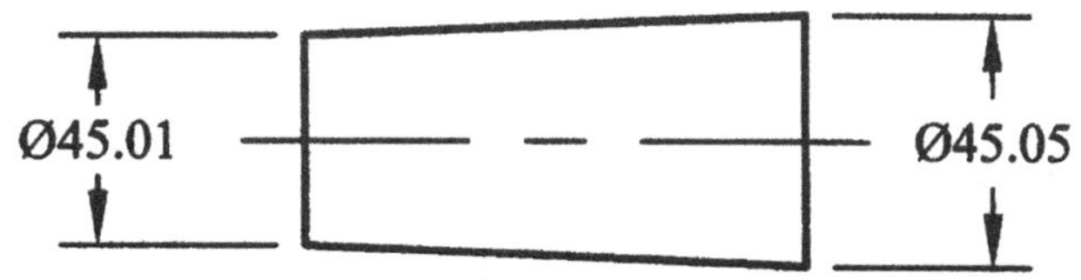

15. Circularity & Circular Runout: Answer the questions for each of the produced parts.

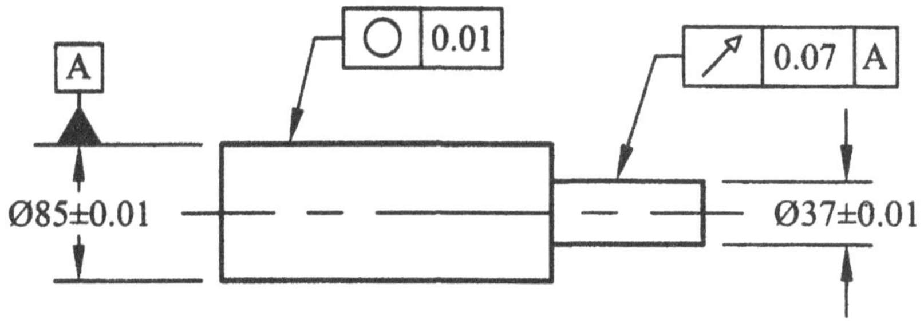

as produced:
1st part

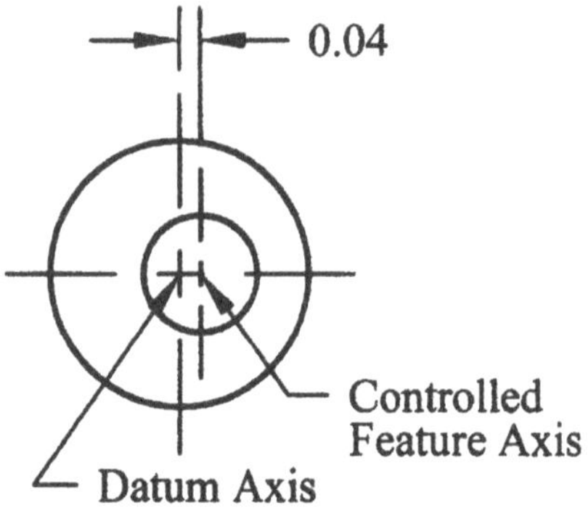

Is the part good?
Why, or why not?

2nd part

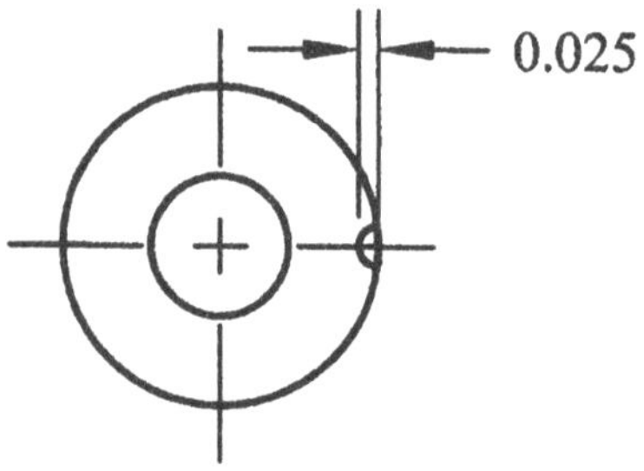

Is the part good?
Why, or why not?

3rd part

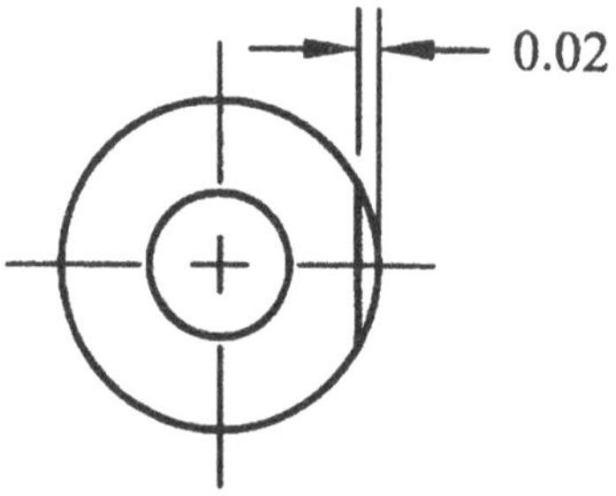

Is the part good?
Why, or why not?

16. Circularity, Straightness & Total Runout

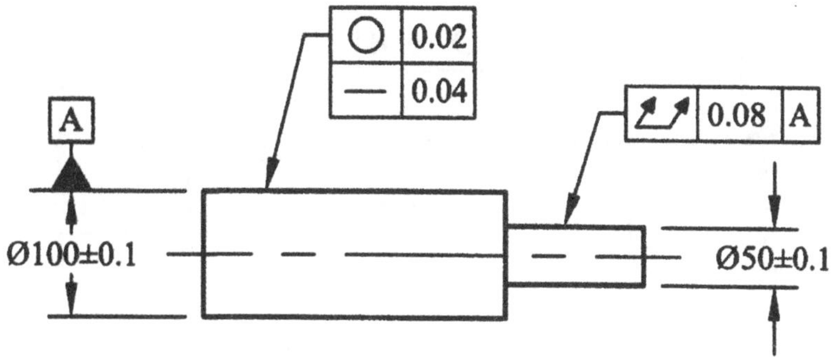

as produced:

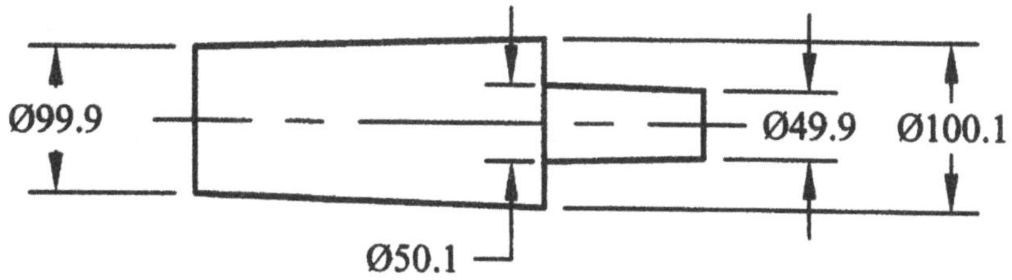

Do any of the "as produced" measurements shown indicate that the part should be rejected? Why, or why not?

17. Cylindricity & Circular Runout

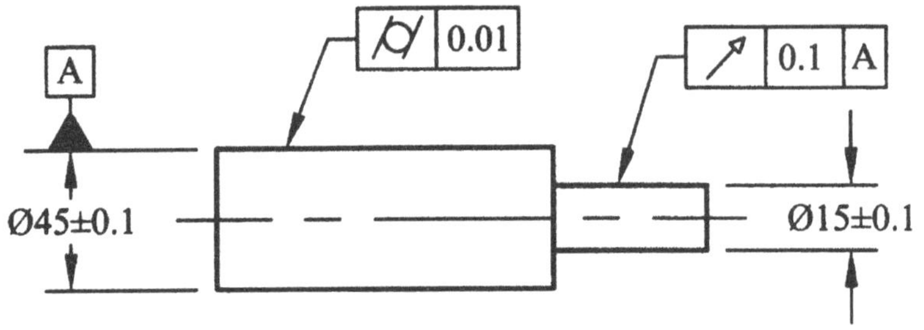

as produced:

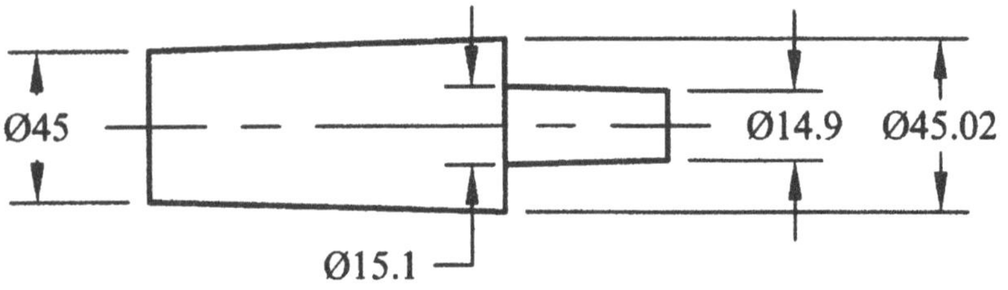

Do any of the "as produced" measurements shown indicate that the part should be rejected? Why, or why not?

18. Cylindricity & Circular Runout

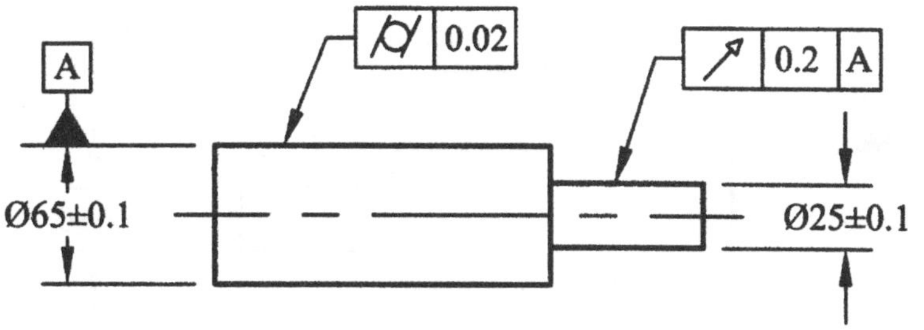

as produced:

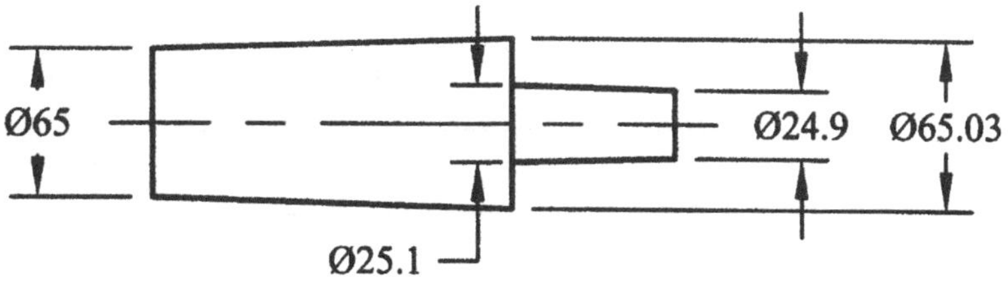

Do any of the "as produced" measurements shown indicate that the part should be rejected? Why, or why not?

19. Cylindricity & Total Runout

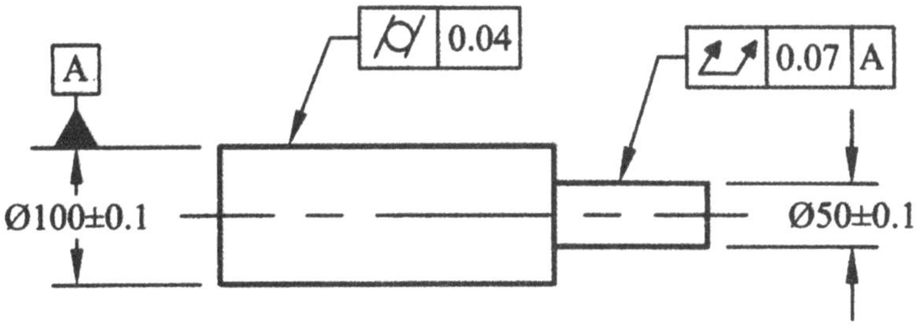

as produced:

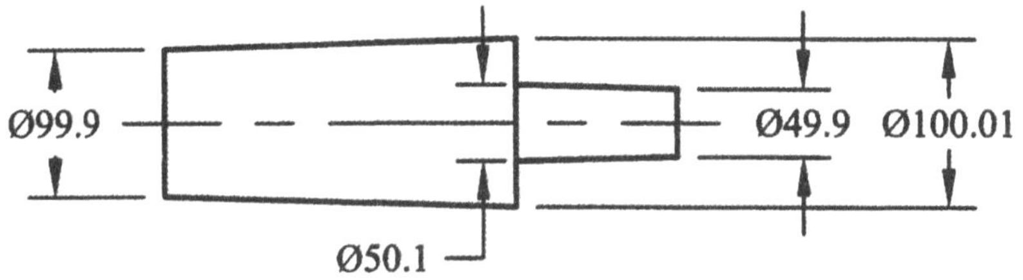

Do any of the "as produced" measurements shown indicate that the part should be rejected? Why, or why not?

20. Cylindricity & Total Runout: Answer the questions below.

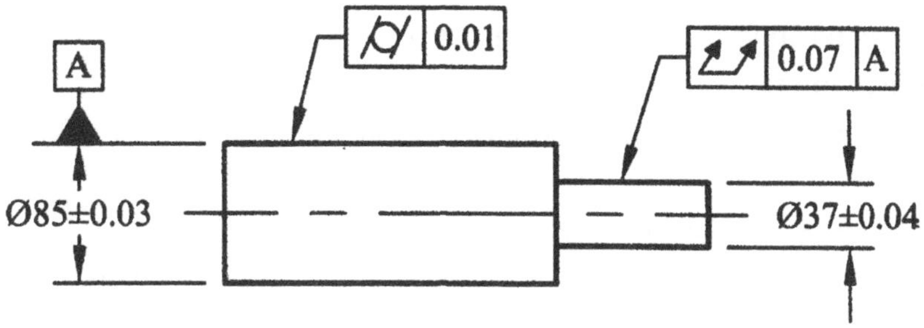

as produced:
1st part

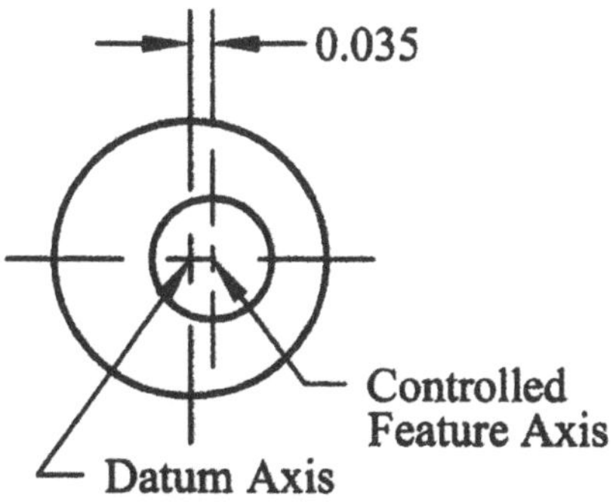

Is the part good?
Why, or why not?

2nd part

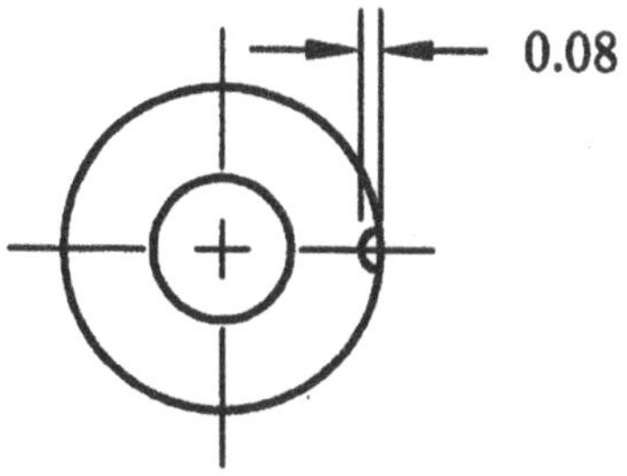

Is the part good?
Why, or why not?

3rd part

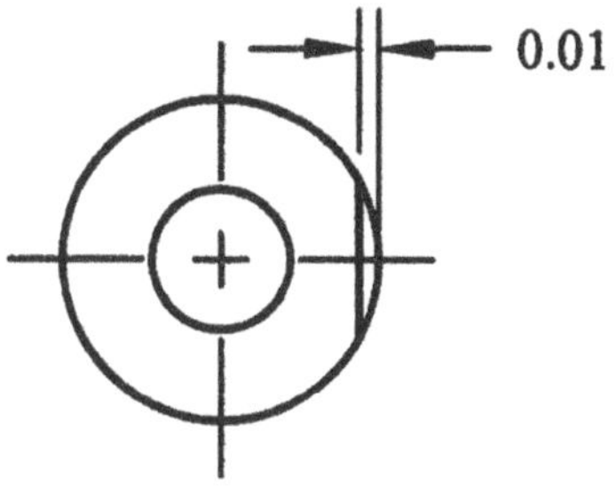

Is the part good?
Why, or why not?

21. Part 12135 – Bushing: Using the following variables data, assess the part on the following page and make assessments pertaining to its compliance to the tolerances depicted.

Available collected measurement data:

- Outside diameter measures .525 at its largest and .519 at its smallest.
- Part outside diameter is round to within .0002.
- FIM of I.D. while centered on and rotating about datum axis D is .002 at each cross section.
- Inside diameter is .2550 at its largest and .2549 at its smallest.
- Part thickness (shown in the left side view) varies between .196 and .195. It is rigid.
- Part thickness (shown in the left side view) has a slight uniform bow. Its derived median plane is out of straightness .003.

21. (continued)

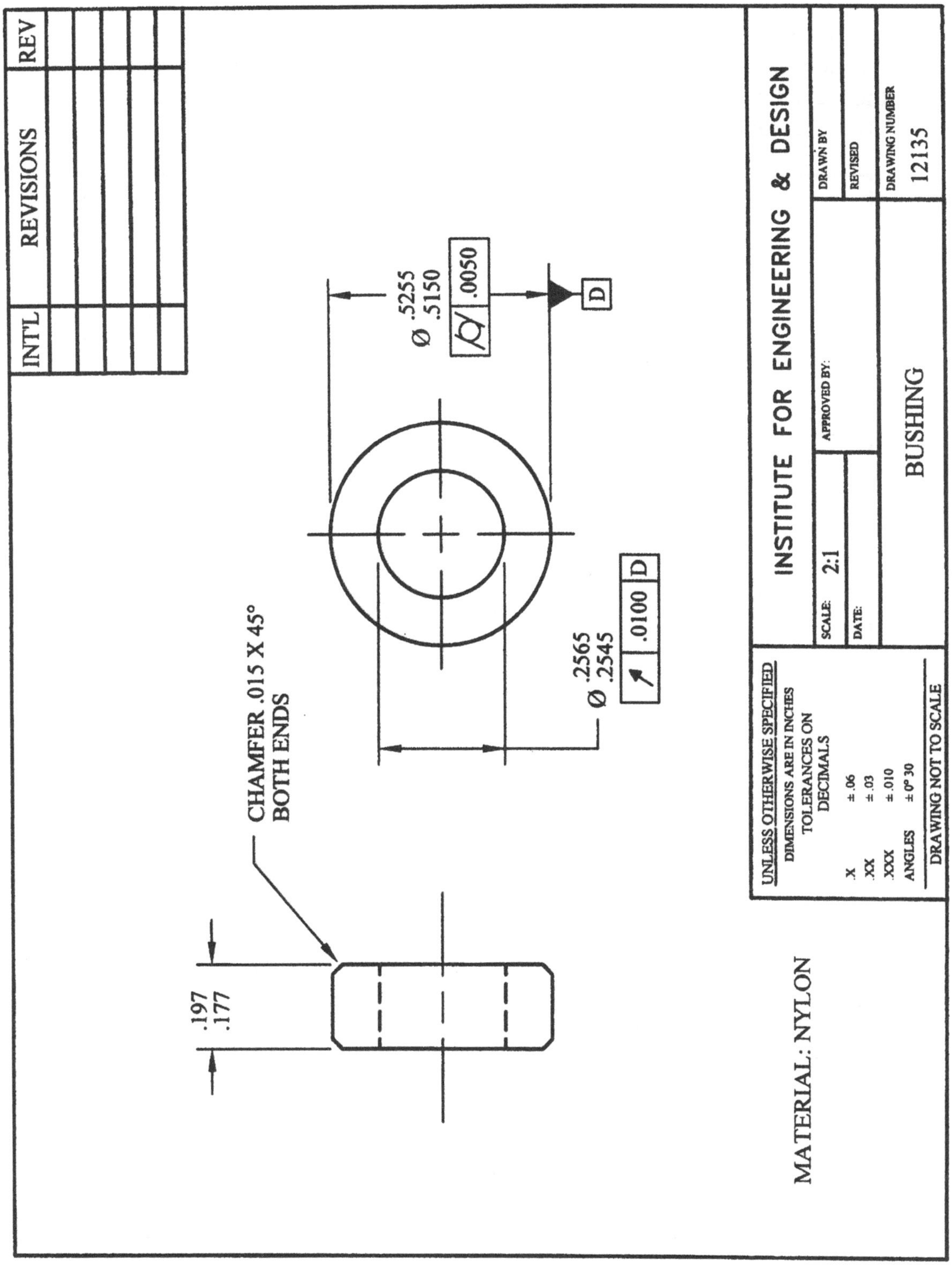

22. Please tolerance the two mating parts on this page using the following steps:
    a) assign a primary datum feature to each and give each a flatness control of 0.03.
    b) assign a secondary datum feature to each and give each a perpendicularity control.
    c) assign a tertiary datum feature to the second part (but not the first part).
    d) position the 19.8-20.0mm diameter on the first part and the 19.6-19.8mm diameter on the second part.
    e) give each part an all around unilateral profile of a surface tolerance.

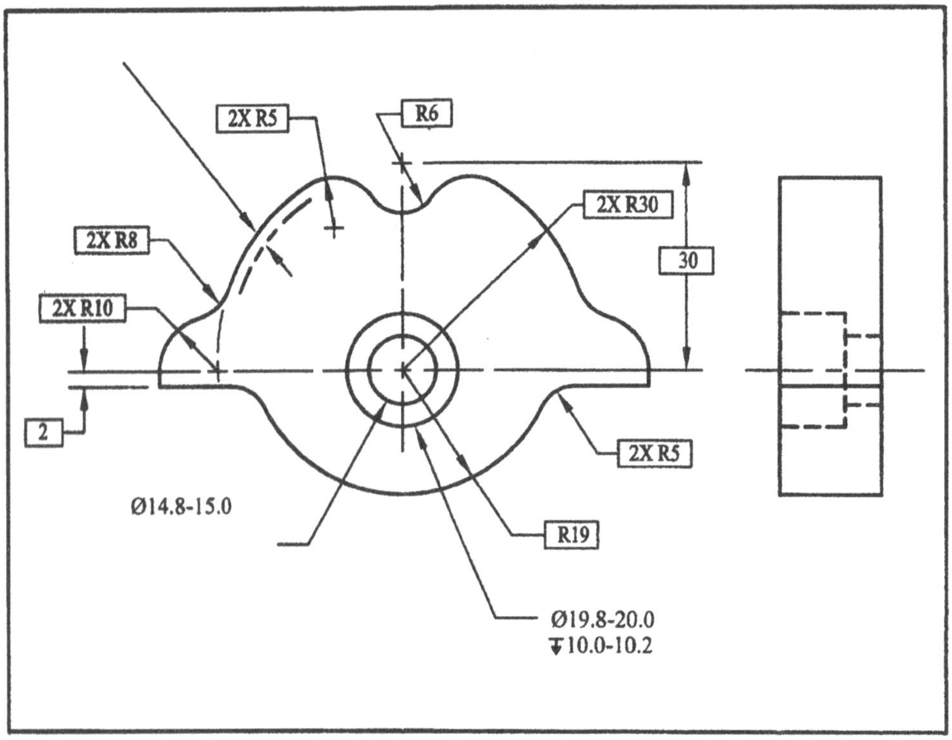

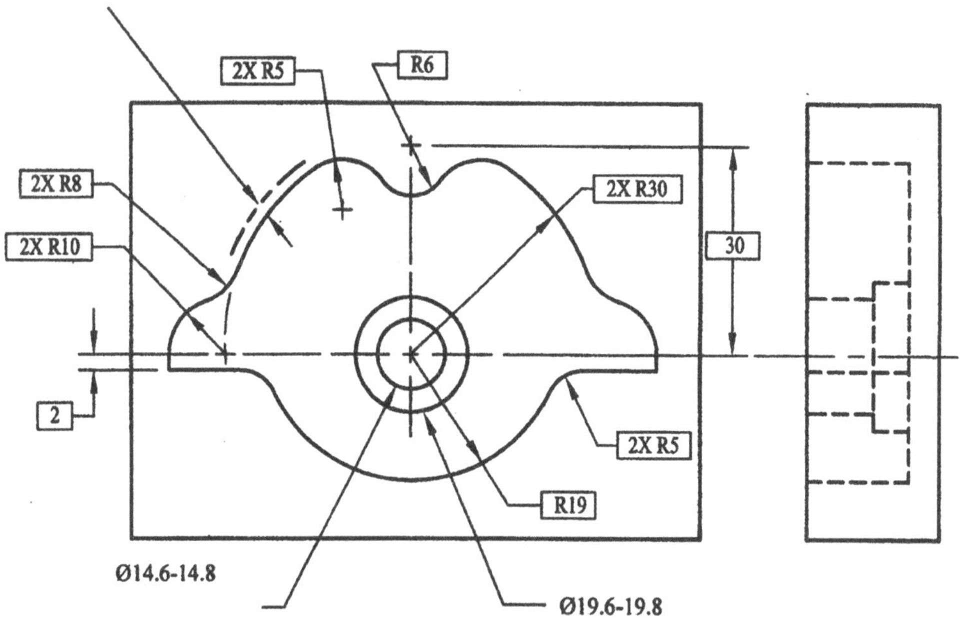

23. Convert this plus and minus toleranced part to a composite profile toleranced part.

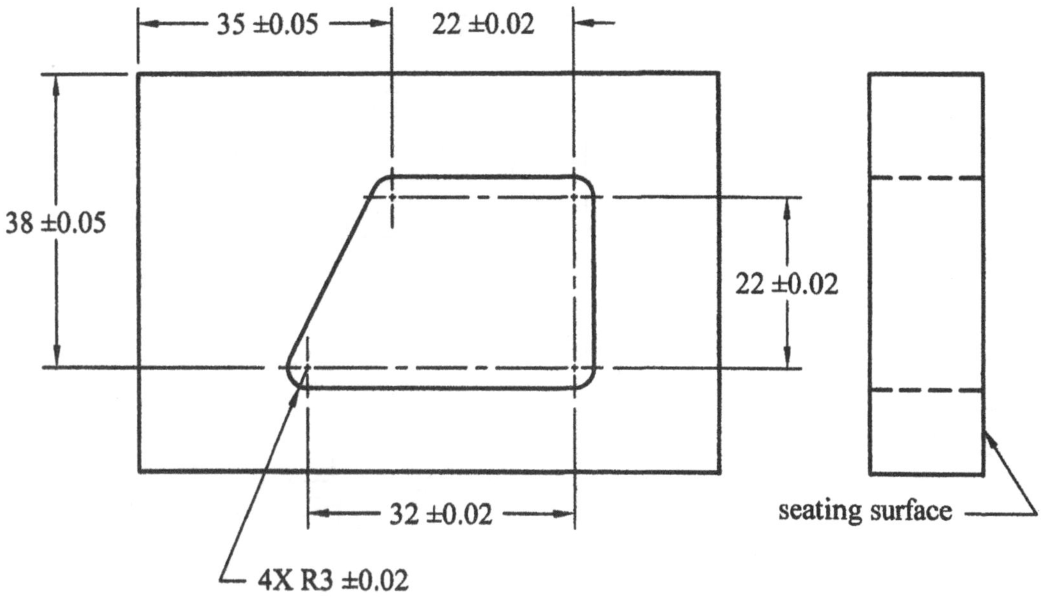

24. Answer the following questions.

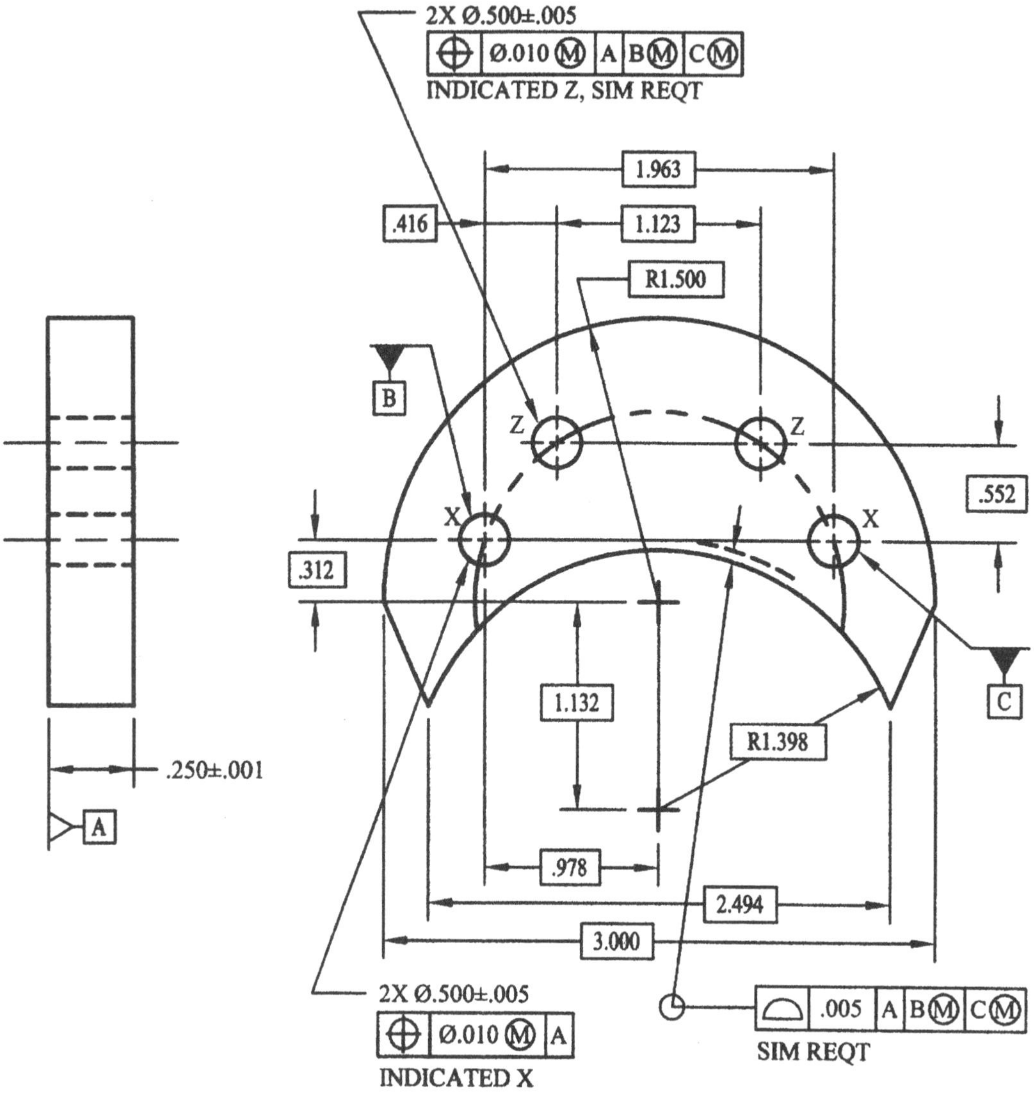

1) If the part thickness is measured with a micrometer at .249 and the median plane bows .002, is the part within its size tolerance?

2) If the two holes 'INDICATED X' are produced at a size of ∅.500 and with their axes a distance apart of 1.982, are they acceptable?

3) If the profile size shown as a basic 3.000 is produced at 2.980, does it exceed its tolerance?

24. (continued)

4) If the two Z holes are produced at ∅.505 and their axes at a distance of 1.099 from each other, are they within tolerance?

5) If the holes are produced at ∅.500 and out-of-perpendicularity to datum A a ∅.015, are they out of tolerance?

6) If the surface profile has a bump on the upper left radius that exceeds the basic radius shown by .005, is the surface to be rejected as out of tolerance?

7) If the holes measure a ∅.499 at each circular cross section and can receive a gage pin (without regard to orientation or location) no larger than a ∅.492, do they exceed their size tolerance?

8) If the largest gage pin that can be inserted into datum feature B measures a ∅.505, the largest gage pin that can be inserted into datum feature C measures a ∅.495, and a vernier caliper measurement from the outermost points on the gage pin diameters is a distance of 2.458, would that indicate that they exceed their positional tolerance?

25. Answer the questions 1-11 below for the illustration shown on Page 54. Question 12 refers to the illustration on Page 55.

    1) How flat must datum feature A be when produced?

    2) For the four holes that become datum pattern D, what are they being positioned to besides datum feature A?

    3) How many holes does the profile control tie the profile to?

    4) What acts as the origin of measurement for location for the three hole pattern?

    5) What characteristic of geometry does datum A tolerance when used in each feature control frame?

    6) Is the profile control a total wide zone of .01 or is it ±.01?

    7) What is the virtual condition of the holes in the three hole pattern?

    8) How many gages would be necessary to gage the MMC of all seven holes, the two positional controls and the profile control (if only the outer tolerance zone boundary of the profile control was to be gaged)?

    9) If the two outermost .51-.52 holes are produced at a diameter of .514 and their axes a distance apart of 1.126, are they within tolerance?

    10) If the surface profile has a bump on the upper left radius that exceeds the basic radius shown by .005, is the surface in violation of the tolerance zone under all circumstances?

    11) If the largest gage pin that can be inserted into the middle .510-.520 hole measures a diameter of .518 and the hole just to its left can receive a gage pin no larger than .512 and a vernier caliper measurement from the outermost points on the gage pin diameters is a distance of 1.080, would that indicate that they exceed their positional tolerance?

    12) How does the gaging you visualized when answering Question 8 differ from the gaging that would be used for the part on Page 55?

25. (continued)

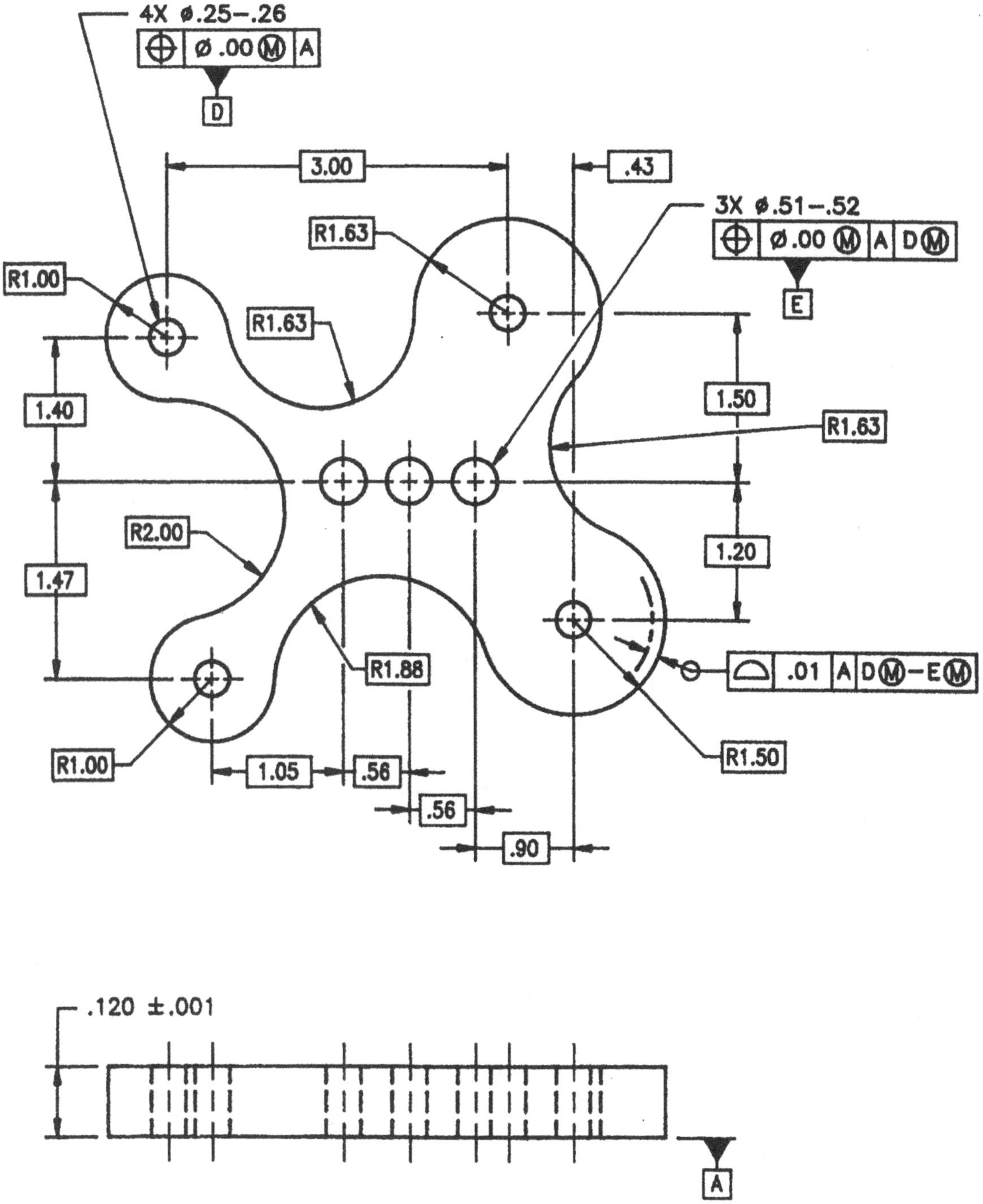

25. (continued)

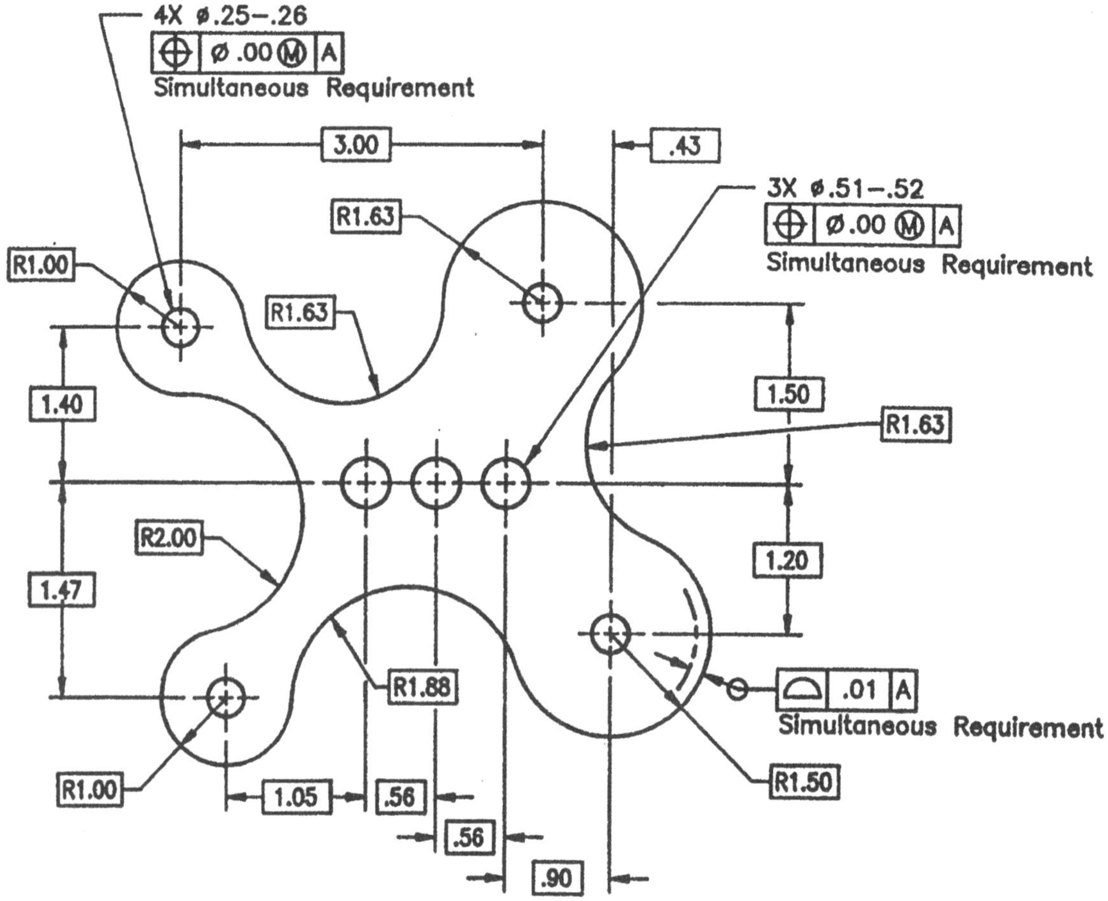

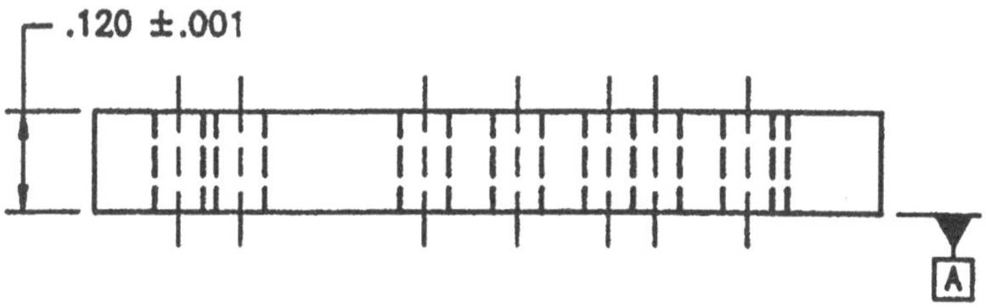

26. Position to Control Symmetry: Answer the questions about the part below as produced.

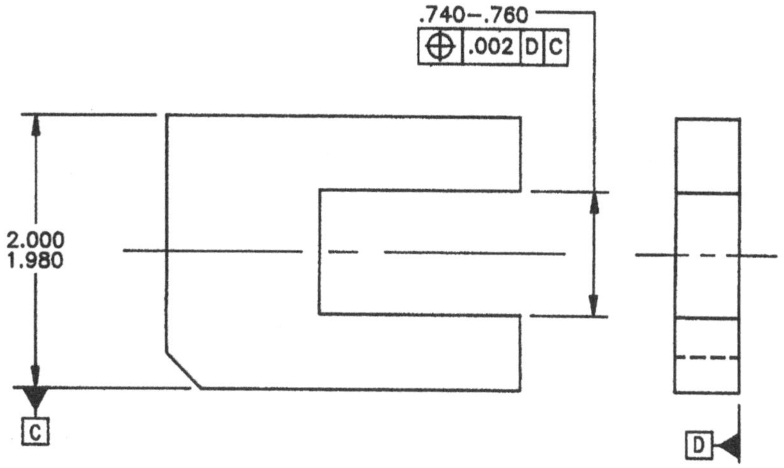

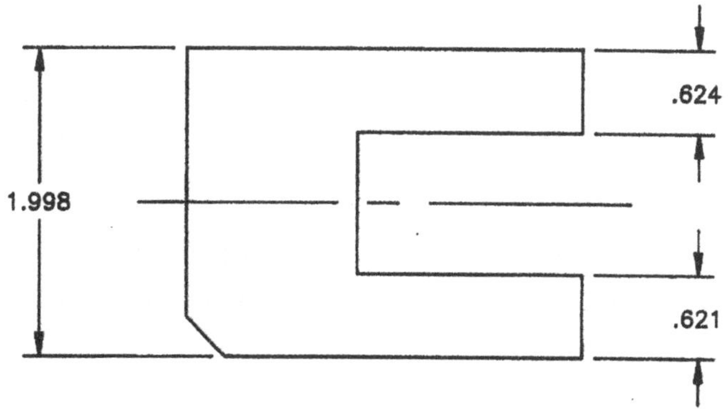

1) What dimension represents the center of the 1.998 width of the produced part? (Hint: divide the 1.998 by 2.)
2) What is the size of the part, not including the slot size?
3) What is the slot size?
4) What is half the slot size?
5) How far is the slot centerplane from the top of the part?
6) How far is the slot centerplane from the bottom of the part?
7) Starting with the answer to Question 1, as a mean dimension, give the limits of the tolerance zone. __________ ± ________ as plus and minus
__________ - ________ as a limit dimension
8) Is the slot centerplane within tolerance?
9) If not, what is the minimum that can be removed from the top of the slot to be made acceptable per the position control? Use your answers to Questions 1 through 7 above to prove your answer.

27. Please answer the following questions as they pertain to the illustration on Page 58.

    1) If the holes on the 1.600 diameter bolt circle have been produced at a distance from datum axis B of .820, under what conditions might the holes meet their positional tolerance?

    2) If the 2 holes opposite each other on the bolt circle are produced at a distance apart of 1.630 and the holes measure a diameter of .578, do they meet their positional tolerance?

    3) If the center hole is produced at a diameter of .586 and is off center of datum axis B a radius of .017, it is within its positional tolerance?

    4) If the part thickness measures .626 with a vernier caliper and curves like a banana .006 in the right side view, does it meet its size requirements?

    5) If datum feature B is produced at a diameter of 2.976 and is perfectly perpendicular to datum A, how much pattern shift would be allowed for the 5-hole pattern?

    6) If the outside diameter is produced at a size that will fit into a hole that is 2.980 and seat on datum feature A with three high points of contact, how much perpendicularity tolerance is the O.D. allowed?

    7) If none of the allowed perpendicularity tolerance in question #6 was used by the O.D., what could the tolerance be used for instead?

    8) If gage pins were inserted into two opposite holes on the 1.600 bolt circle and a measurement taken over the pins was 2.230, could the holes be within their positional tolerance? Under what circumstances?

    9) If two holes on the bolt circle were produced at a diameter of .580, what would their maximum allowed axial separation be?

    10) What is the maximum amount of pattern shift for the five hole pattern that may ever occur on this part?

27. (continued)

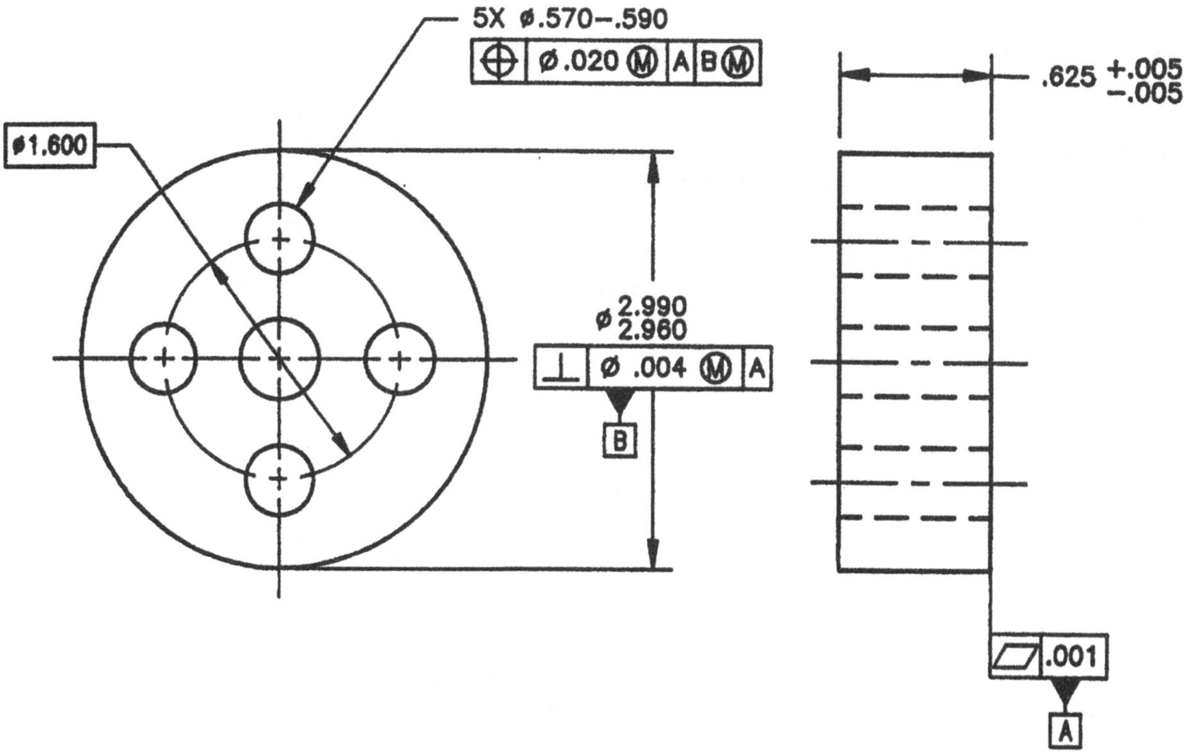

28. Fixed Fastener Assembly: Tolerance the adaptor depicted in the detail drawing callout, saving a portion of the positional tolerance for the next part in the assembly shown in the assembly drawing. To calculate the positional tolerance, use an 8mm screw.

Extra information: The bore that mates with the adaptor's 49.95-50.00 diameter has a virtual condition of 50.00.

Sequence reminder:

Step 1 – assign a primary datum feature and give it a flatness tolerance of 0.03.

Step 2 – assign a secondary datum feature and give it a perpendicularity control. Calculate the tolerance using the information given above.

Step 3 – position the 4-hole pattern to the datum reference frame created in Steps 1 and 2.

Assembly

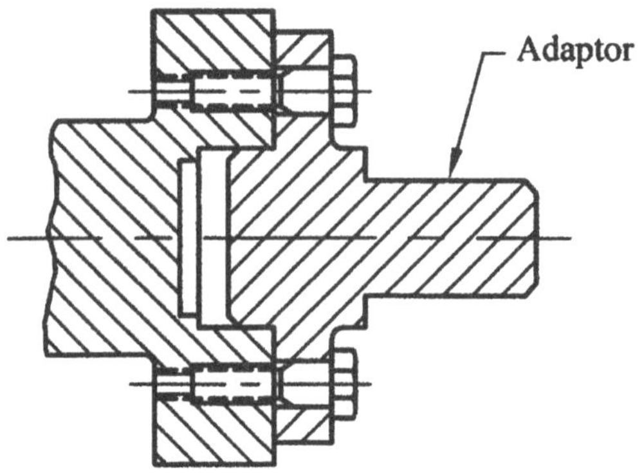

Detail drawing callout

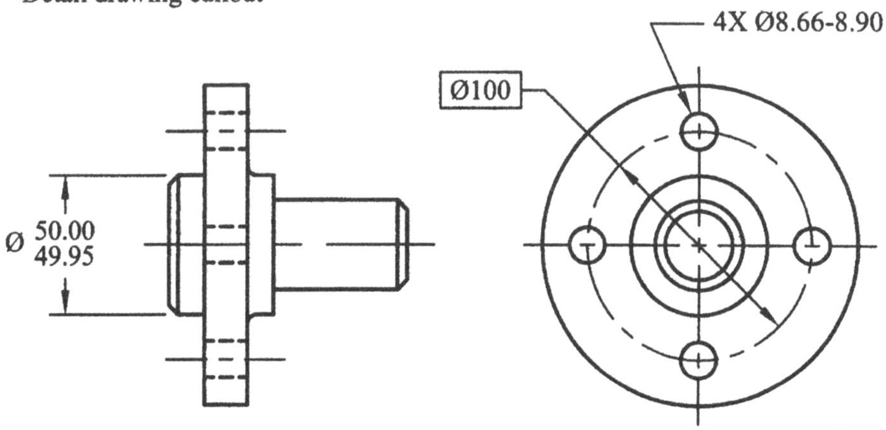

29. Center Datum Feature of Size Mating Part Design:
    1) Consider a flatness control for datum features A and C of 0.01.
    2) Using the datum features shown, tolerance these parts to mate at assembly. Tolerance only those features showing size dimensions. Give the location datum features orientation controls only.

Assembly | **Detail drawing callouts**

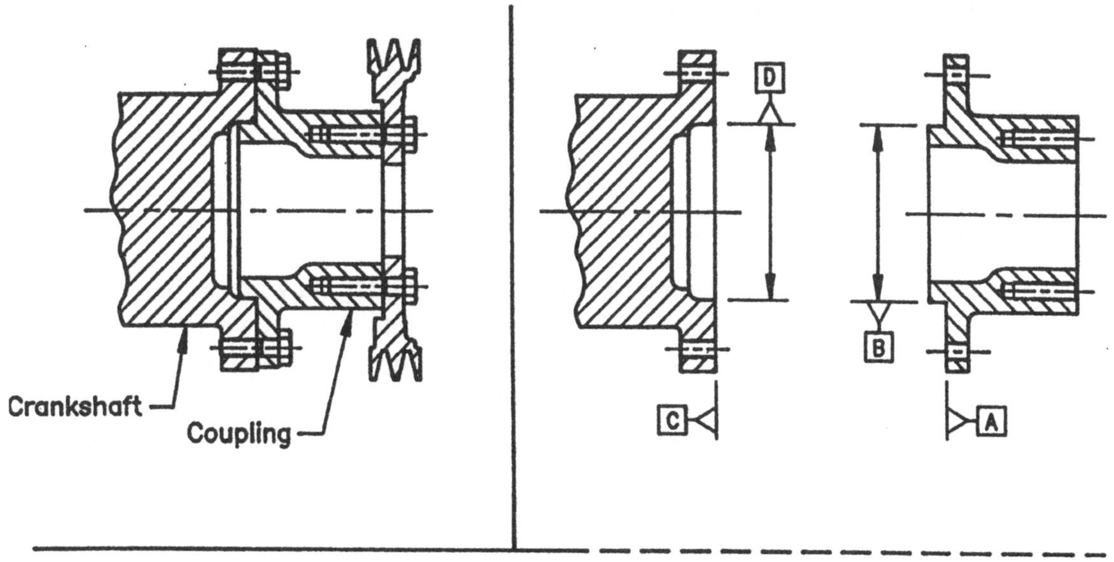

Detail drawing callouts

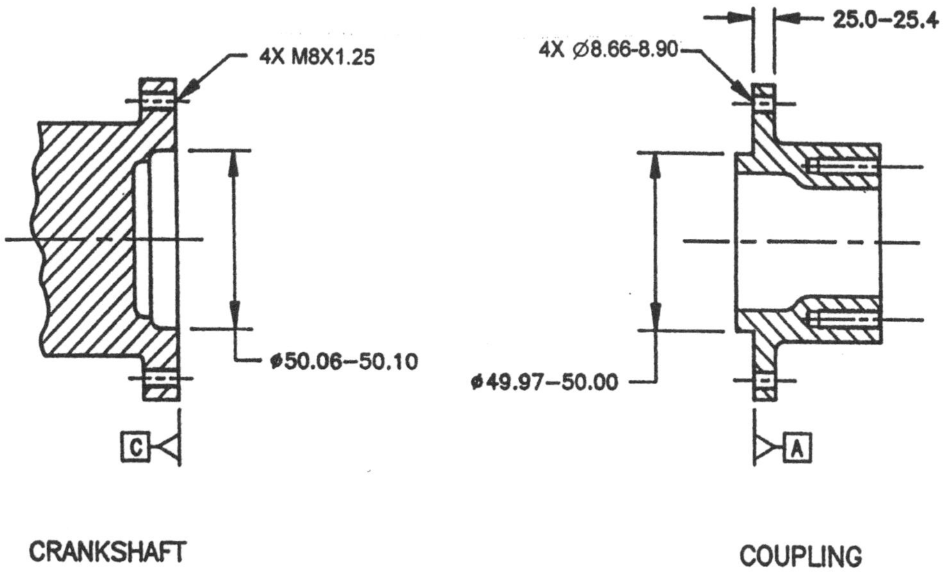

30. Multiple Patterns Made One: The following instructions are to be applied to the detail drawing callout on the next page. The purpose of this approach is to tie all ten holes together as one pattern so that they may be gaged with one functional gage to inspect their positional tolerances.

Multiple patterns of features located from the same datum references, using the same material condition symbols after those datum features referenced, are considered one pattern and, consequently, are gaged with the same gage. This is to be done on both hole patterns so all ten holes are considered one pattern.

1) Give the primary datum feature a flatness tolerance of 0.03.

2) Consider parallelism of 0.05 between surfaces into which holes are produced.

3) The virtual condition of the mating hole for the 99.95-100.00 shaft is 100.00. Assign this shaft a perpendicularity tolerance.

4) Locate the ten holes with geometric controls that dictate they are to be considered one pattern of features. Calculate appropriate positional tolerances where applicable. **Save a portion of the available tolerance for the mating parts depicted in the assembly drawing.** The threaded holes mate with holes of a diameter of 8.6-8.8 on a part where the thickness of that mating part is 50.0-51.0. Use your own judgment for missing screw sizes.

30. (continued)

Assembly

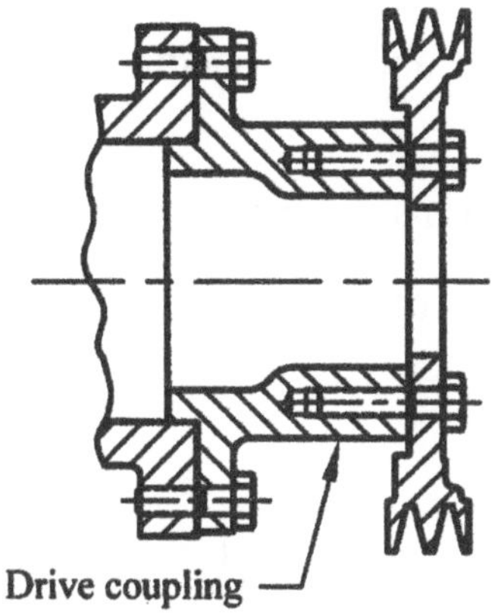

Detail part and gage for checking position of M8X1.25 tapped holes

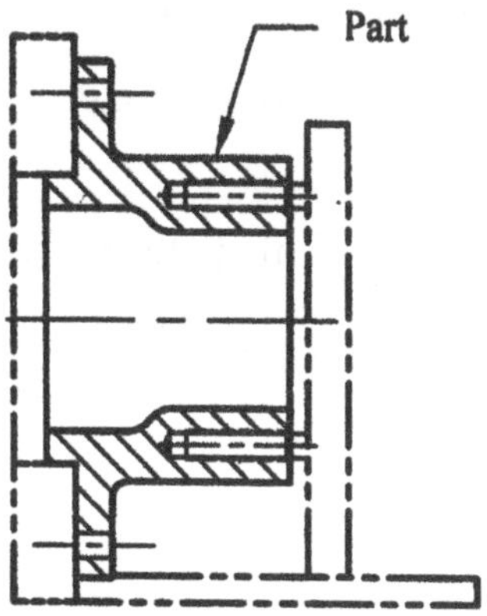

Detail drawing callouts

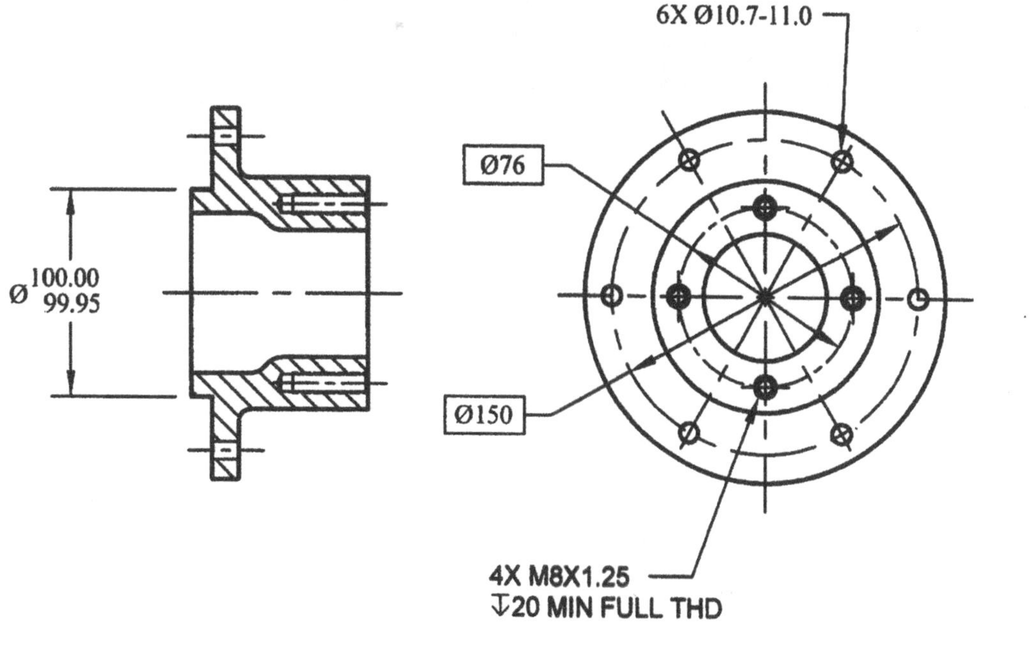

31. Center Datum Features of Size & Mating Part Tolerancing in a Fixed Fastener Assembly: Consider these parts cylindrical. Tolerance them to mate at assembly. The assembly drawing is shown below. All work is to be done on the next page on the detail drawing callouts (Part 1 and Part 2).

    1) Assign each part a primary datum feature and a flatness tolerance of 0.03.
    2) Assign each part a secondary datum feature of size and a perpendicularity tolerance that you calculate.
    3) Assign each hole pattern a positional tolerance with appropriate datum references.
    4) Assign each outside diameter a positional tolerance of 0.8 at MMC

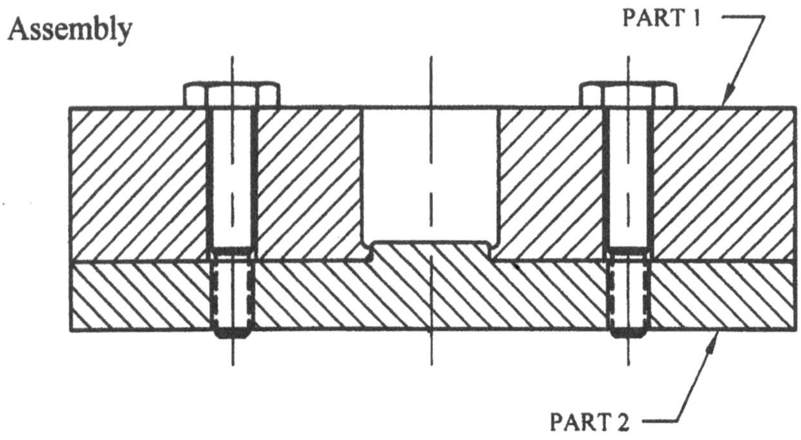

31. (continued)

PART 1

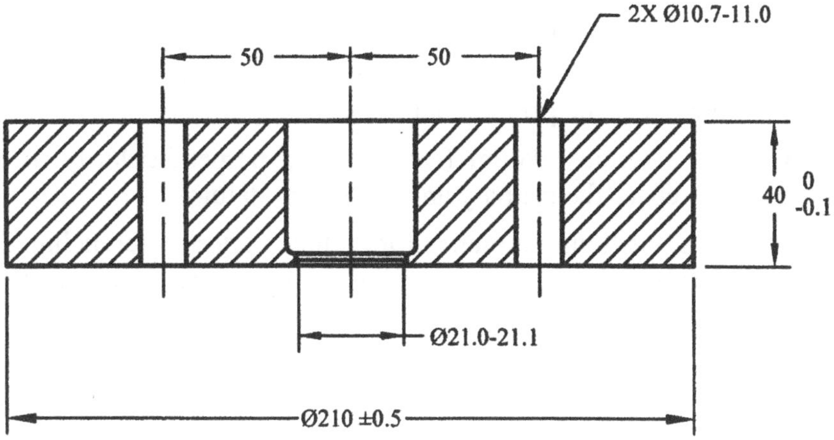

PART 2

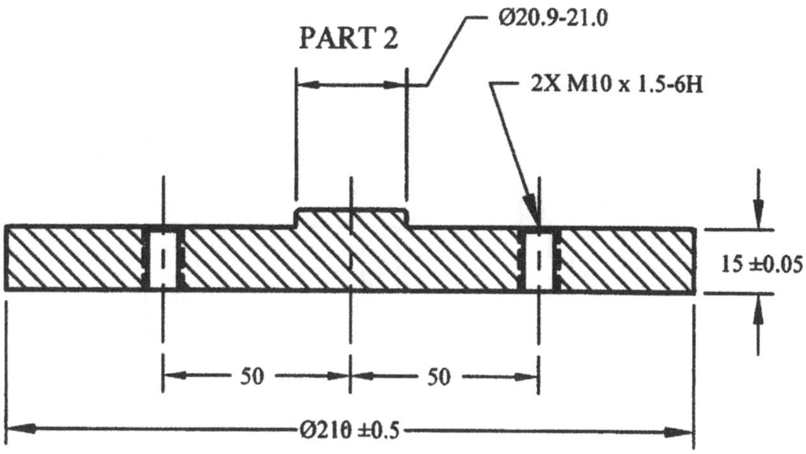

32. Centerplane Datums Fixed Fastener Assembly: Tolerance the following rectangular parts to mate at assembly. The assembly drawing is shown below. All work is to be done on the next page on the detail drawing callouts (Part 1 and Part 2).
    1) Assign each part a primary datum feature and a flatness tolerance of 0.03.
    2) Assign each part a secondary datum feature and a perpendicularity of the centerplane tolerance that references the primary datum.
    3) Assign each part a tertiary datum feature and a perpendicularity of the centerplane tolerance that references the primary and secondary datums.
    4) Assign each part a positional tolerance for its hole pattern referencing all three datums. Use maximum material condition symbols everywhere it is appropriate. Use a projected tolerance zone on the threaded holes.
    5) Assign any remaining widths positional tolerances to within 1mm regardless of feature size. Reference only two datum features in each of these controls. Use no maximum material condition symbols in this step.

<u>Assembly Drawing</u>

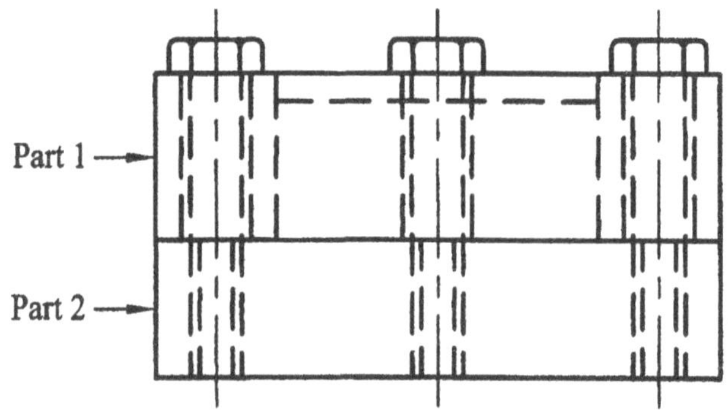

32. (continued)

Part 1

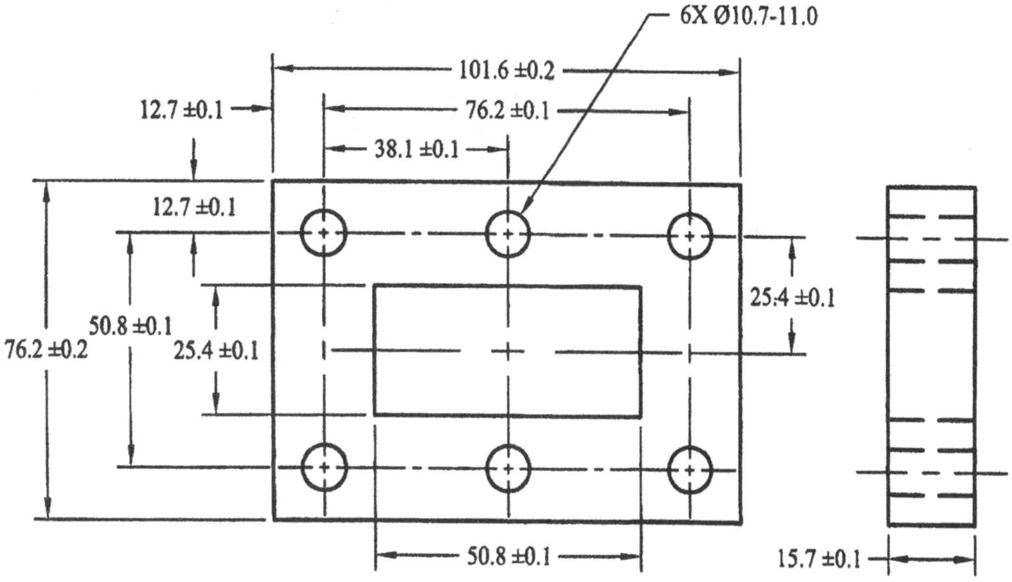

UNLESS OTHERWISE SPECIFIED ALL ANGLES ARE ±1°

Part 2

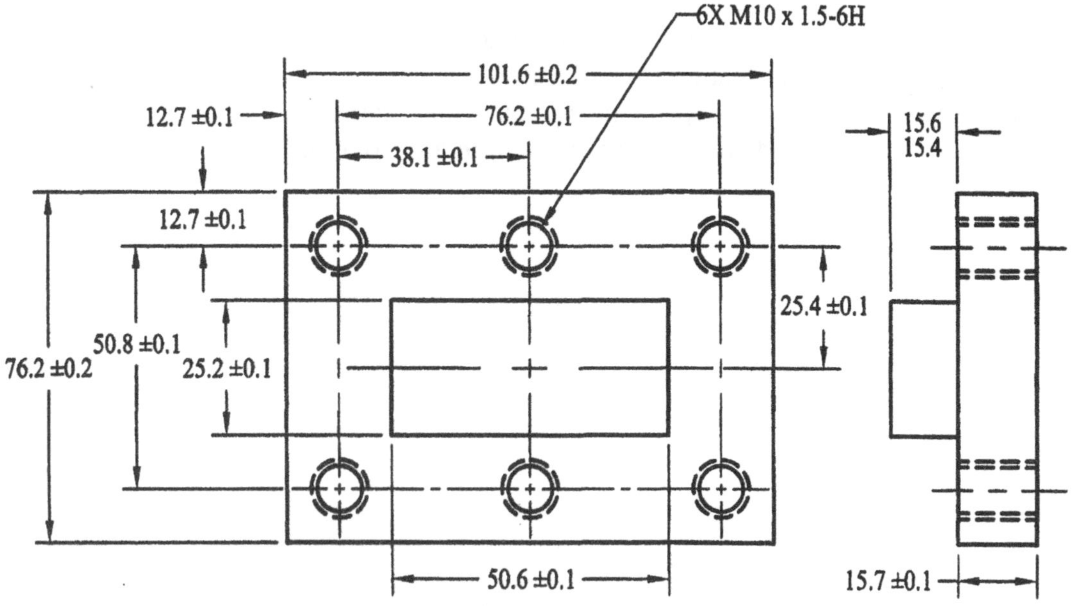

UNLESS OTHERWISE SPECIFIED ALL ANGLES ARE ±1°

33. Punch Pattern (Positioning Rectangular Holes): The virtual condition of the shaft that will mate with the ∅.250-.260 holes is ∅.250. The virtual condition of the rectangular shafts that mate with these rectangular holes is .690 x .890. Use exactly the same datum feature references for the .700-.710 widths as you use for the .900 -.910 widths. The illustration in this exercise appears on the following page.

   1) Assign this part a primary, secondary and tertiary datum feature as the edges of the part. Use the surface marked 'mating surface' and the two edges from which the 1.00 dimensions originate.

   2) Position the ∅.250-.260 hole within a positional tolerance that is the diameter of a circumscribed circle about a .120 x .120 square tolerance zone. To calculate this, use the formula that says the diameter of the circumscribed circle around a square is equal to one side of the square (.120) times the square root of 2 (the square root of 2 is approximately 1.4142, but for this calculation, use 1.4).

   3) Assign the ∅.250-.260 hole a refinement of perpendicularity to the primary datum. Calculate this tolerance using the formula: MMC of the hole minus the virtual condition of the mating shaft equals the geometric tolerance for the hole. Make this hole a datum feature.

   4) Assign the 6 x .700 -.710 widths a positional tolerance to
      a) the primary datum feature (for perpendicularity)
      b) the hole described in steps 2 and 3 above (for location)
      c) one side of the part (for angular orientation/anti-rotation).

   5) Assign the 6 x .900 -.910 widths a positional tolerance to
      a) the primary datum feature (for perpendicularity)
      b) the hole described in steps 2 and 3 above (for location)
      c) one side of the part (for angular orientation/anti-rotation).

33. (continued)

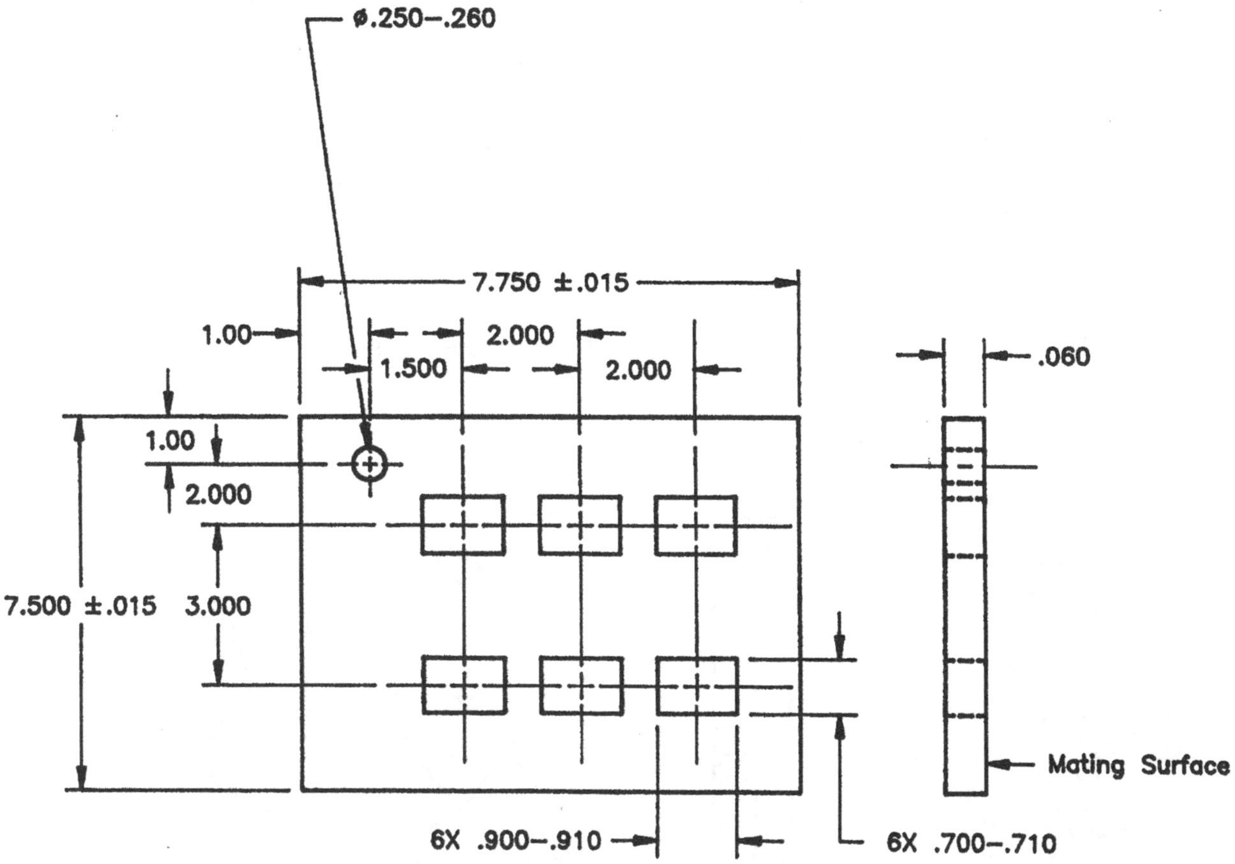

.XX = ±.06
.XXX = ±.005
∠ = ±1°

34. Complete the following:
    1) Assign a primary datum feature and give it a flatness tolerance of 0.03.
    2) Assign a secondary datum feature and relate it back to the primary datum.
    3) Assign a tertiary datum feature and relate it back to the primary and secondary datums.
    4) Position the four-hole pattern to this datum reference frame.

[**NOTE:** The MMC of the mating key is 4mm. The virtual condition of the shaft that fits into the center bore on this part is ∅50mm. Assume the four-hole pattern will be bolted to the next part in the assembly using 8mm screws. Save some of the available positional tolerance for the mating part.]

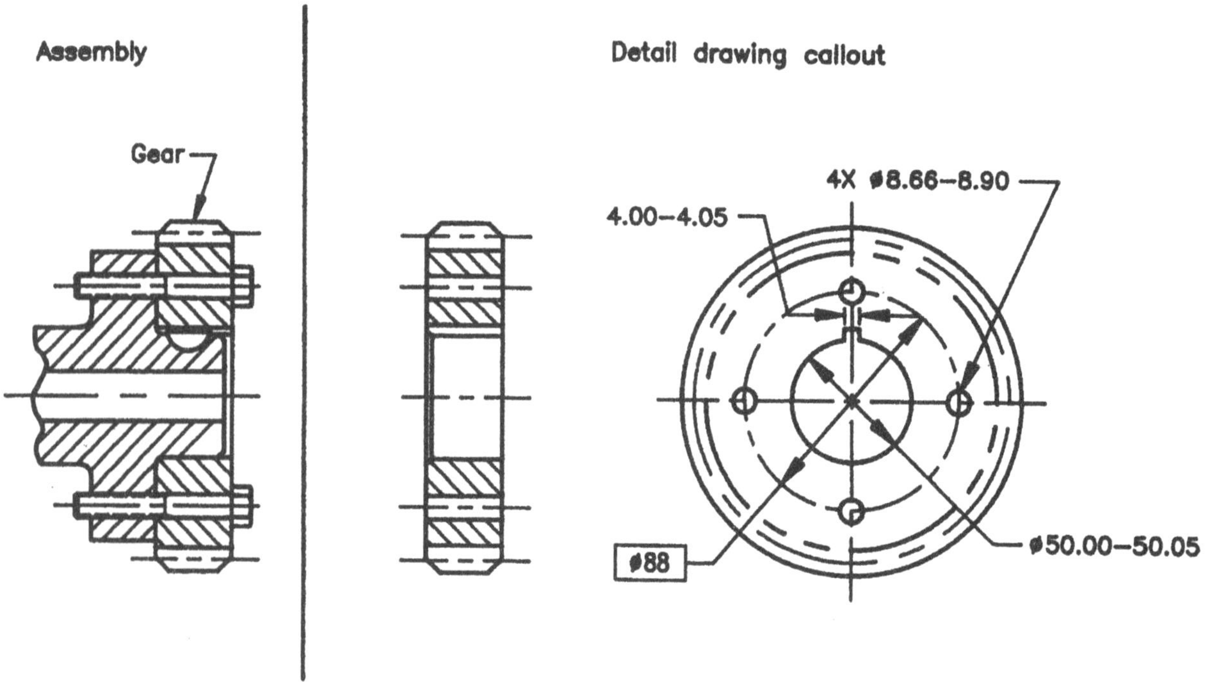

35. Combined Positional and Perpendicularity vs. Composite Positional Tolerancing
    1) What would be the difference between the composite positional control given to datum feature D and a control that, instead, combined position to within a diameter of .060 at MMC to A, B and C on the top and perpendicular to A to within .000 at MMC on the bottom? Which would be most appropriate?
    2) In controls such as the perpendicularity of C to A, the word 'Total' was used in the distant past inside of some feature control frames. What was meant by the word 'Total'?
    3) What purpose does datum feature C serve in the 3-hole pattern's positioning?
    4) What would be the result of leaving datum feature C out of the control?

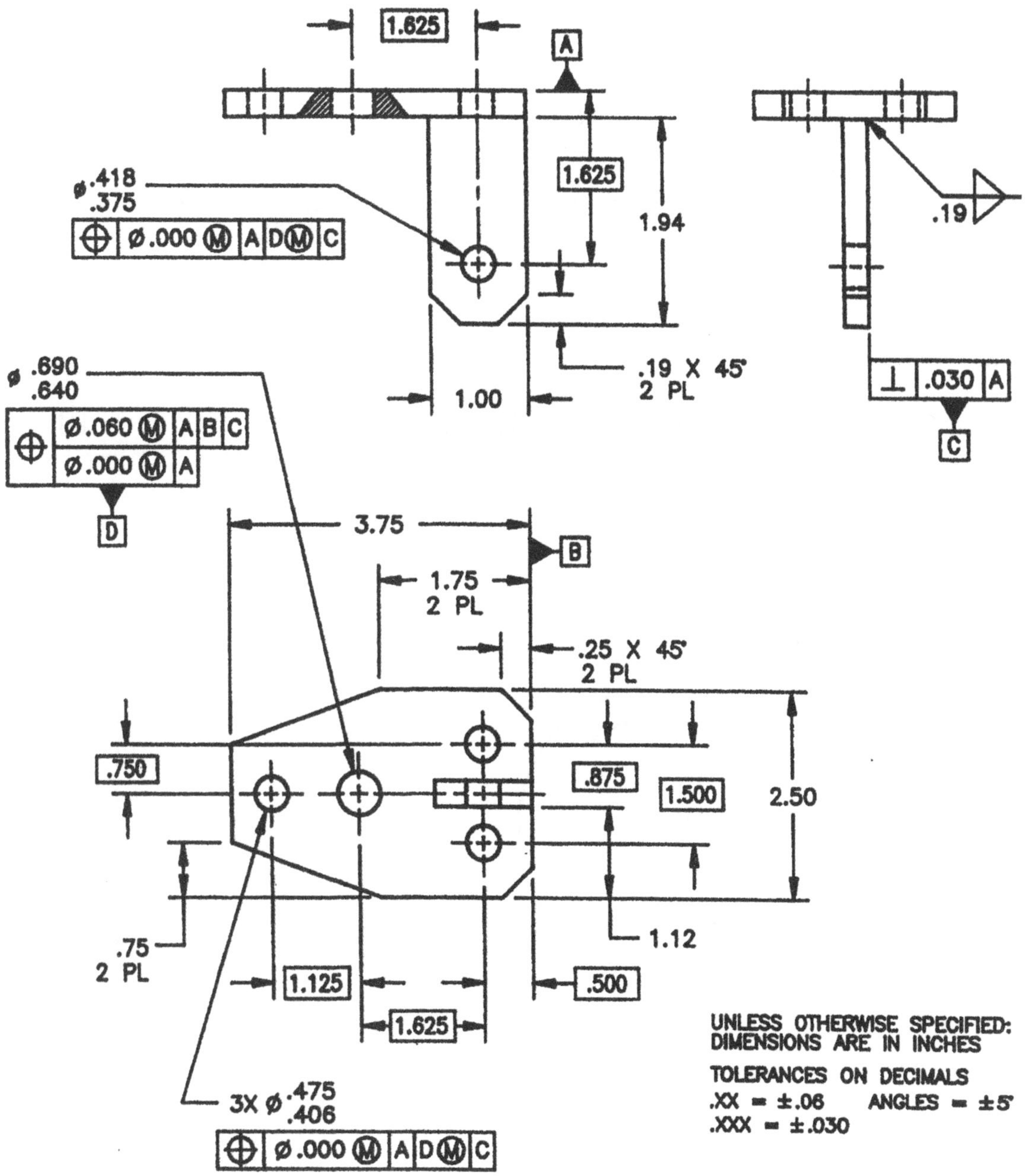

36. Square Shaft Positioning; Mating Part Tolerancing: Tolerance the part on the bottom of the page using the following steps:
    1) Assign a flatness control of .001 and primary datum feature.
    2) Make all maximum material conditions the same as those given the part on the top of the page. Make the least material condition of each feature larger than the MMC by the same size tolerance given to the part on the top of the page.
    3) Assign a secondary datum feature and a perpendicularity tolerance.
    4) Assign a positional tolerance in each appropriate direction for the square hole. Assign the necessary basic dimensions.

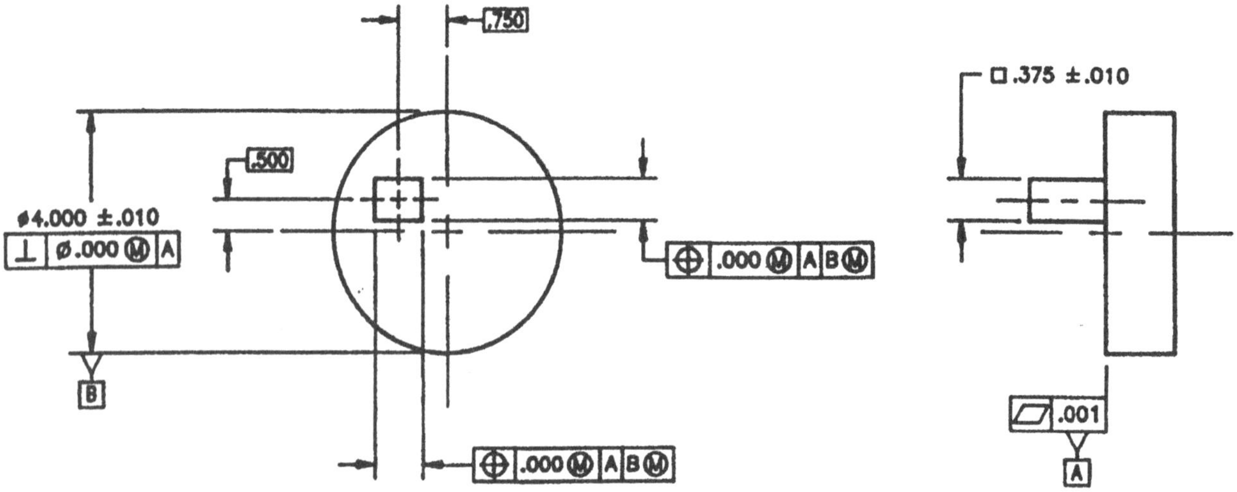

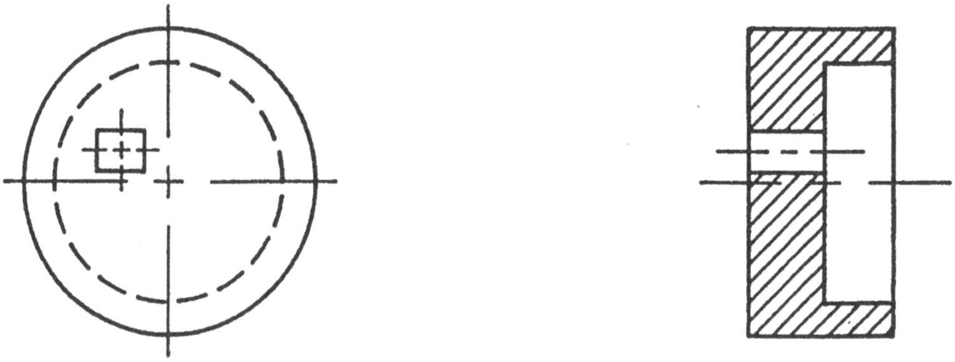

37. Problem #1 Floating Fastener Assembly Condition   The two part assembly shown below must mate with the following requirements:
    1) The inner shaft must go into the outer shaft.
    2) The planar end of the ∅37.95-38.00 inner shaft must bottom out in the ∅38.05 – 38.12 outer shaft's hole.
    3) The ∅6.25 clearance holes in both parts must align well enough to insert the ∅6.0 bolt. This bolt and its accompanying nut are used to bind the parts together. Since the bolt is not fixed at any angle by either part's ∅6.25 holes, this is considered a floating fastener assembly condition. Use a single line of position tolerance for each part's ∅6.25 holes.
    4) The title block tolerance for angles is $\pm 1^{\circ}$.

a) Assembly

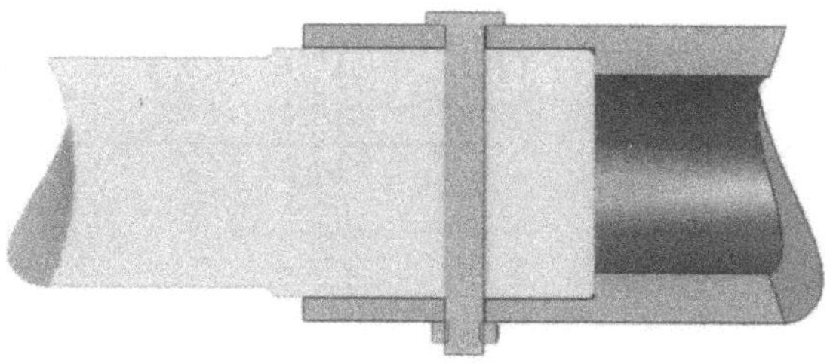

b) Section view of Assembly

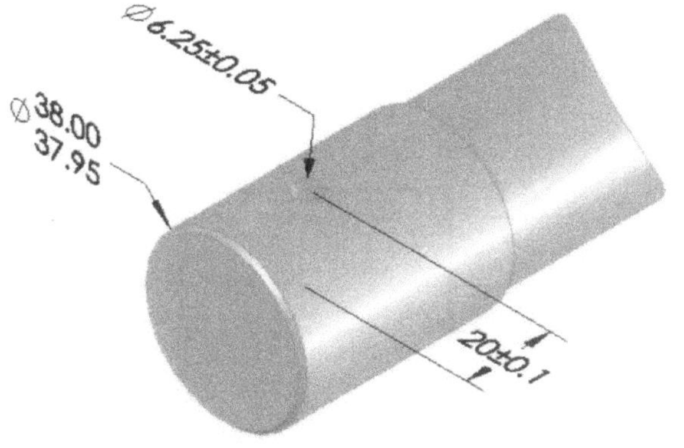

c) Inner shaft plus and minus tolerance scheme

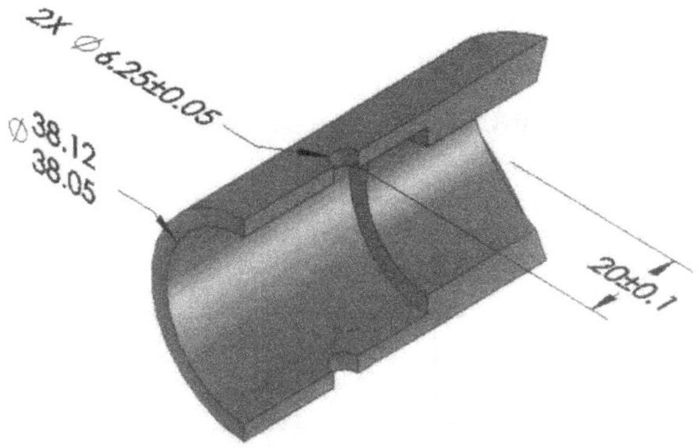

d) Outer shaft plus and minus tolerance scheme

SolidWorks Corporation

37. Problem #2 Mating Part Design; Holes Perpendicular to Part Axis

1) Assign the primary datum feature on each part as the large mating diameters.
2) Assign a secondary datum feature on each part as the end of the part.
3) Assign a two-single segment position tolerance on the 6.2-6.3 holes on each part. In the upper level segment, use a 1mm position tolerance to both datums. In the lower level segment, calculate the position tolerance using the floating fastener formula and reference only the functional diameter datum feature at MMC.

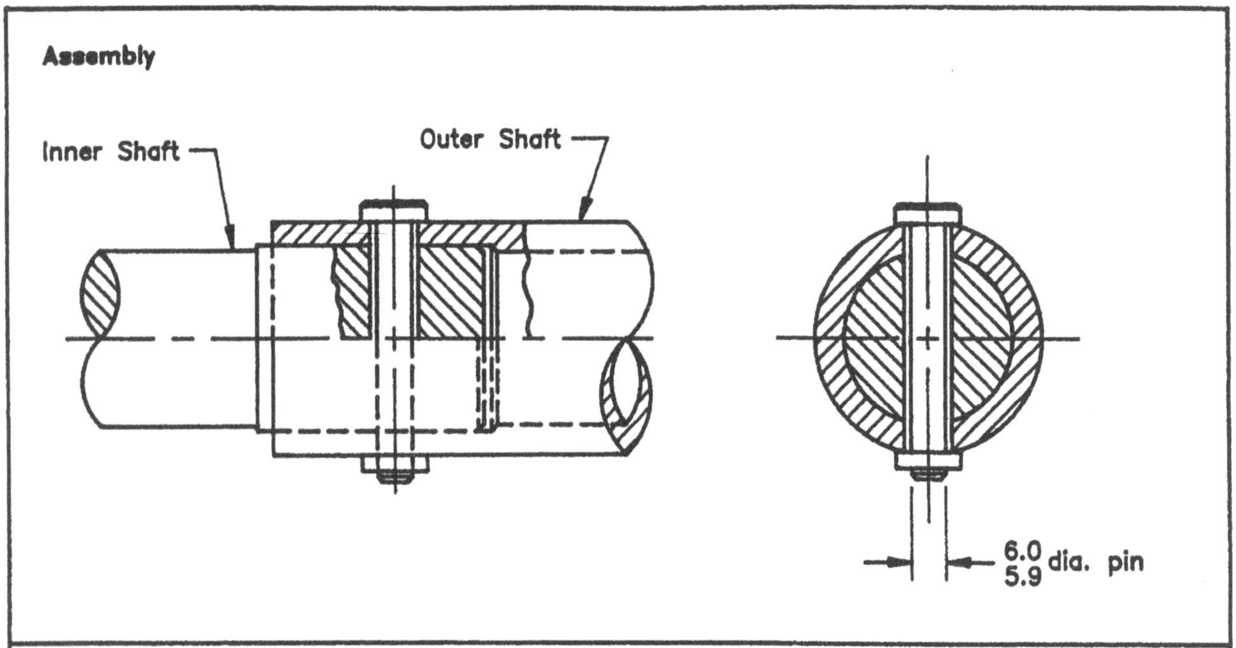

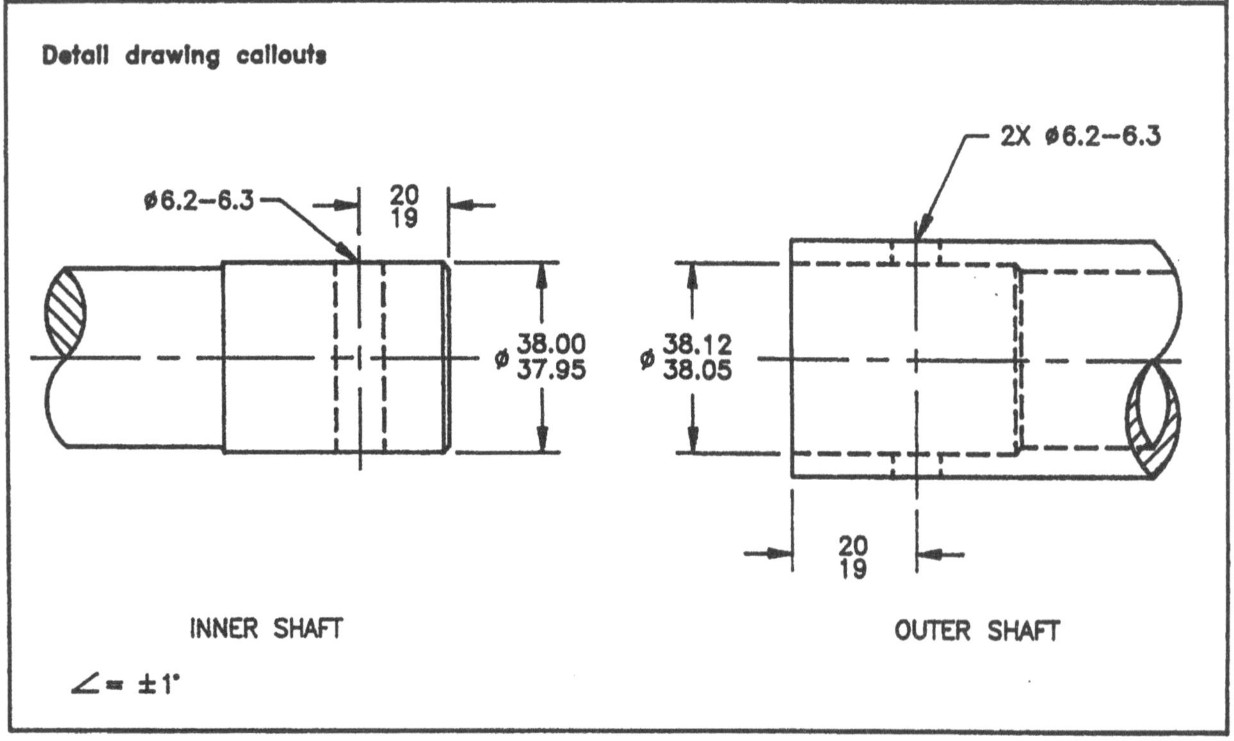

38. Multiple Datum Reference Frame Complete the detail drawings on the next page using the following information:

1) Give a flatness tolerance to datum feature A of 0.03mm. Give a perpendicularity tolerance of 0.03mm to datum feature C. Give datum feature D a combined control of position to within 1mm at MMC followed by a perpendicularity tolerance requirement to be calculated. The mating shaft has a virtual condition of ∅50mm.
2) Assign a two-single segment position tolerance for the ∅8.04-8.07 holes. Use 1mm of position tolerance to three datums in the upper level control. In the lower level control, calculate the position tolerance between the holes and to the primary datum using the fixed fastener formula. These holes use 8mm screws.
3) Give the three threaded holes a two-single segment position tolerance. In the upper level control, reference three datum features and use a 1mm at MMC position tolerance. In the lower level control, reference two datum features and calculate the position tolerance using the fixed fastener formula. Assume 8mm screws are to be used. Save some of the tolerance for the mating ∅8.04-8.09 holes. The screws must pass through 25.4mm of material before reaching the threaded holes, so use a projected tolerance zone in both levels of control.

[**NOTE:** The assembly drawing appears below and the detail drawings appear on the next page.]

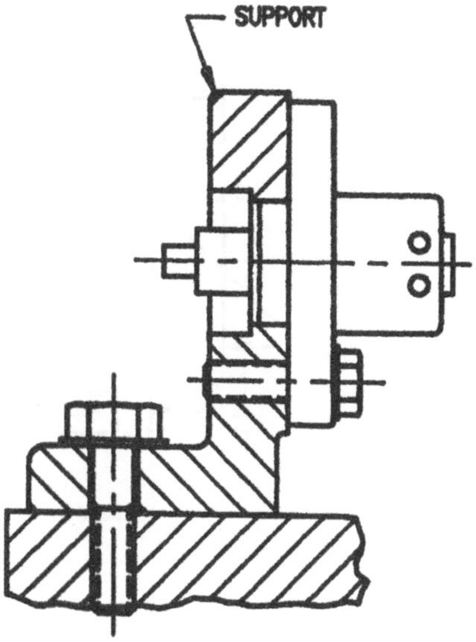

38. (continued)

Detail drawing callout

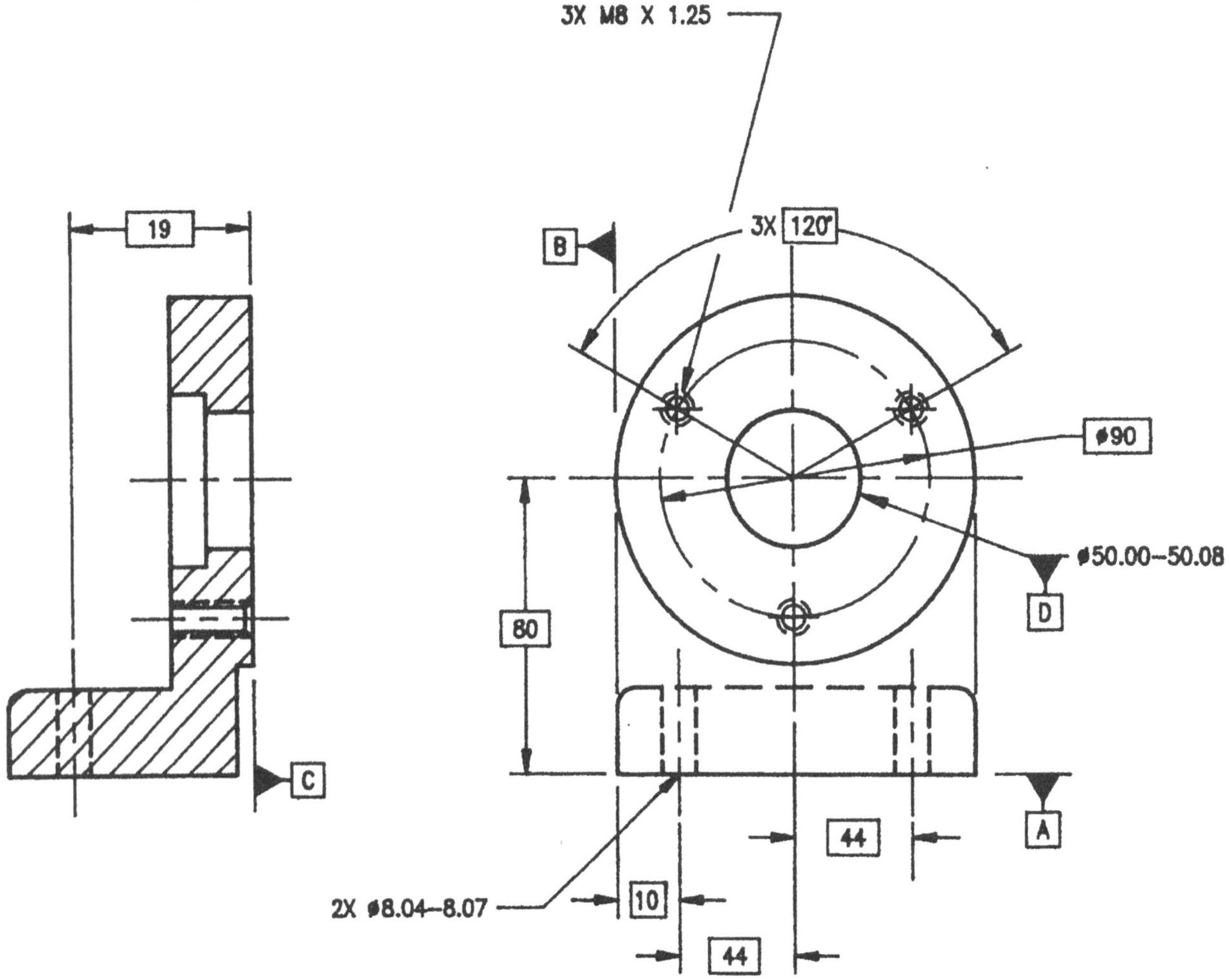

39. Secondary and Tertiary Datum Features of Size
    1) Make the bottom of the part the primary datum feature and give it a flatness tolerance of 0.03mm.
    2) Make the four surfaces co-planar to one another and parallel to the primary datum to within 0.1mm.
    3) Position two of the ∅15.6-16.0 holes to each other and the primary datum. The mating pins (for all four of these holes) have a virtual condition of ∅15.4. Make these two holes the secondary and tertiary datum features. Indicate them with the letter 'X'.
    4) Position the two remaining ∅15.6-16.0 holes to the established datum reference frame.
    5) Position the 3X ∅8.6-9.0 holes to the established datum reference frame. The virtual condition of the mating pins is ∅8.4.
    6) Profile the outer periphery of the top view to the established datum reference frame to within 0.5mm.

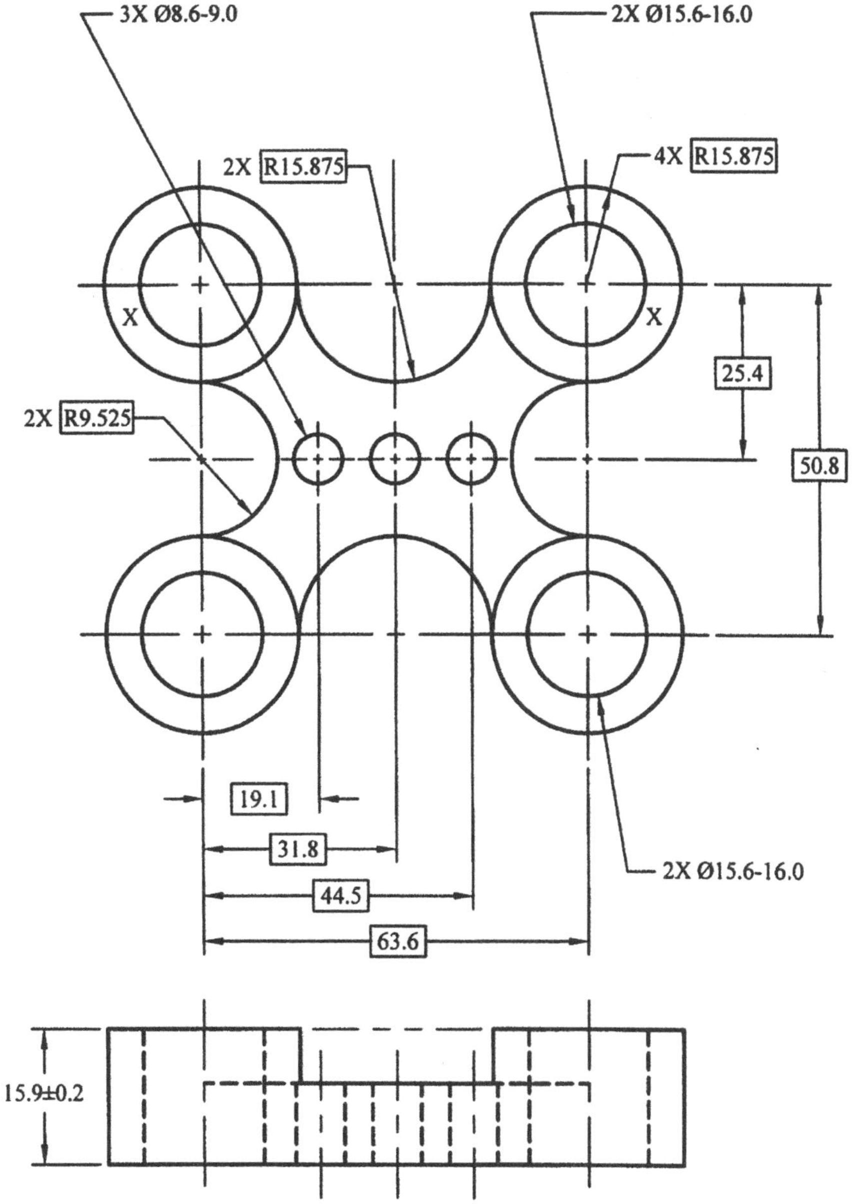

40. Sheet Metal Part #9 Using the sheet metal drawing on the next page:
    1) Assign a primary datum feature. Control its form to within .010.
    2) Assign a secondary and a tertiary datum feature as two of the six holes. Relate them to each other and the primary datum by giving the secondary datum feature a perpendicularity control and the tertiary datum feature a position control. [**NOTE:** These six holes bolt the part to six threaded holes which have yet to be toleranced. The assembly will use #10-24 screws (MMC of ∅.190).]
    3) Locate the four holes that remain to the datum reference frame. Use the MMC concept where possible.
    4) Locate the .06 x 5.22 slot to the datum reference frame. Make it part of the same inspection set-up used for the features in item 3 above. Give it a (MMC concept) virtual condition of .03 x 5.19. Use a local note stating BOUNDARY beneath the position control.
    5) Form, orient, locate and size the profile in the front view to within .02 per surface. Use the datum reference frame already established. Make it a simultaneous requirement with the features in items 3 and 4 above.

40. (continued)

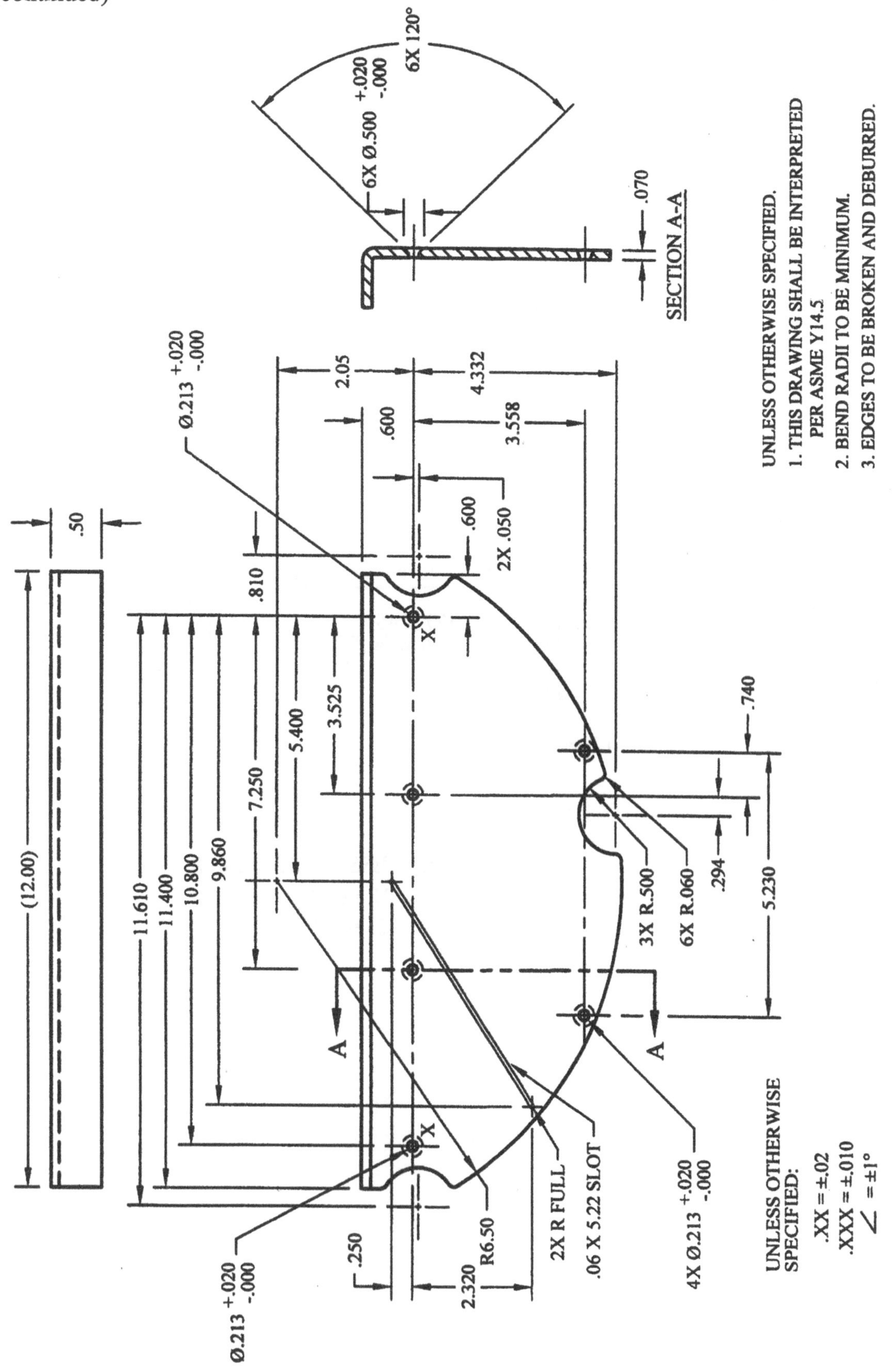

41. Sheet Metal Part #8 Using the sheet metal drawing below:
    1) Convert this form from plus and minus tolerancing to position and profile. It is a fixed fastener situation in that these three clearance holes bolt to three threaded holes. The virtual condition of the threaded holes using a projected tolerance zone is ∅.128.
    2) The workpiece is to be inspected while restrained.

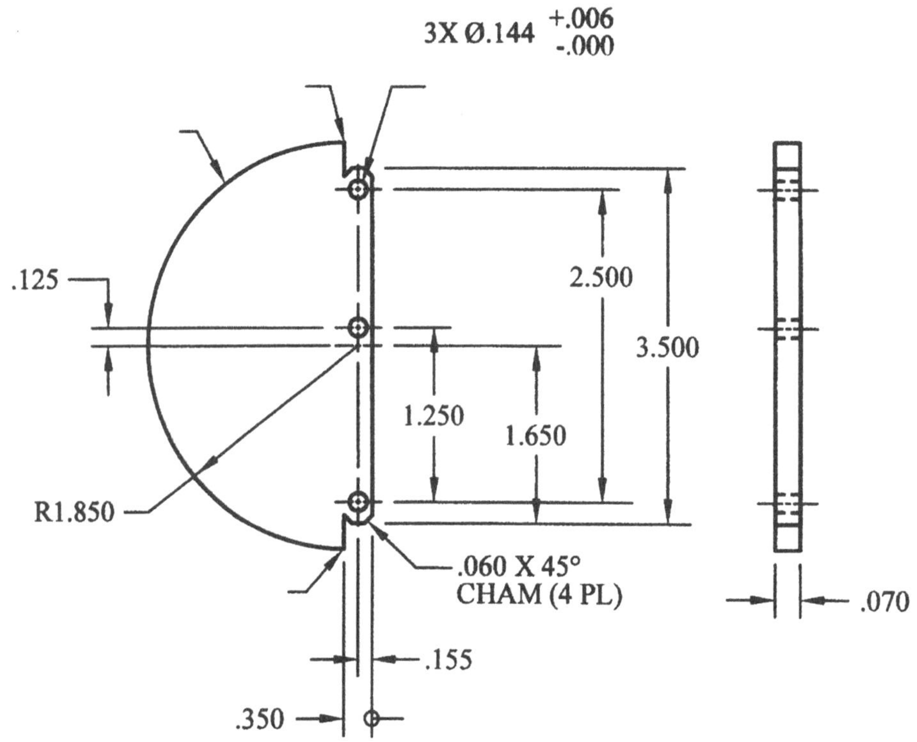

UNLESS OTHERWISE SPECIFIED:
.XXX = ±.010
∠ = ±1°

42. Sheet Metal Part #6 Using the sheet metal drawing on the next page:

1) Assign eight datum target areas to establish a primary datum. Make the areas 25 x 25mm.
2) Position the four elongated holes (using the 'BOUNDARY' method) to each other and the primary datum. The four holes should have a virtual condition (MMC concept) of 20 x 50mm. Make the width of each hole (the 20mm) a datum feature.
3) Profile the rest of the part simultaneously in all views to the primary datum and the common plane formed between two of the 20mm widths for the secondary datum and the common plane formed between the other two 20mm widths for the tertiary datum to within 1mm.
4) Add a general note to indicate workpiece is to be restrained during inspection.

42. (continued) Sheet Metal Part #6

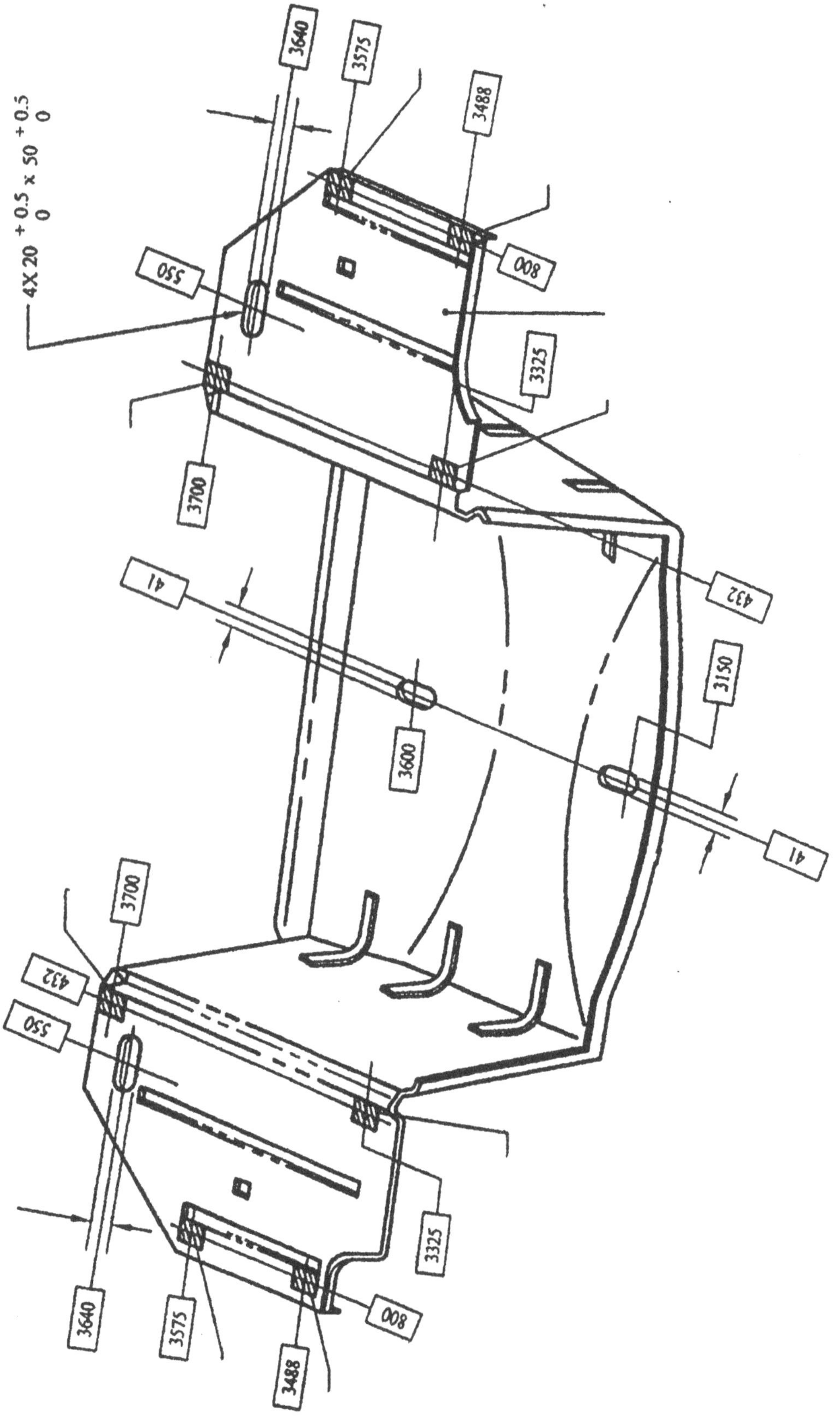

43. Sheet Metal Part #7 Using the sheet metal drawing on the next page:

1) Assign four 25 x 25mm datum target areas.
2) Relate the 30mm diameter hole to the primary datum to create a (MMC concept) virtual condition of ∅29.5. Designate this hole as the secondary datum feature.
3) Relate the 30 x 60mm elongated hole to the primary and secondary datums to create a (MMC concept) virtual condition 'BOUNDARY' of 29 x 59mm. Designate the 30mm width as the tertiary datum feature.
4) Assign the following controls:
   a. Position the Z holes to within a ∅1.6.
   b. Profile the entire mating surfaces to the front floor pan to within 1mm.
   c. Profile the local mating surface to the front seat mounting bracket to within 1mm.
   d. Profile the entire mating surface to the front seat to within 1.4mm.
   e. Profile the three entire mating flanges to the sill to within 1mm.

43. (continued) Sheet Metal Part #7

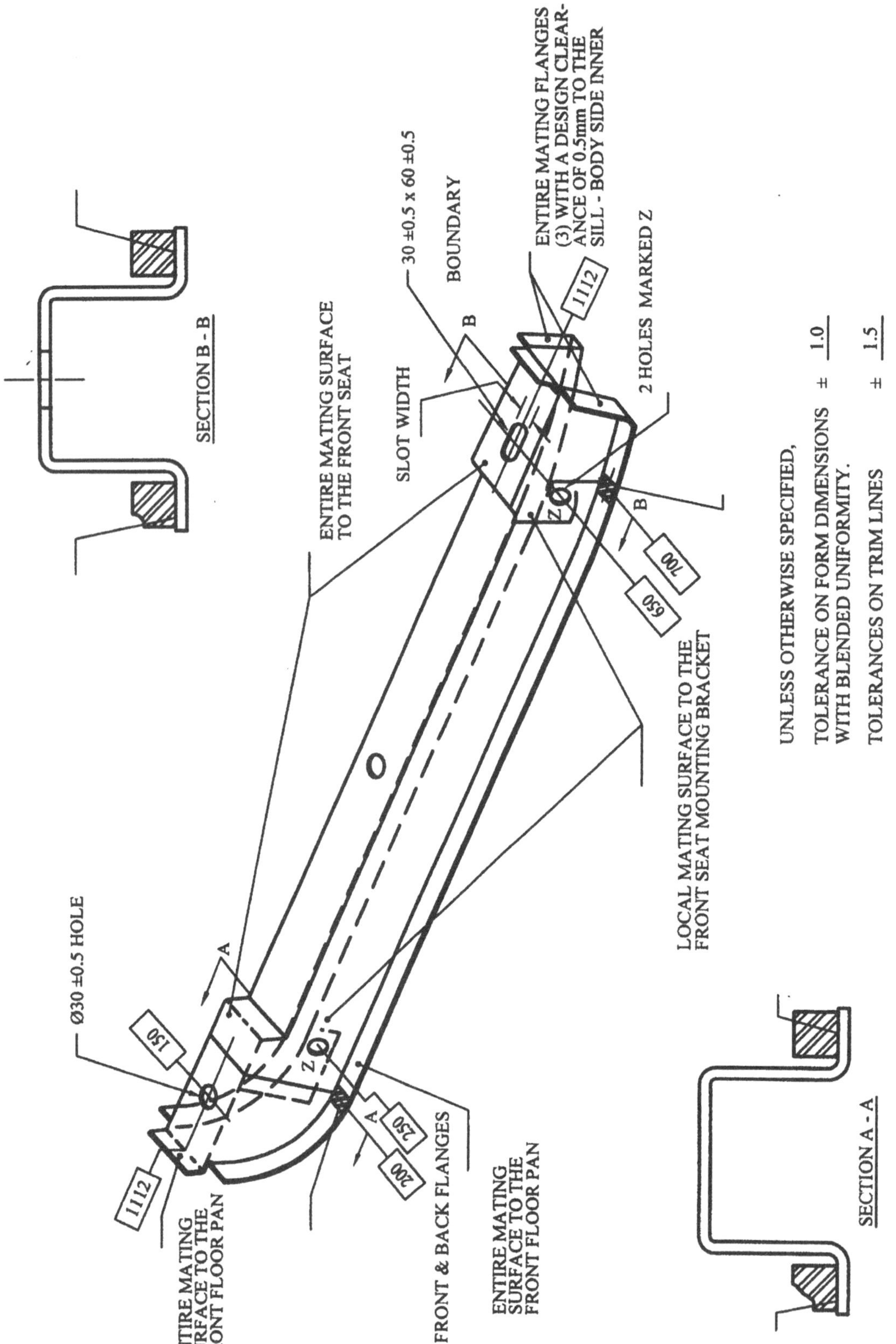

44. For the hose on the next page:
    1) Make the 20mm and 30mm internal diameter lengths datum targets. Position them to each other to within 2mm at MMC. A note may be added to the bottom of the drawing to clarify the meaning of this position control.
    2) Position the ⌀100mm outside diameter of the entire hose to create an outer boundary of 103mm. Reference datum feature A at MMC. Notes may be added to explain the extent of this positional boundary control.
    3) Add a note to the drawing explaining that unless otherwise specified, all CAD dimensions are to be considered basic dimensions.

44. (continued)

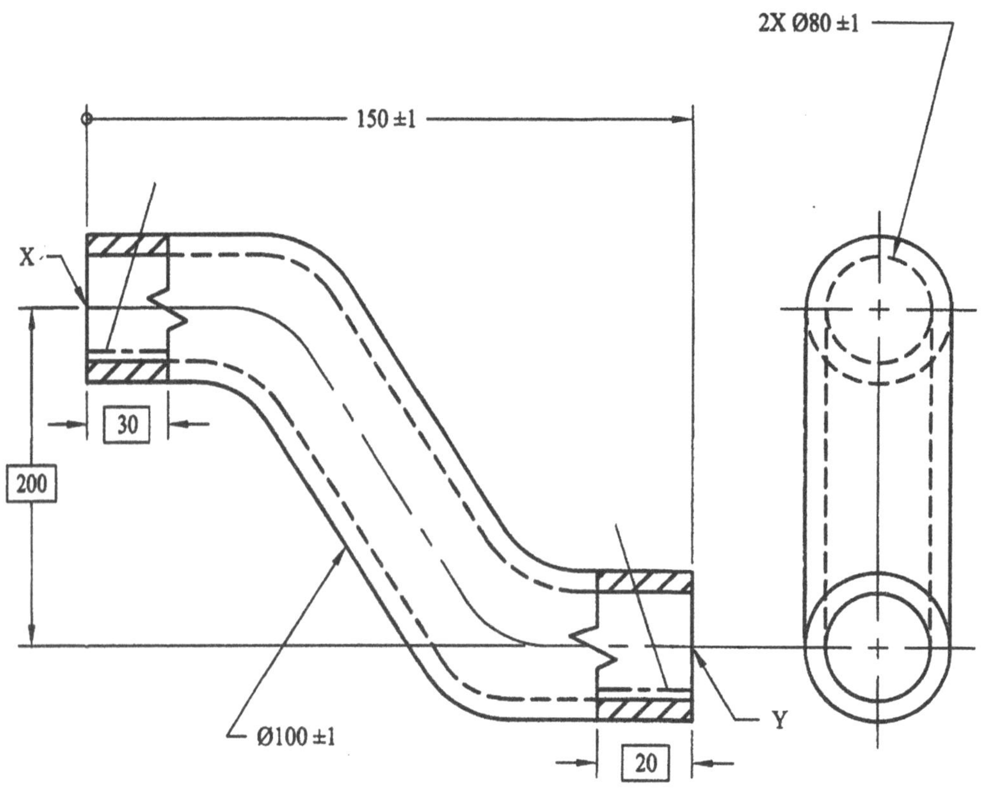

NOTE 1: FOR COMPLETE DEFINITION OF THIS PART REFER TO THE CAD MODEL.

45. For the gage on the next page:
    1) Assign a primary datum feature and give it a flatness control that is 10% of the part tolerance.
    2) Assign the oddly configured shaft as a secondary datum feature. Dimension all its radii. Assign it a profile of the surface control that uses 10% of the profile tolerance of the part feature it represents. Reference the primary datum.
    3) Dimension and tolerance the four gage pins for size and position. Use 10% of the part tolerance.

45. (continued)

Workpiece

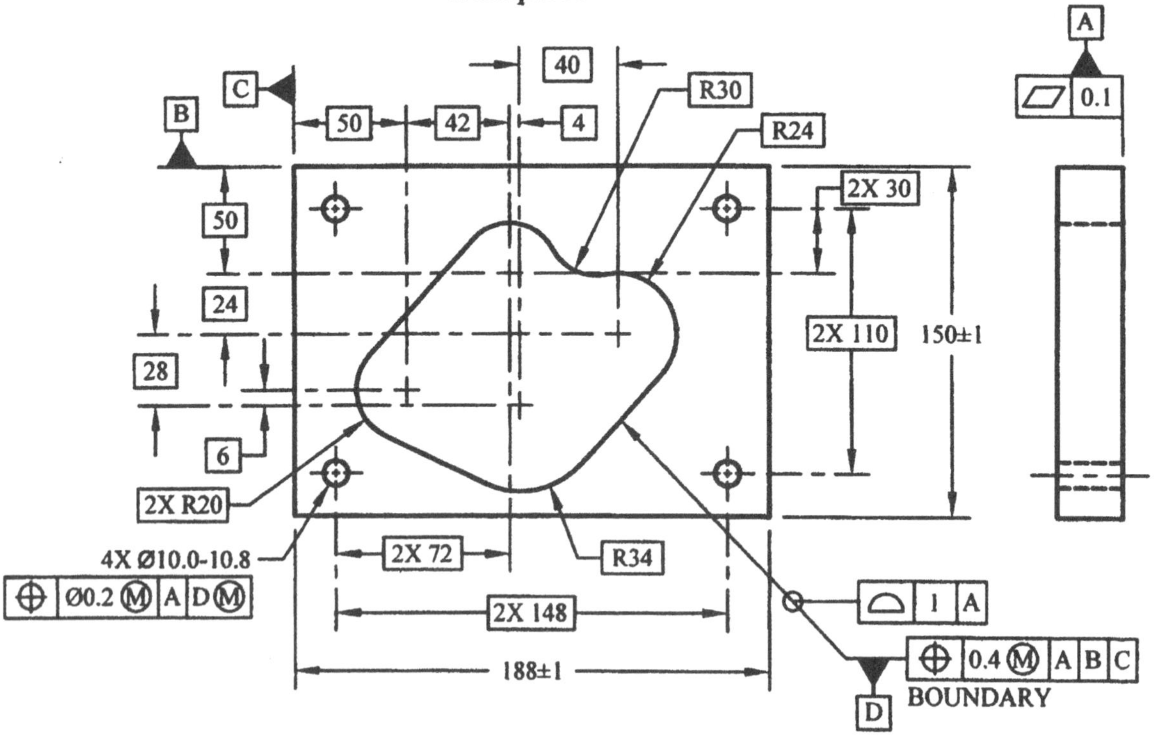

Gage

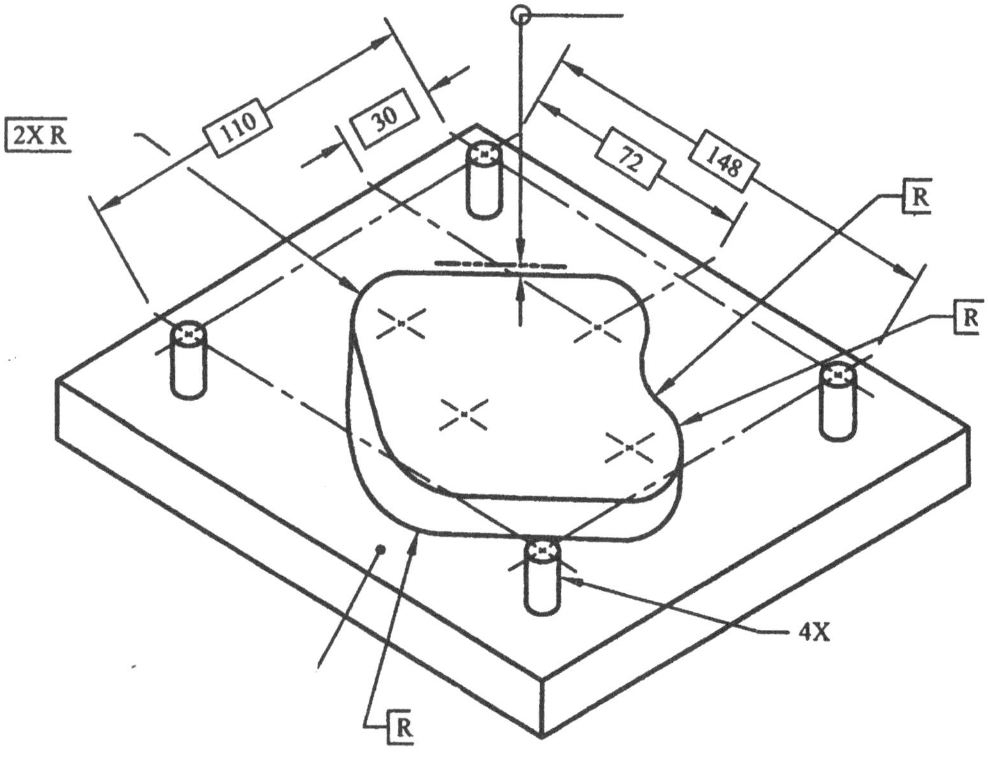

46. Use composite position tolerancing to locate all five holes on this part to the datums within 2mm and to each other to within 0.5mm. These are mating features, so reference all geometric tolerances at MMC.

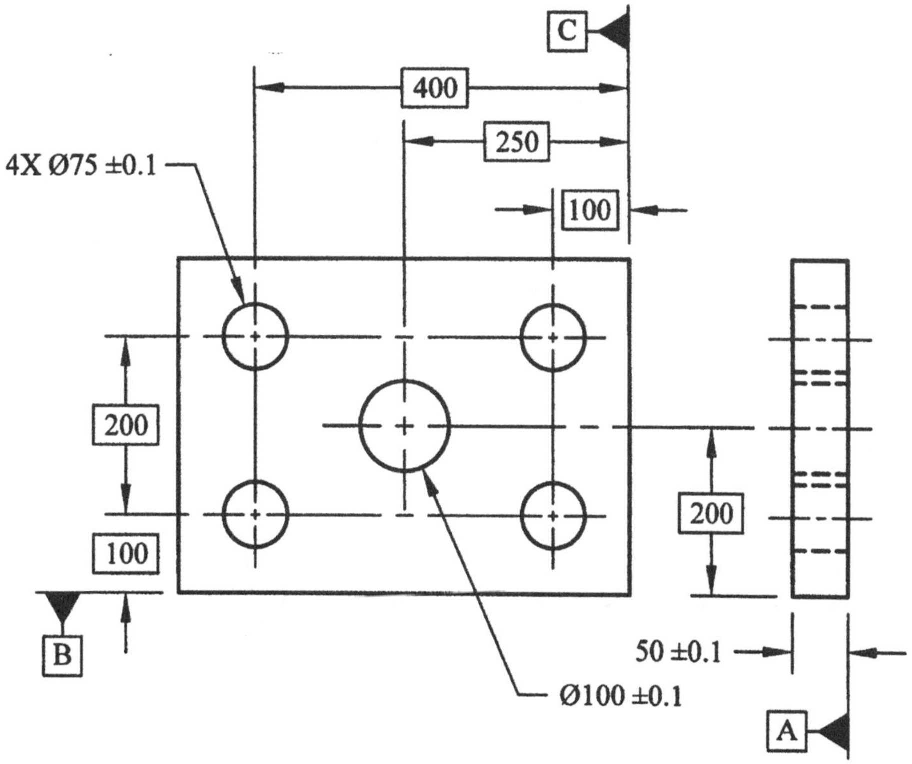

47. Converting from Plus and Minus to Composite Position Tolerancing: Converting one part from plus and minus tolerancing to composite position tolerancing, then redesigning its mating part(s) around its virtual condition boundaries, is sometimes desirable. However, it is recommended parts be designed initially using the appropriate geometric controls. Convert this drawing from plus and minus to composite positional tolerancing. Consider the following information:
    1) Assign the part three datum features.
    2) To calculate the pattern locating tolerance to all three datums, circumscribe a circle around the existing square tolerance zone using the tolerances on both 1.000 inch dimensions.
    3) To calculate the feature-relating tolerance (from hole to hole), consider this part as one piece in a floating fastener assembly, using a screw that has a ∅.212 MMC. Reference only the primary datum in this control.

[**NOTE:** Don't forget to convert all location dimensions to basic dimensions.]

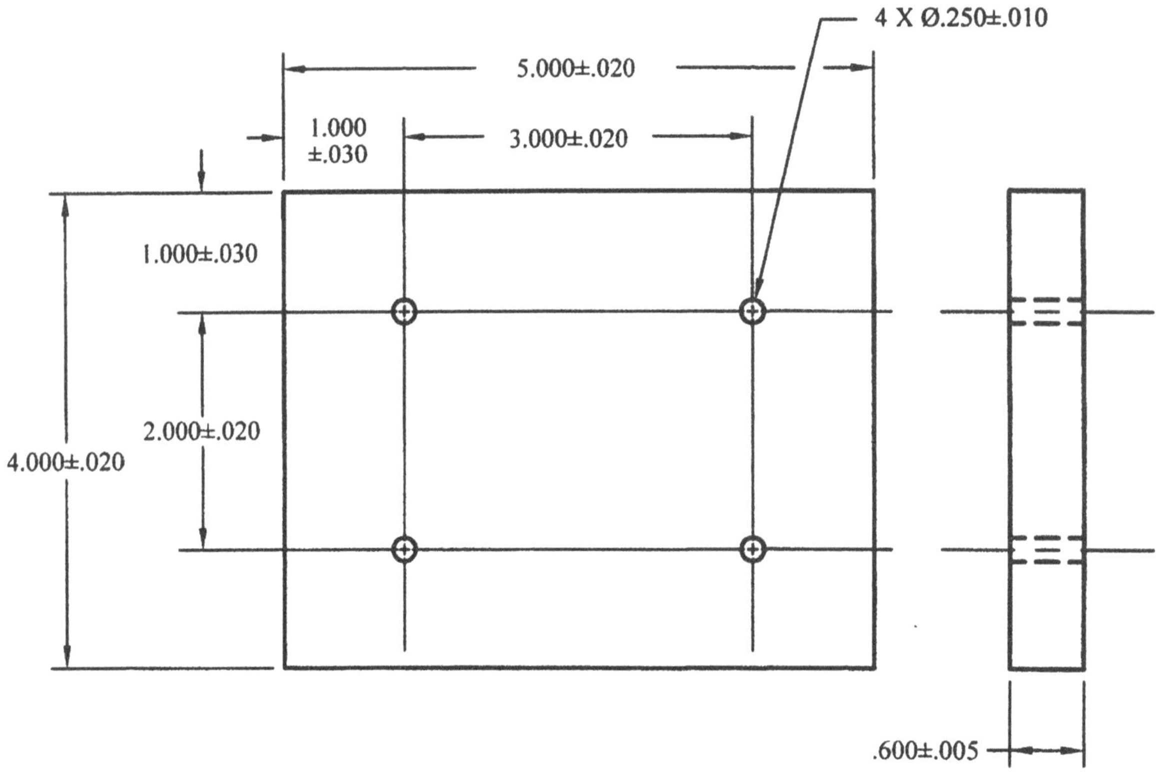

48. Convert this part from plus and minus tolerancing to composite position tolerancing.
    1) Assign the part three datum features.
    2) To calculate the pattern locating tolerance to all three datums, circumscribe a circle around the existing square tolerance zone using the tolerances on both 25mm dimensions.
    3) To calculate the feature-relating tolerance (from hole to hole), consider this part as one piece in a floating fastener assembly, using a screw that has a ∅10mm MMC. Reference only the primary datum in this control.

[**NOTE:** Don't forget to convert all location dimensions to basic dimensions.]

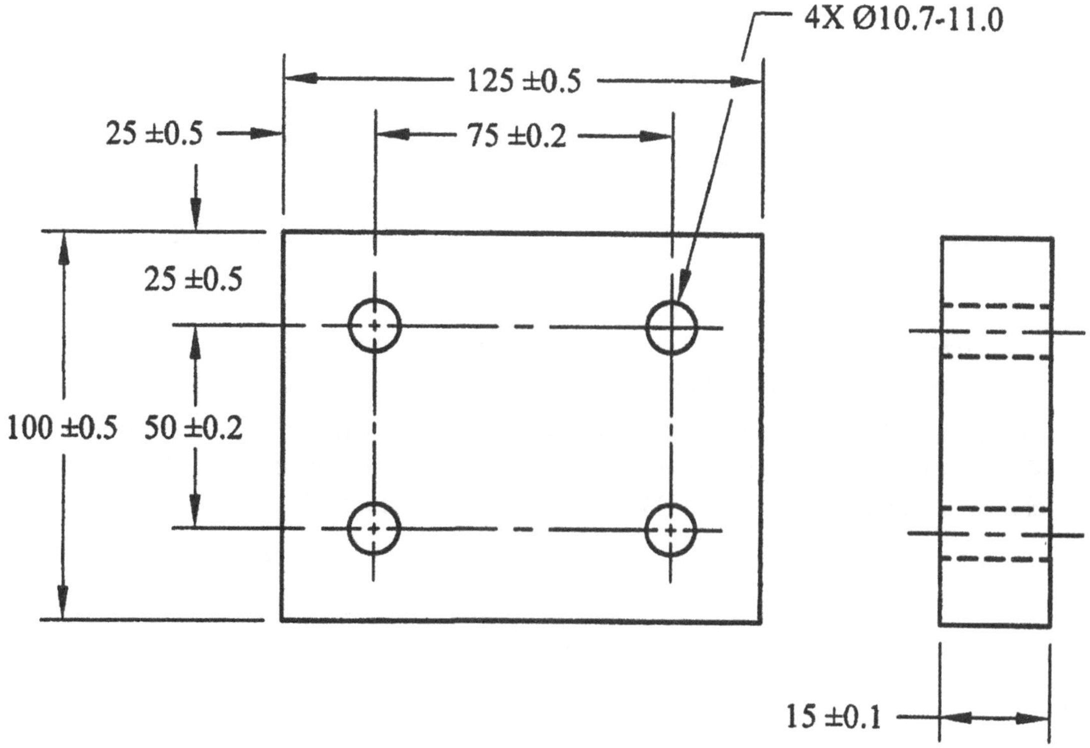

49. The three 8.3-8.5 holes in the drawing on the next page must be held coaxial to within the proper geometric tolerance to receive an 8mm bolt. However, other than that, their placement on the part can be loosely controlled.

1) Tolerance the part using a composite position control. Use only the datums shown.
2) To calculate the hole-to-hole coaxiality tolerance, use the floating fastener formula.
3) To calculate the pattern-locating tolerance, assume a minimum wall thickness at X of 1.1mm. The following formula may be used to calculate the position tolerance (Unknown Position Tol. at MMC listed below):

   LMC of Holes =
   \+ Unknown Position Tol. at MMC =
   \+ Bonus Tol. =
   Outer Boundary of Holes

   MMC of Datum Feature B =
   \- Outer Boundary of the Holes =
   Material on Both Sides of Hole

   Material on Both Sides of Hole ÷ 2 = Wall Thickness without Pits

   Wall Thickness without Pits =
   \- Size/Flatness Tolerance on Datum Feature B (Pits) =
   Minimum Wall Thickness = 1.1

4) Could this design be changed (while retaining the same virtual conditions) so as to allow inspection to accept more parts that would mate at assembly? If so, make the change. **Hint:** use a zero at MMC tolerance to hold coaxiality.

49. (continued)

Assembly

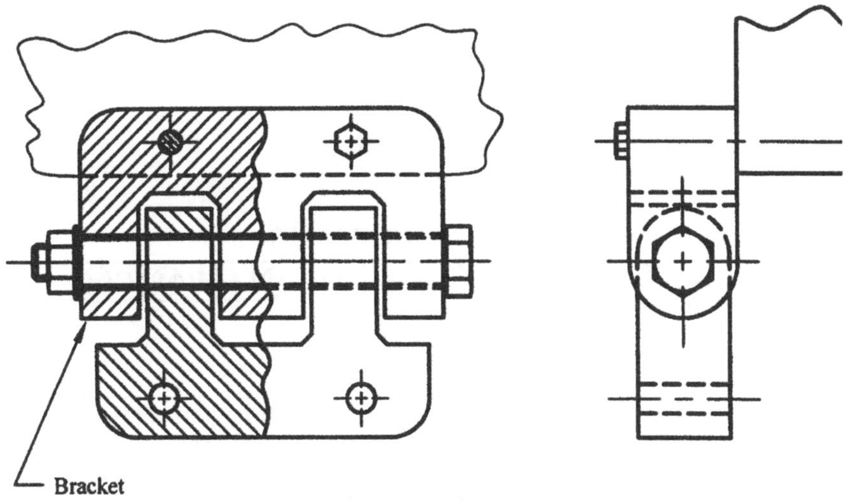

Detail Drawing Callout

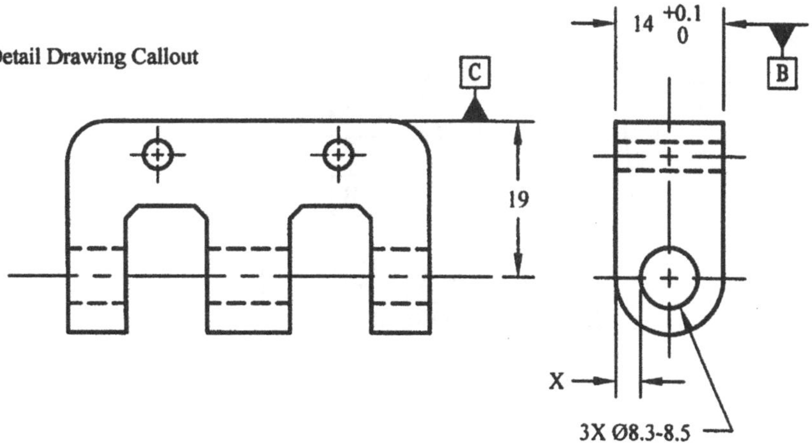

50. Please dimension and tolerance the gage shown on the next page for the four-hole pattern shown on the workpiece below.
    1) The gage should use the Practical Absolute tolerancing method with 10% of the part tolerance.
    2) Use the MMC symbol after geometric tolerances where appropriate.
    3) All datum features to be implied at RFS.

Workpiece

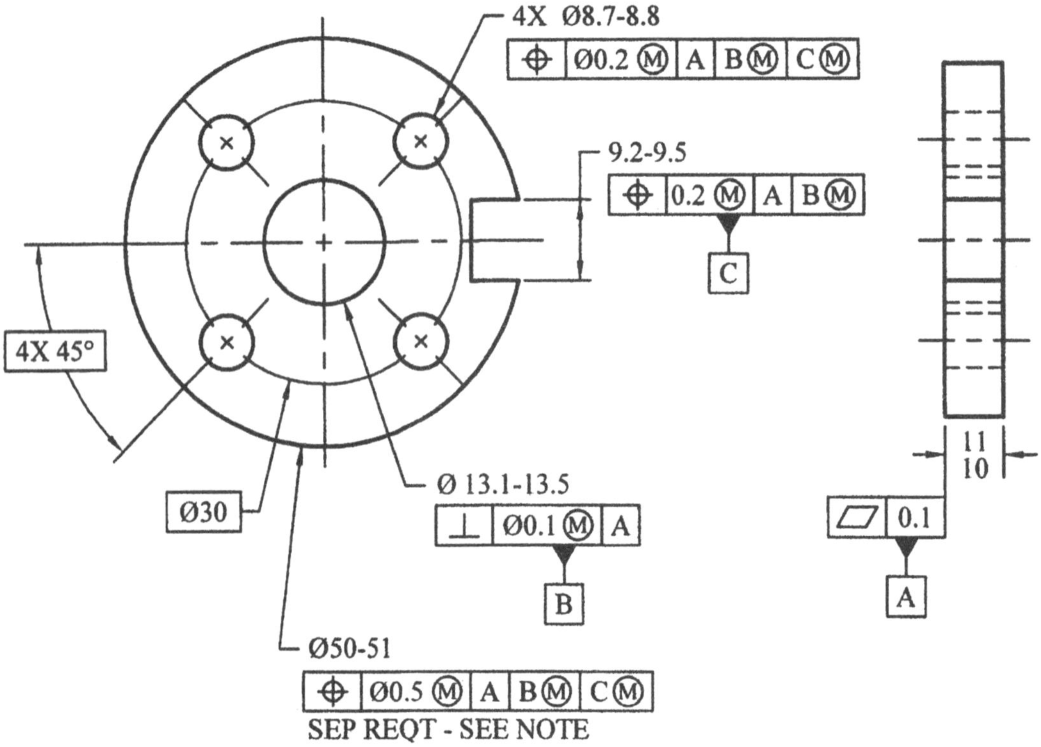

NOTE: THIS POSITION CONTROL IS A SEPARATE GAGING REQUIREMENT. IT WILL NOT BE GAGED WITH THE SAME GAGE AS THE 4 HOLE PATTERN.

50. (continued)

GAGE

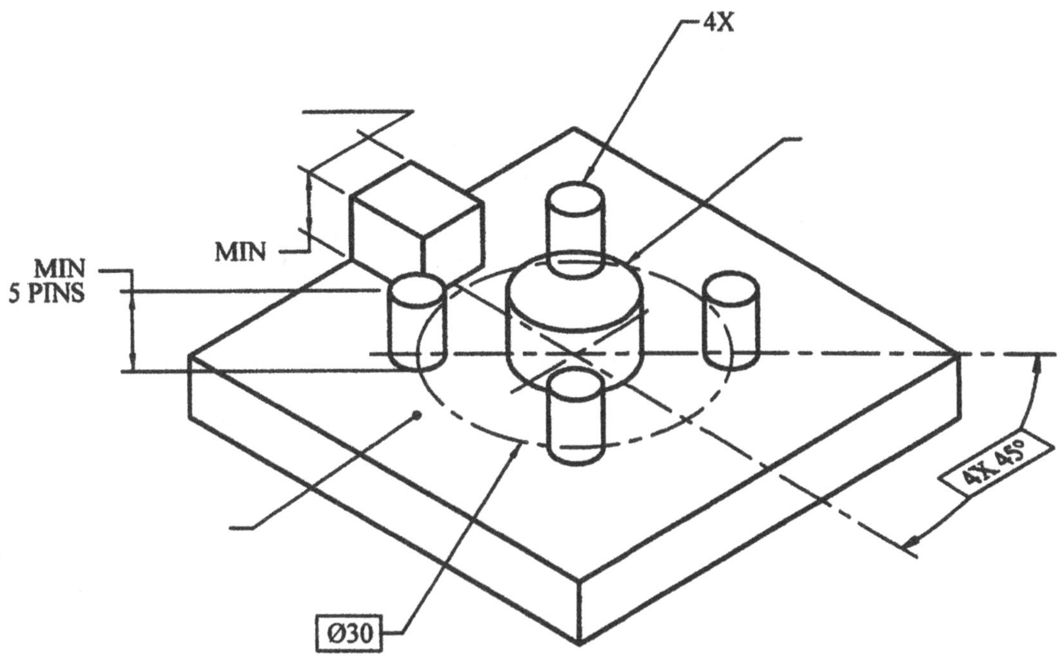

WORKPIECE APPLIED TO GAGE

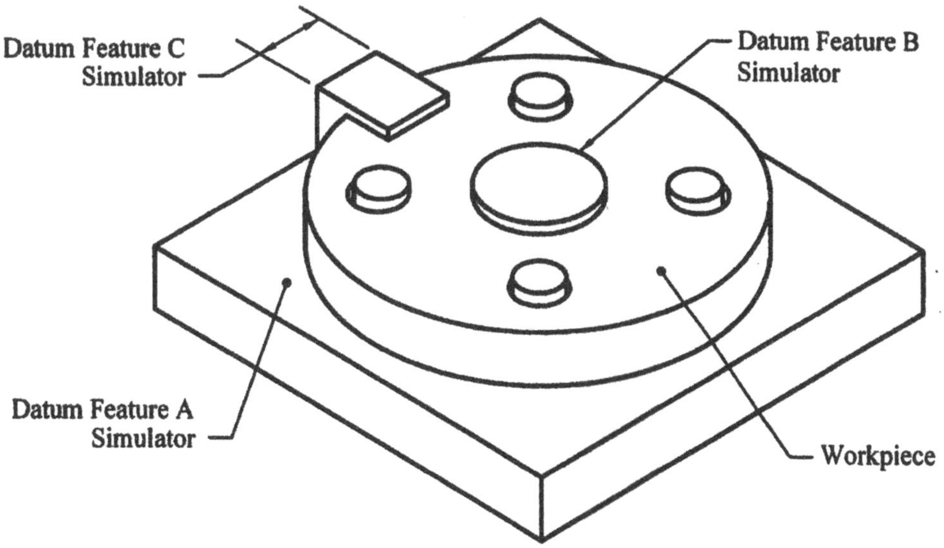

51. Functional Gage Design   Design, dimension and tolerance a functional gage to verify each positional requirement for the four-hole pattern.
    1) Use the Practical Absolute gage tolerancing method.
    2) Positional tolerances on the gage will use zero at MMC.

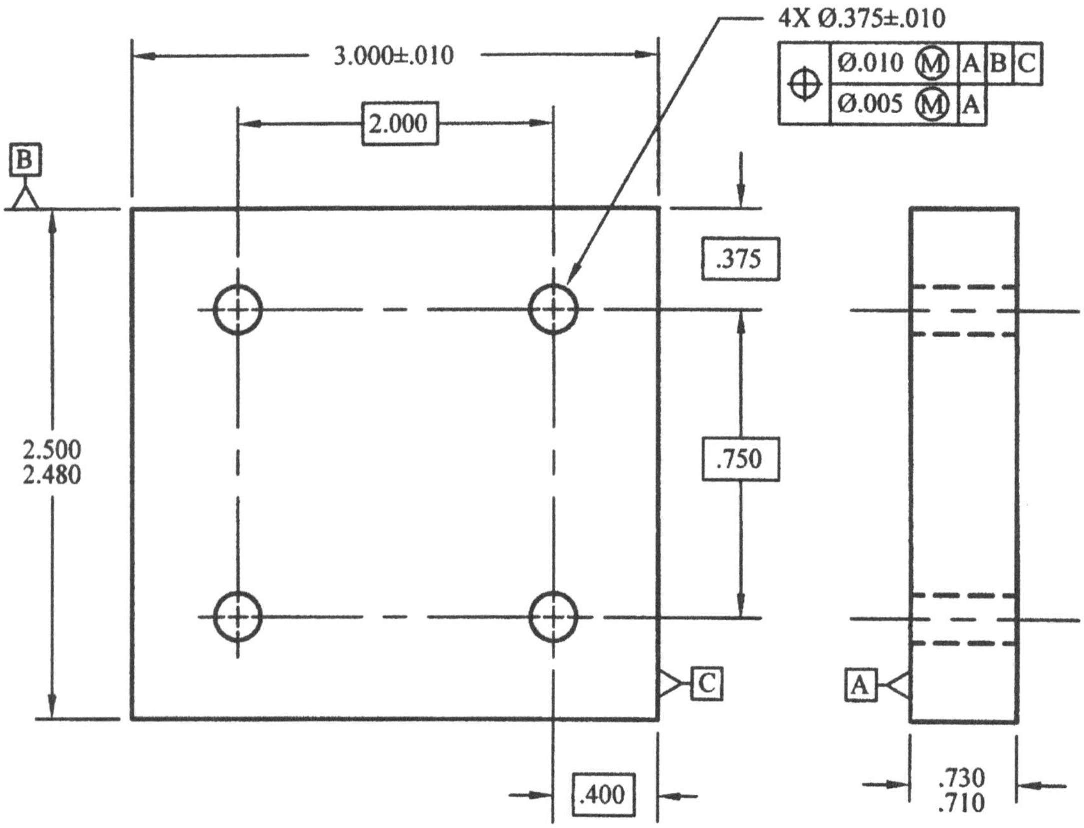

52. Please use the calculations boxes and the Loop Analysis Chart provided on this page and the next to calculate the minimum and maximum gap depicted in the assembly drawing for all eleven parts when assembled.

Assembly Drawing

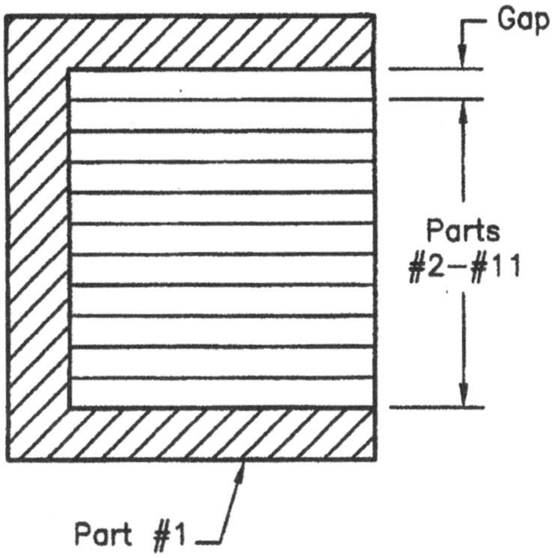

Part #1

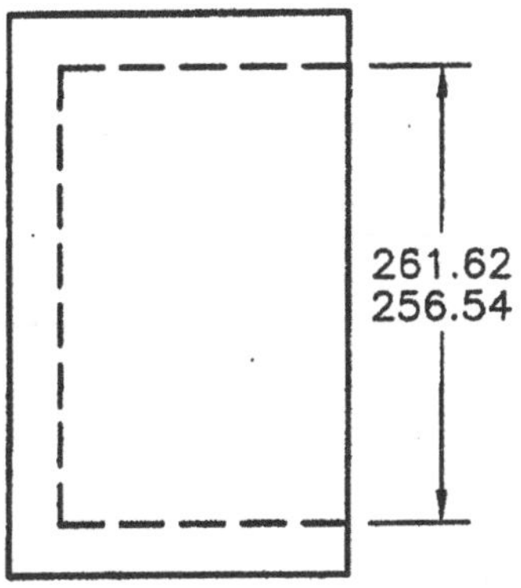

Part #2–#11

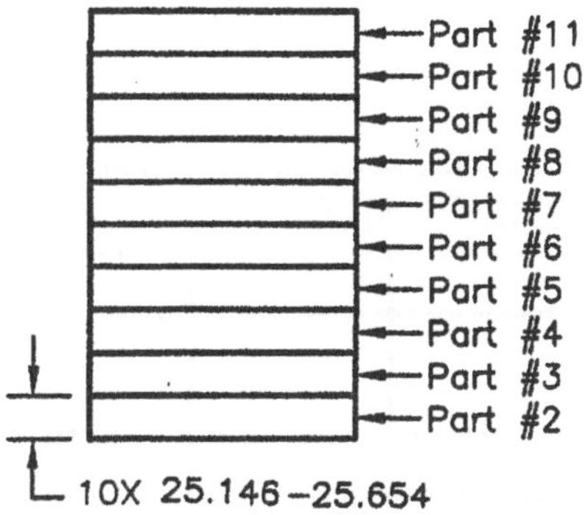

Calculations for Part #1

Calculations for Parts #2–#11

52. (continued)

Loop Diagram

<table>
<tr><td colspan="3">DIMENSIONS ±TOL</td></tr>
<tr><td></td><td></td><td></td></tr>
<tr><td></td><td></td><td></td></tr>
<tr><td></td><td></td><td>TOTALS</td></tr>
<tr><td>Differences in dimensions</td><td>MAX GAP</td><td>MIN GAP</td></tr>
<tr><td></td><td></td><td></td></tr>
</table>

53. Please use the illustration on this page and the numbers chart and illustrations on the following three pages to calculate and graph the minimum and maximum gaps as depicted.

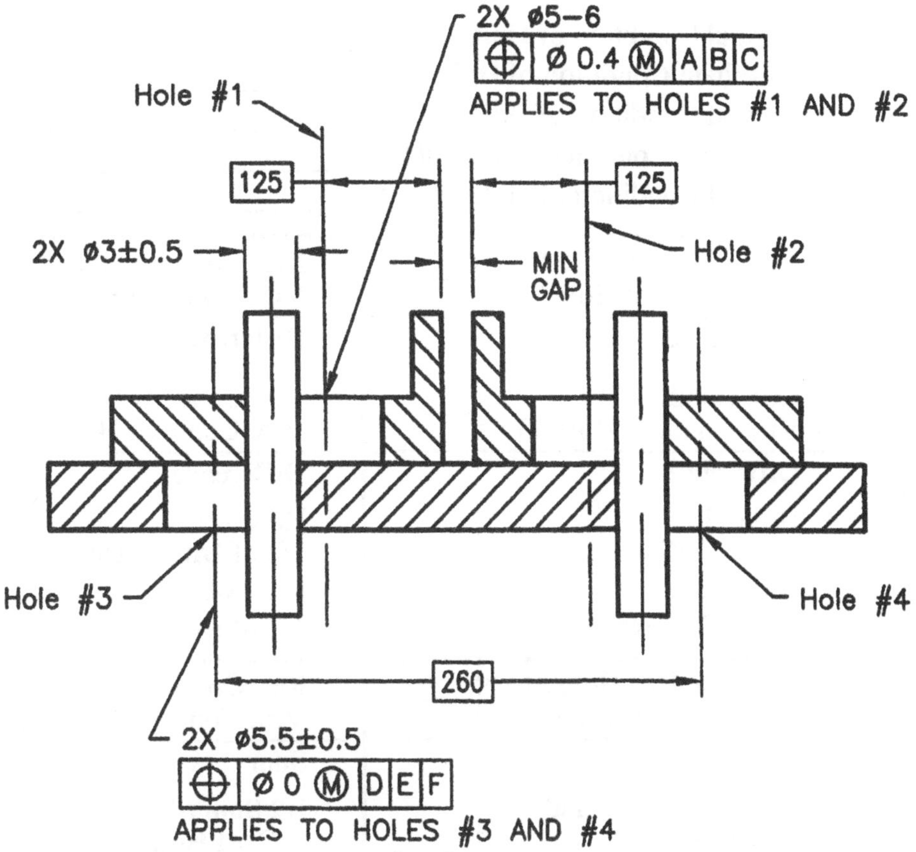

53. (continued)

Resultant Condition of Hole #1 = ______________

Virtual Condition of Hole #1 = ______________

Resultant Condition of Hole #2 = ______________

Virtual Condition of Hole #2 = ______________

Resultant Condition of Hole #3 = ______________

Virtual Condition of Hole #3 = ______________

Resultant Condition of Hole #4 = ______________

Virtual Condition of Hole #4 = ______________

	Res. Cond. of Hole #1 = ______	Res. Cond. of Hole #2 = ______
	- Virt. Cond. of Hole #1 =	- Virt. Cond. of Hole #2 =
	Difference	Difference
	Res. Cond. of Hole #3 = ______	Res. Cond. of Hole #4 = ______
	- Virt. Cond. of Hole #3 =	- Virt. Cond. of Hole #4 =
	Difference	Difference
to get mean tol. →	Hole #1: 1/2 of Difference = ______	Hole #2: 1/2 of Difference = ______
	Hole #3: 1/2 of Difference = ______	Hole #4: 1/2 of Difference = ______

	Res. Cond. of Hole #1 = ______	Res. Cond. of Hole #2 = ______
	+ Virt. Cond. of Hole #1 =	+ Virt. Cond. of Hole #2 =
	Sum	Sum
mean diameter →	1/2 Sum of R.C. and V.C. of Hole #1 = ______	1/2 Sum of R.C. and V.C. of Hole #2 = ______
mean diameter & mean tol. →	1/2 Sum ± 1/2 Difference of Hole #1 = ______	1/2 Sum ± 1/2 Difference of Hole #2 = ______
mean radius ± mean radial tol. →	1/2 of 1/2 Sum ± 1/2 of 1/2 Difference Hole #1 = ______	1/2 of 1/2 Sum ± 1/2 of 1/2 Difference Hole #2 = ______

Res. Cond. of Hole #3 = ______	Res. Cond. of Hole #3 = ______
+ Virt. Cond. of Hole #3 =	+ Virt. Cond. of Hole #3 =
Sum	Sum
1/2 Sum of R.C. and V.C. of Hole #3 = ______	1/2 Sum of R.C. and V.C. of Hole #4 = ______
1/2 Sum ± 1/2 Difference of Hole #3 = ______	1/2 Sum ± 1/2 Difference of Hole #4 = ______
1/2 of 1/2 Sum ± 1/2 of 1/2 Difference Hole #3 = ______	1/2 of 1/2 Sum ± 1/2 of 1/2 Difference Hole #4 = ______

53. (continued) MIN GAP

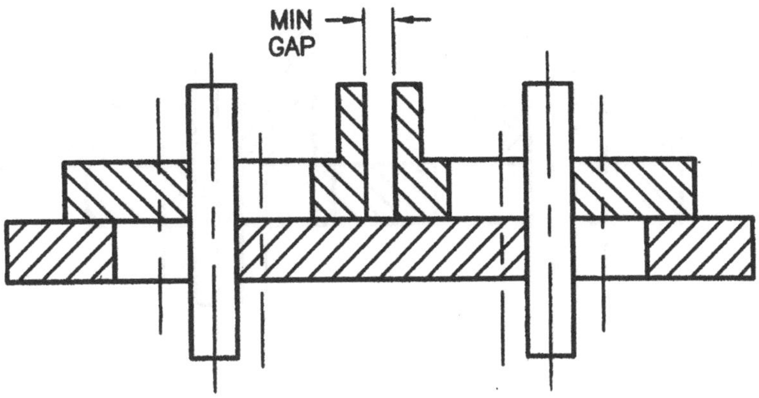

Loop Diagram

Right to Left	Left to Right		
−	+	±Tol	
			Totals
			=MIN GAP

53. (continued) MAX GAP

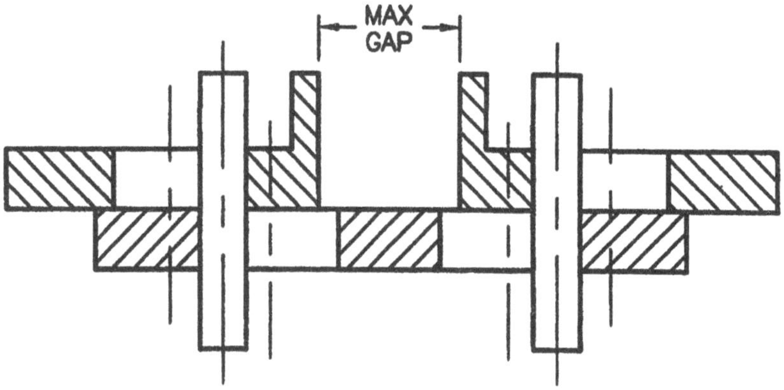

Loop Diagram

Right to Left	Left to Right		
–	+	±Tol	
			Totals
			=MAX GAP

54. Using the illustration from below and the illustrations from Exercise 53, calculate the statistical tolerances to be reintegrated into the assembly for all features used in this MIN GAP calculation. Instead of using the straight RSS formula, use the following RSS formula, with a 1.5 'safety correction' factor. Please indicate all final size tolerances and positional tolerances as ⟨ST⟩.

$$T_A = 1.5\sqrt{T_1^2 + T_2^2 + T_3^2 + T_4^2 + T_5^2 + T_6^2}$$

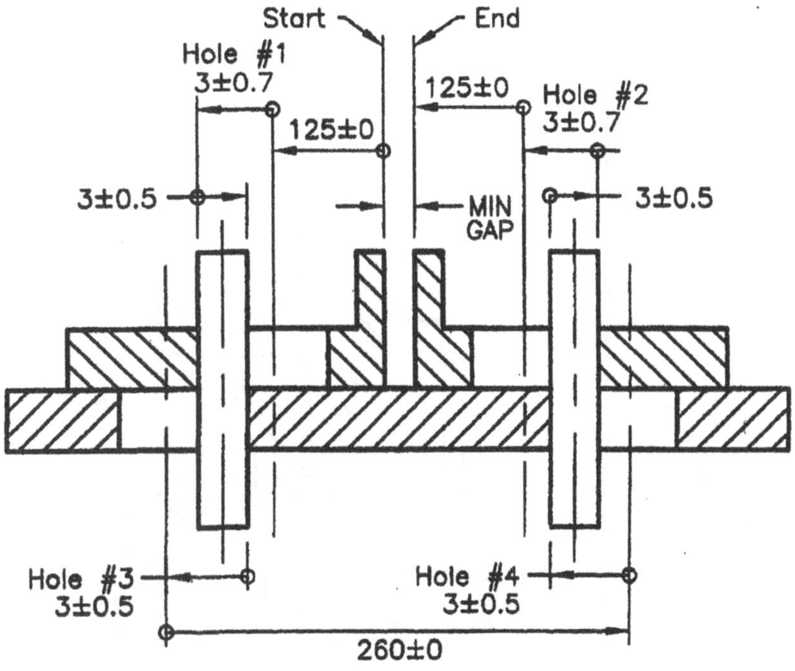

Loop Diagram

Right to Left	Left to Right		
−	+	±Tol	
125		0.0	Basic Dim
3		0.7	Hole #1
	3	0.5	Pin
3		0.5	Hole #3
	260	0.0	Basic Dim
3		0.5	Hole #4
	3	0.5	Pin
3		0.7	Hole #2
125		0.0	Basic Dim
262	266	3.4	Totals
266 −262 4		4.0 −3.4 0.6	MIN GAP

55. PROBLEM: Using all pertinent factors from Part #1 through Part #4, including the central hole on Part #1 as a factor, calculate the minimum clearance between Part #4's outside diameter and Part #1 of the assembly.

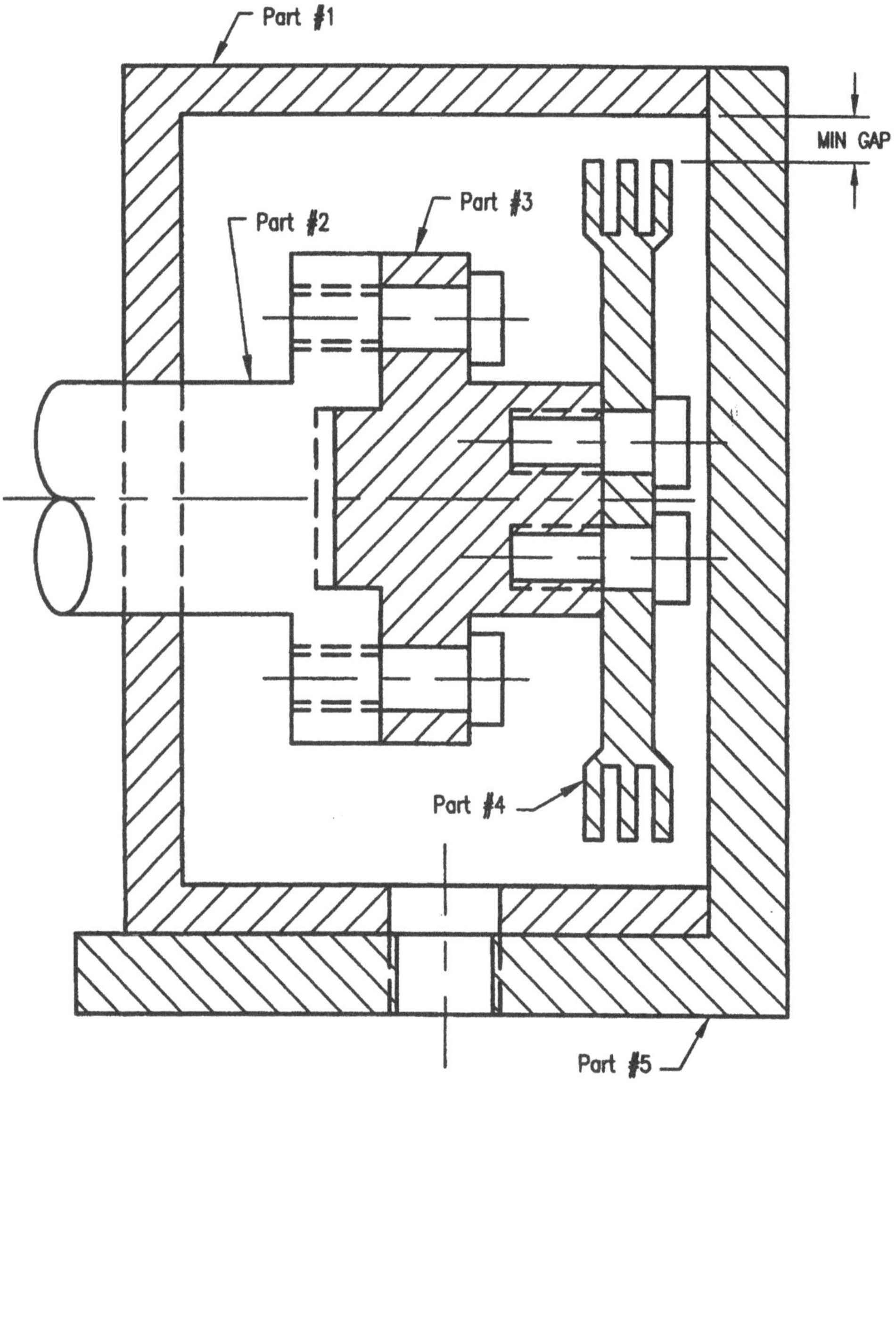

55. (Five-Part Assembly, **Part #1**)

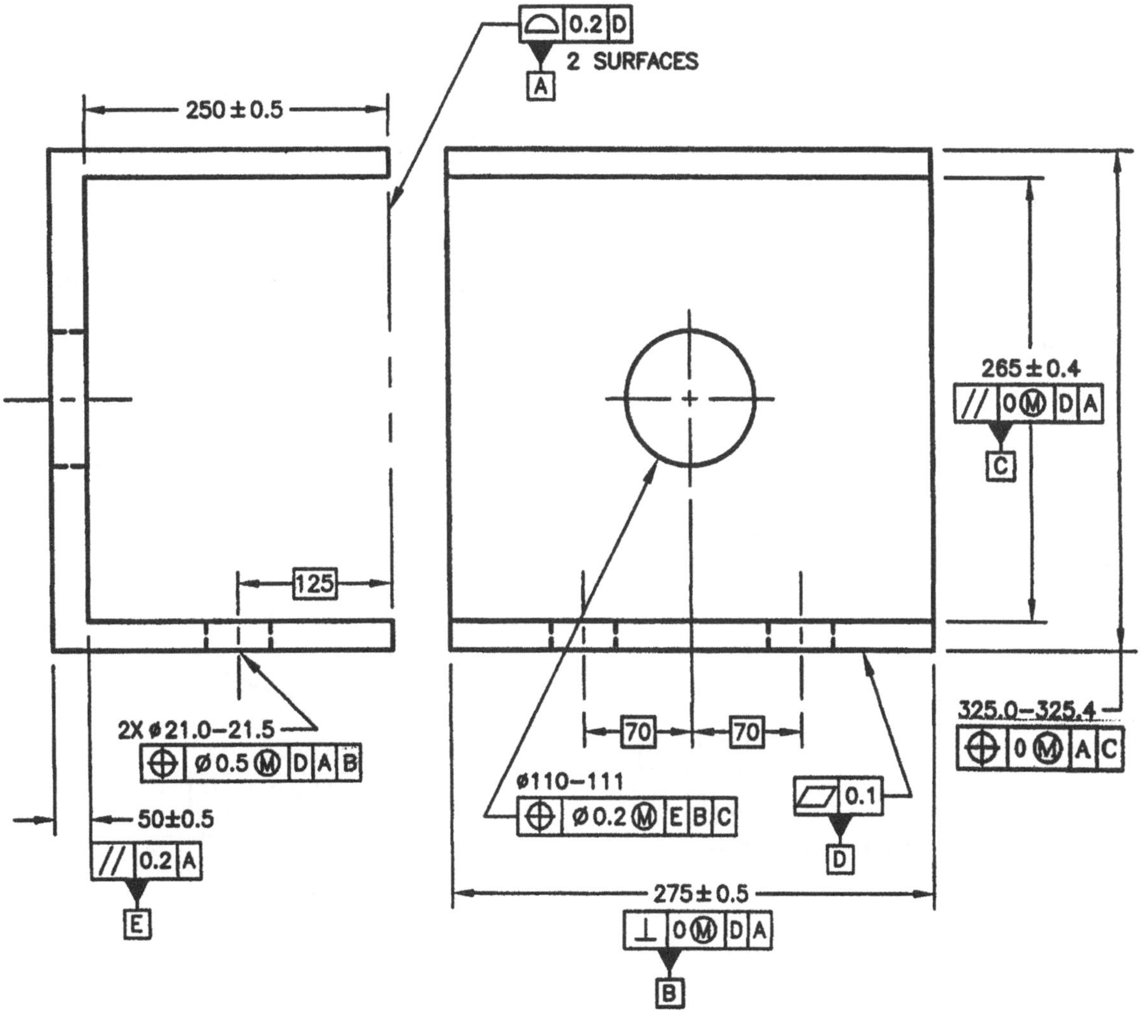

55. (Five-Part Assembly, **Part #2**)

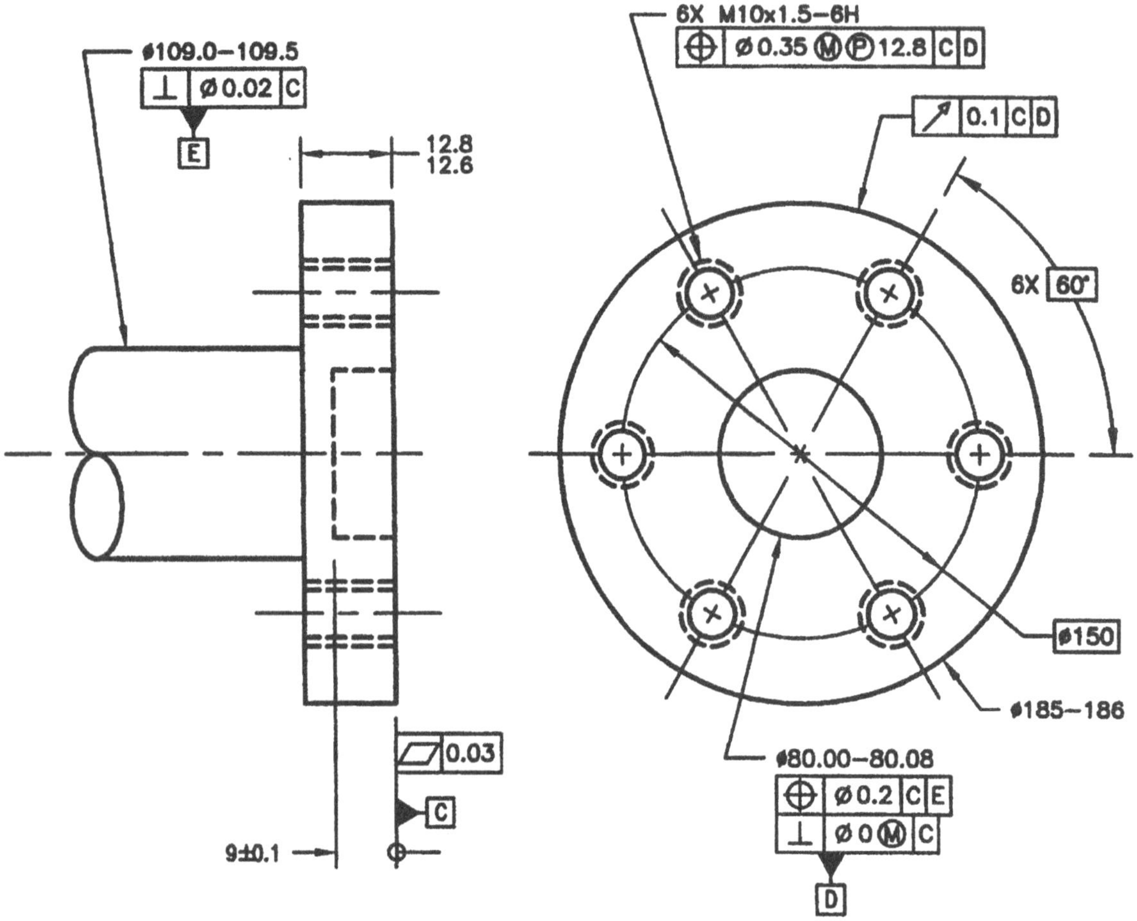

55. (Five-Part Assembly, **Part #3**)

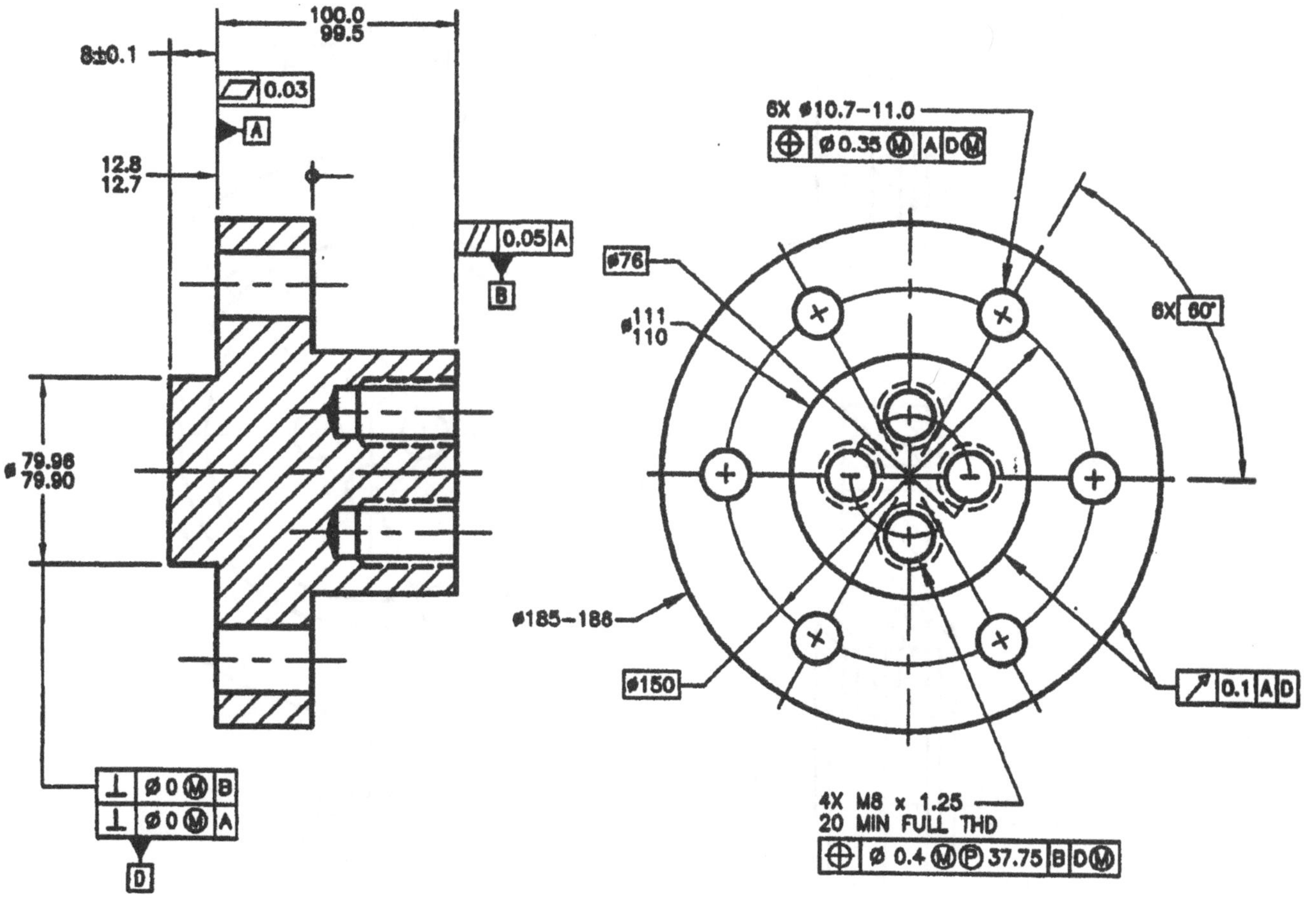

55. (Five-Part Assembly, **Part #4**)

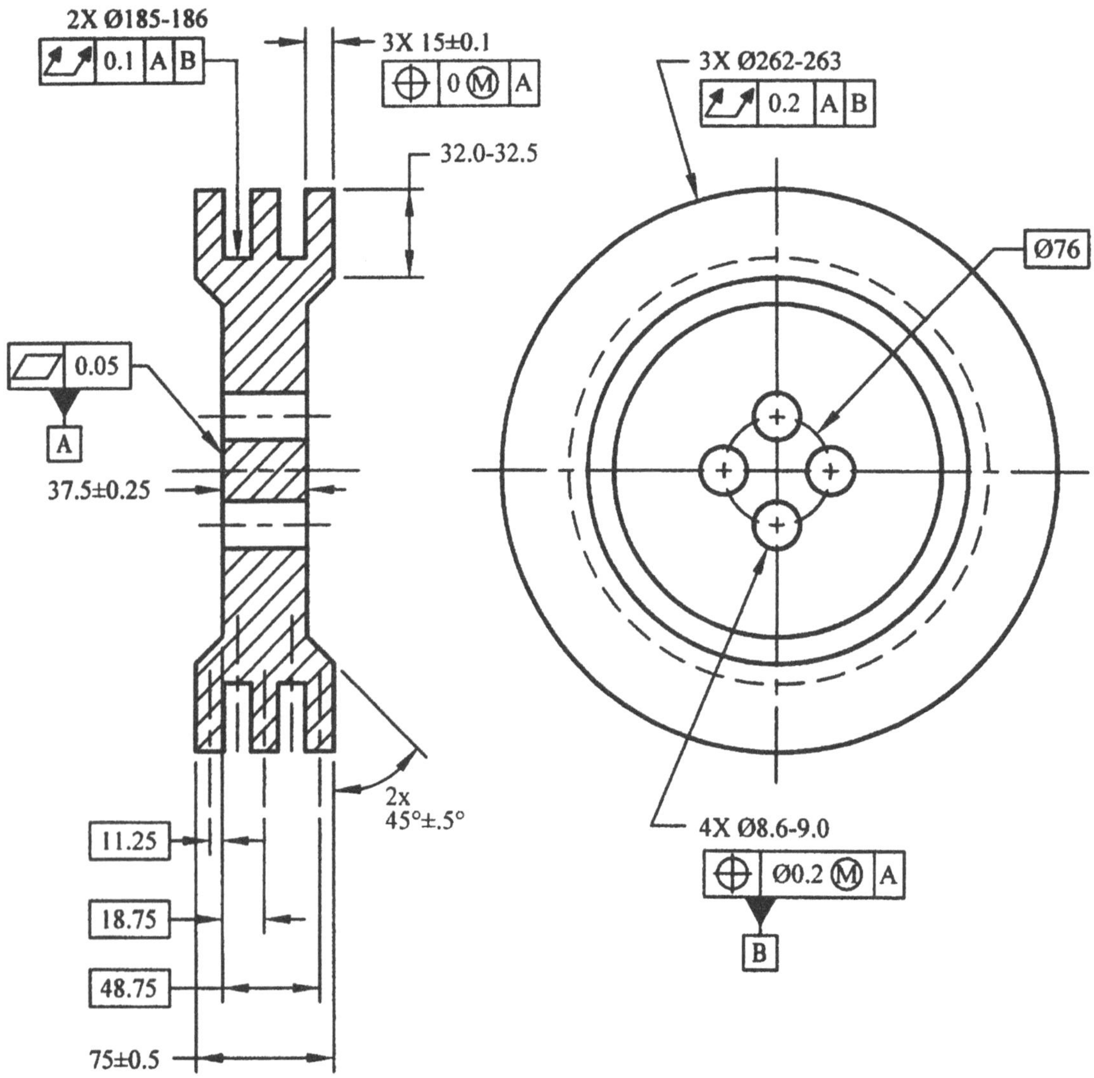

55. (Five-Part Assembly, Part #5)

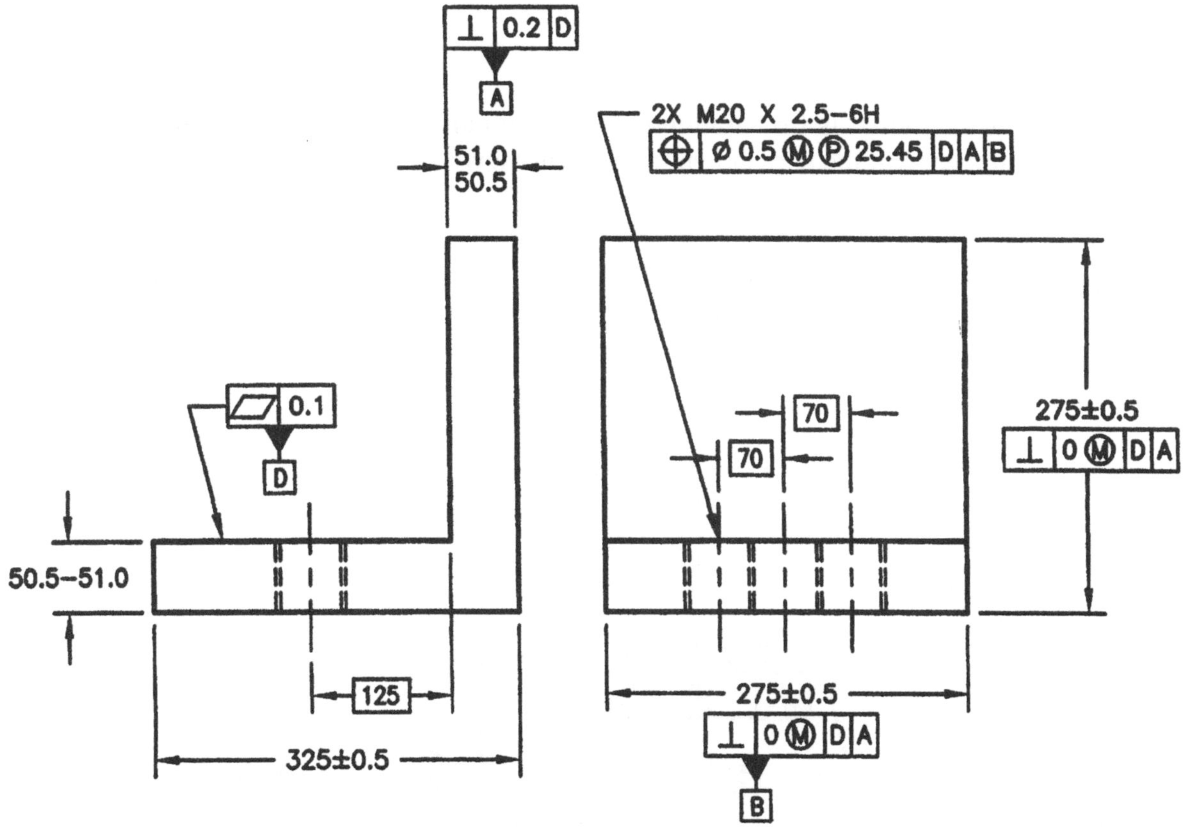

56. Complete the following:

PROBLEM #1: What is the minimum and maximum wall thickness distance between the B and C holes if they are produced at LMC?

PROBLEM #2: What is the minimum and maximum wall thickness distance from the B hole to the top edge of the part shown in the front view?

PROBLEM #3: What is the minimum and maximum wall thickness from the lower right ∅4.0-4.3 hole to the right edge and the bottom edge of the part in the front view?

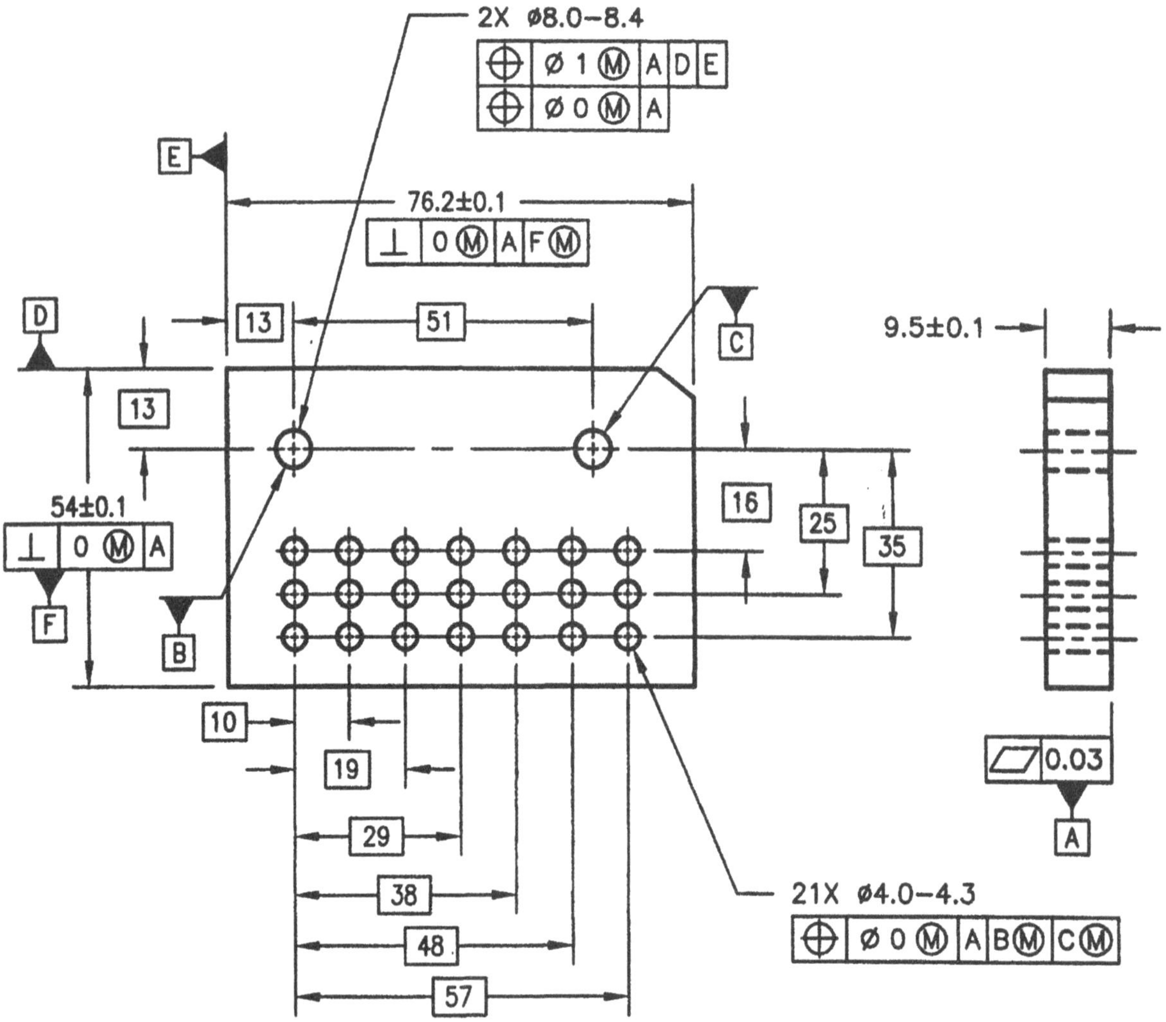

57. Complete the following:

    PROBLEM #1: Calculate MIN and MAX WALLS and axis distances from the left edge of the part to the ∅.125 ± .005 hole closest to it.

    PROBLEM #2: Calculate MIN and MAX WALLS from the bottom edge of the part to the ∅.125 ± .005 hole closest to it.

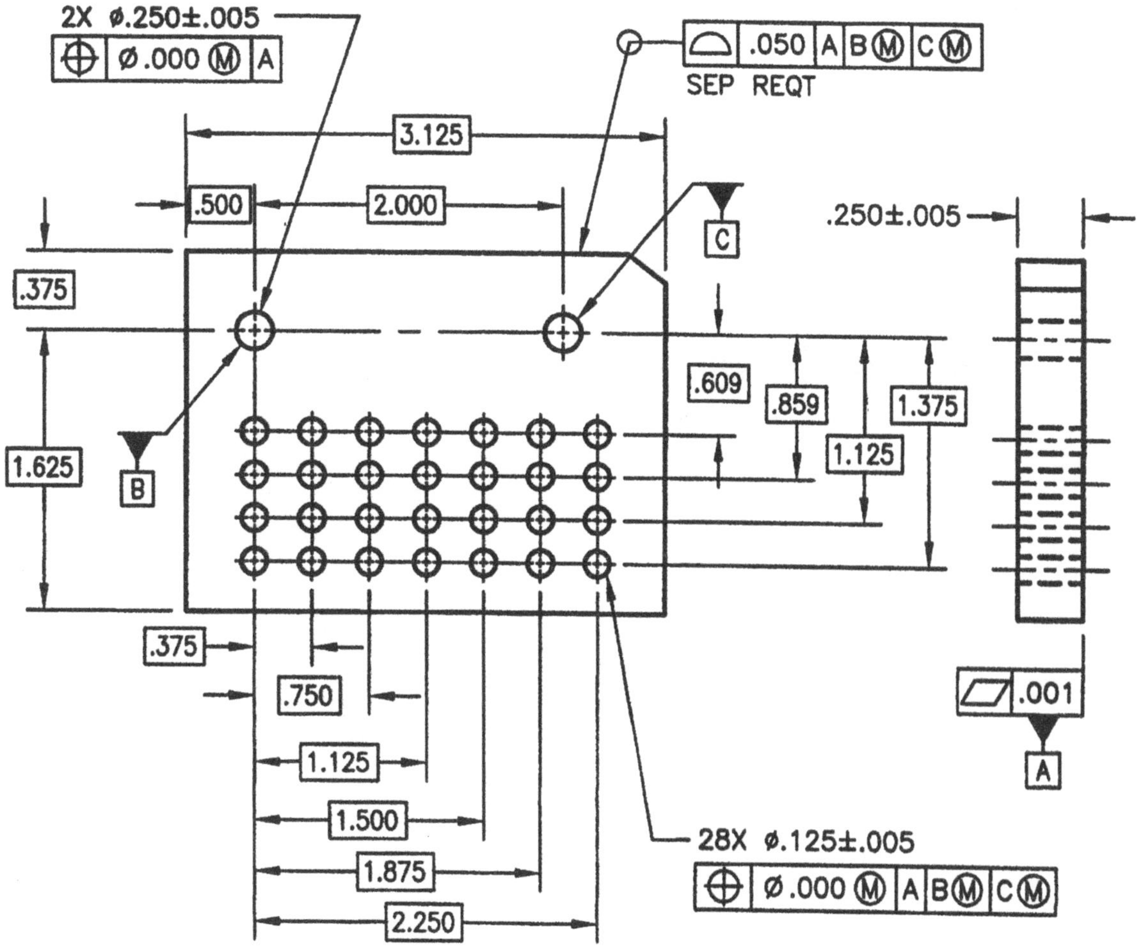

58. Consider this object the shape of a hockey puck that has four holes drilled into it. Calculate the minimum wall thickness between the surface of one hole to the outside diameter of the hockey puck.

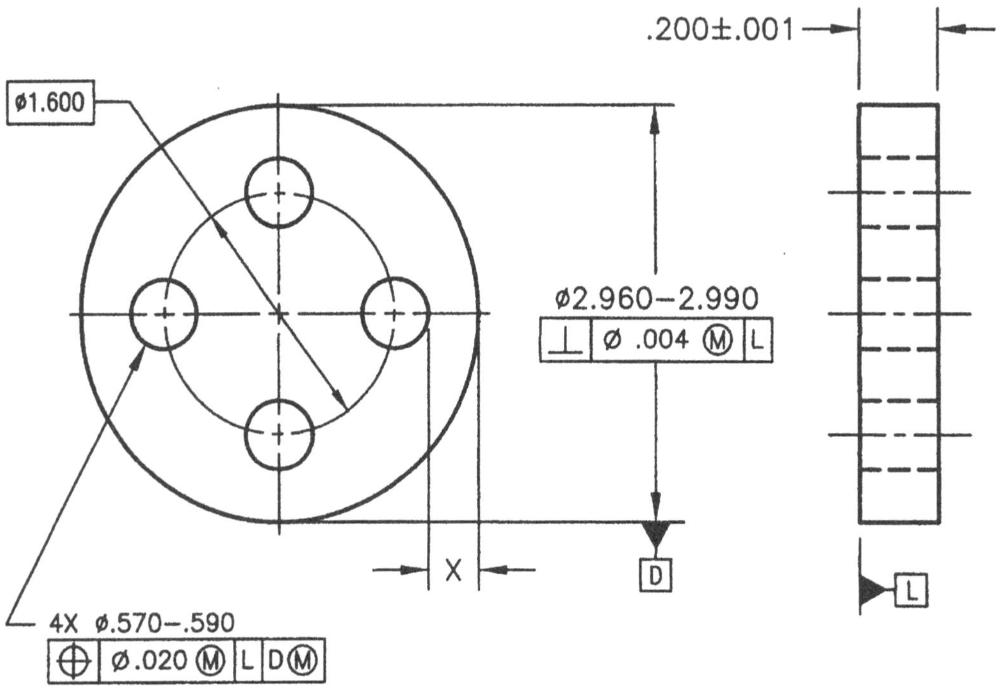

59. What is the minimum wall thickness of the hole pattern to datum feature C?

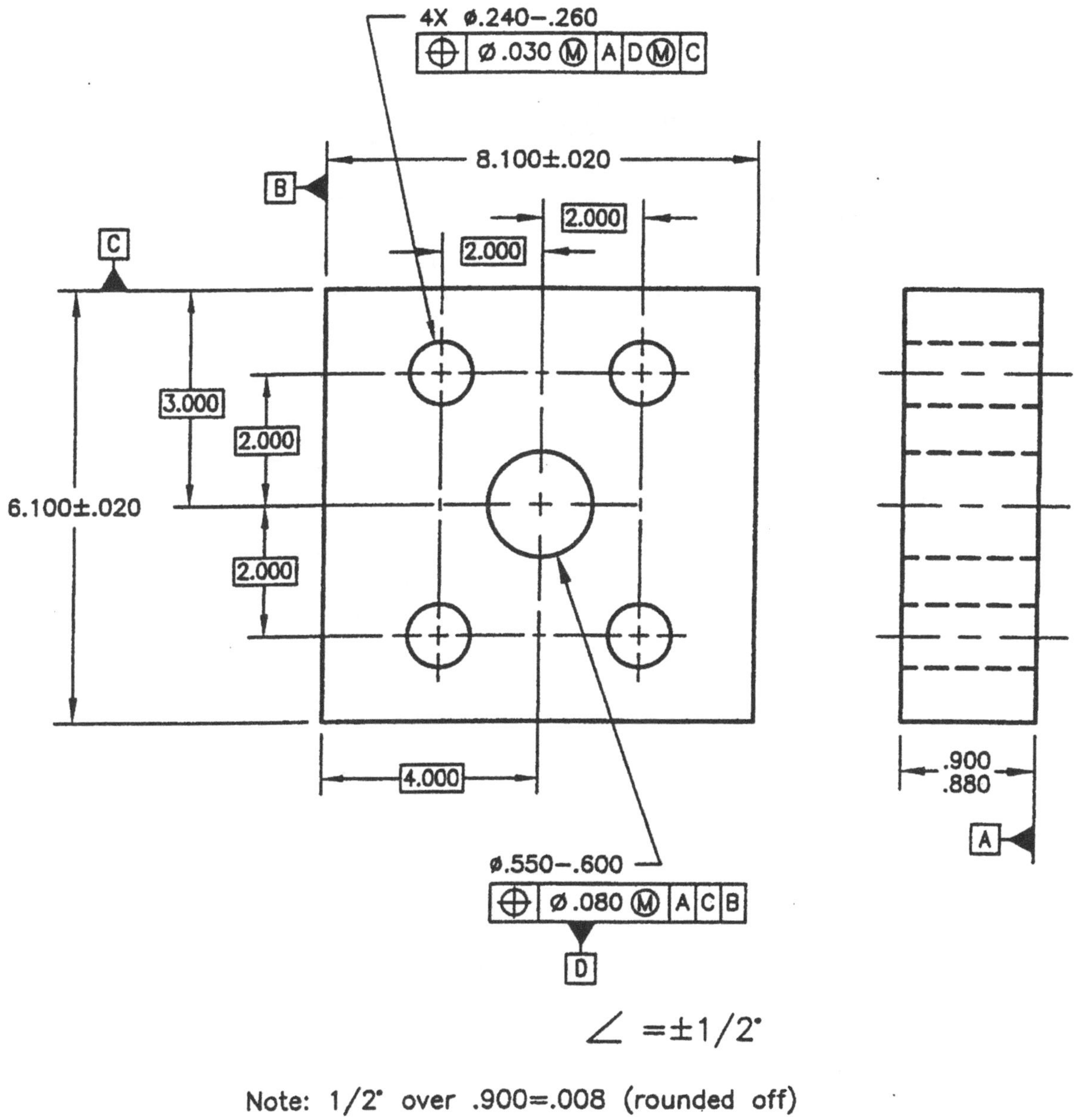

60. Given the 'as-produced' measurements at the bottom of the illustration, is this part good or bad?

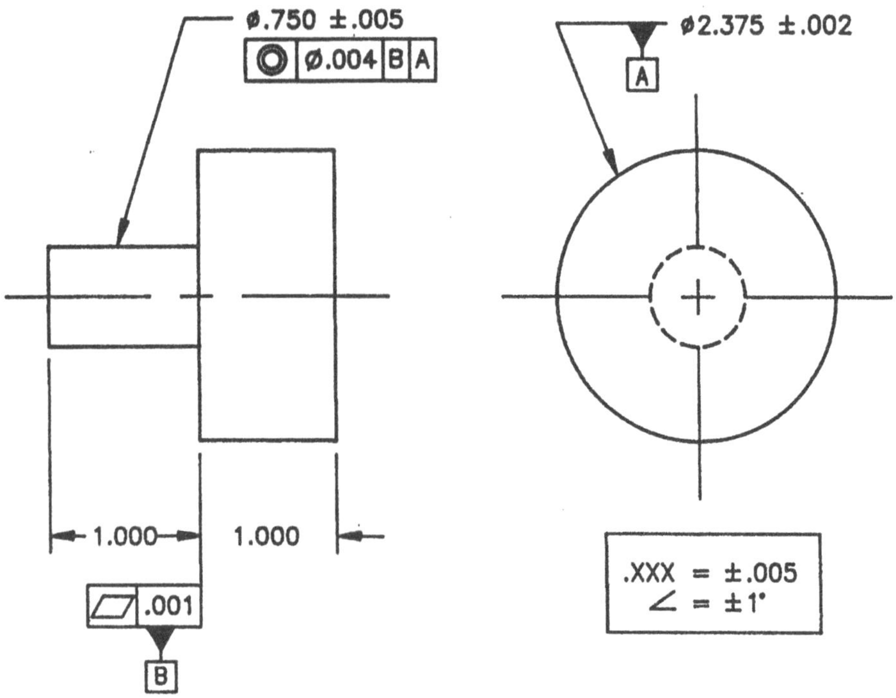

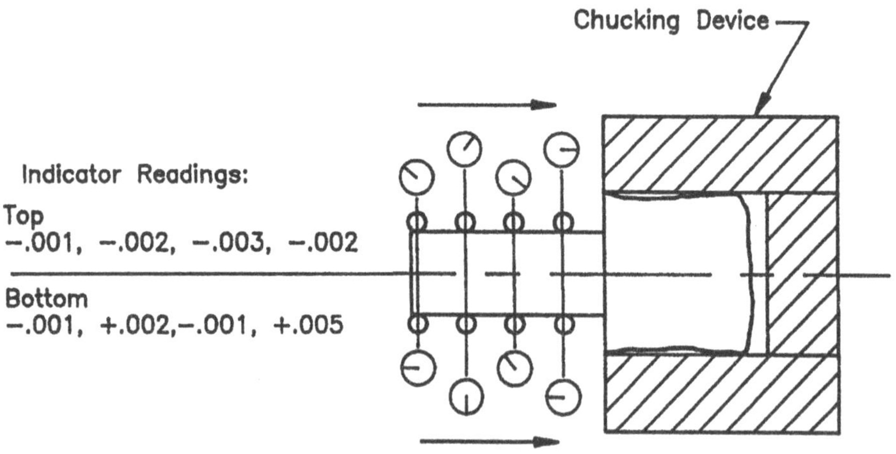

61. Consider each of the following options separately and based on the measurement data, give the status as either good (in tolerance) or bad (out of tolerance).

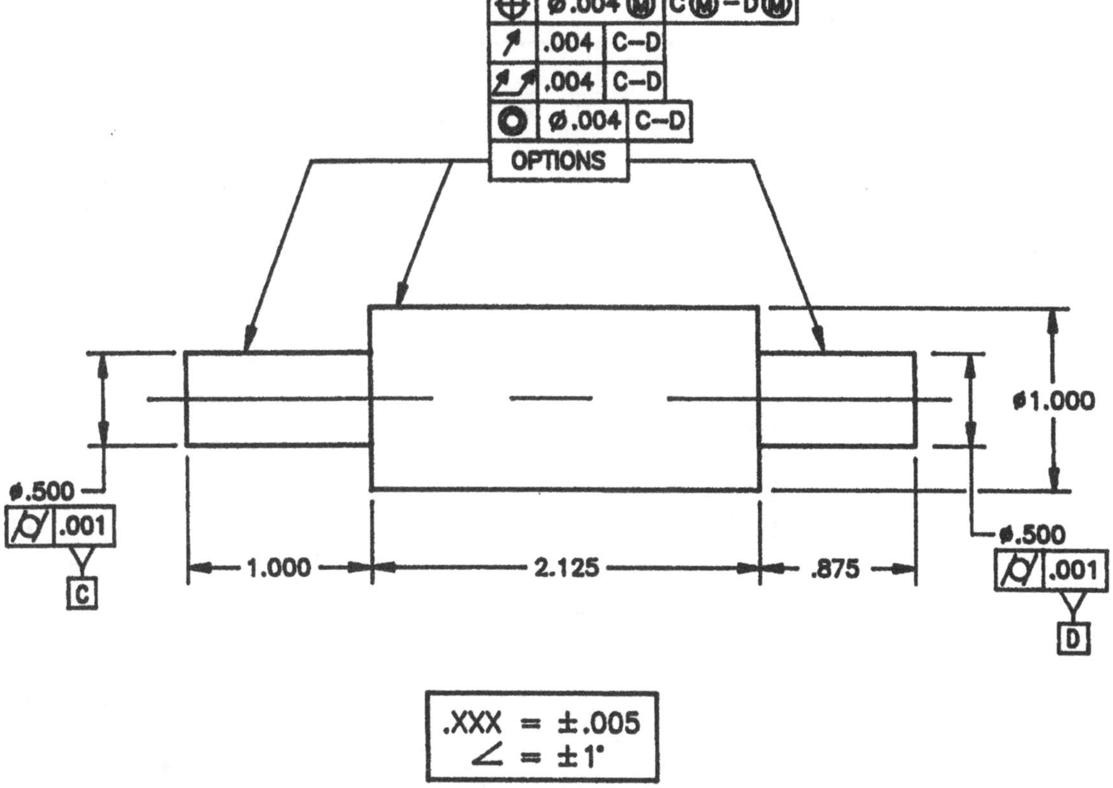

SYMBOL	FIM	Differential Measurements (worst set, part reads zero at some point on surface)	Bonus Tol.	Status
⌖	.004	−.004 − −.006 +.002	.005	
↗	.003	N/A	N/A	
⌰	.005	N/A	N/A	
◎	.005	−.003 − −.006 +.003	N/A	

62. Using this drawing, answer the questions on the following page.

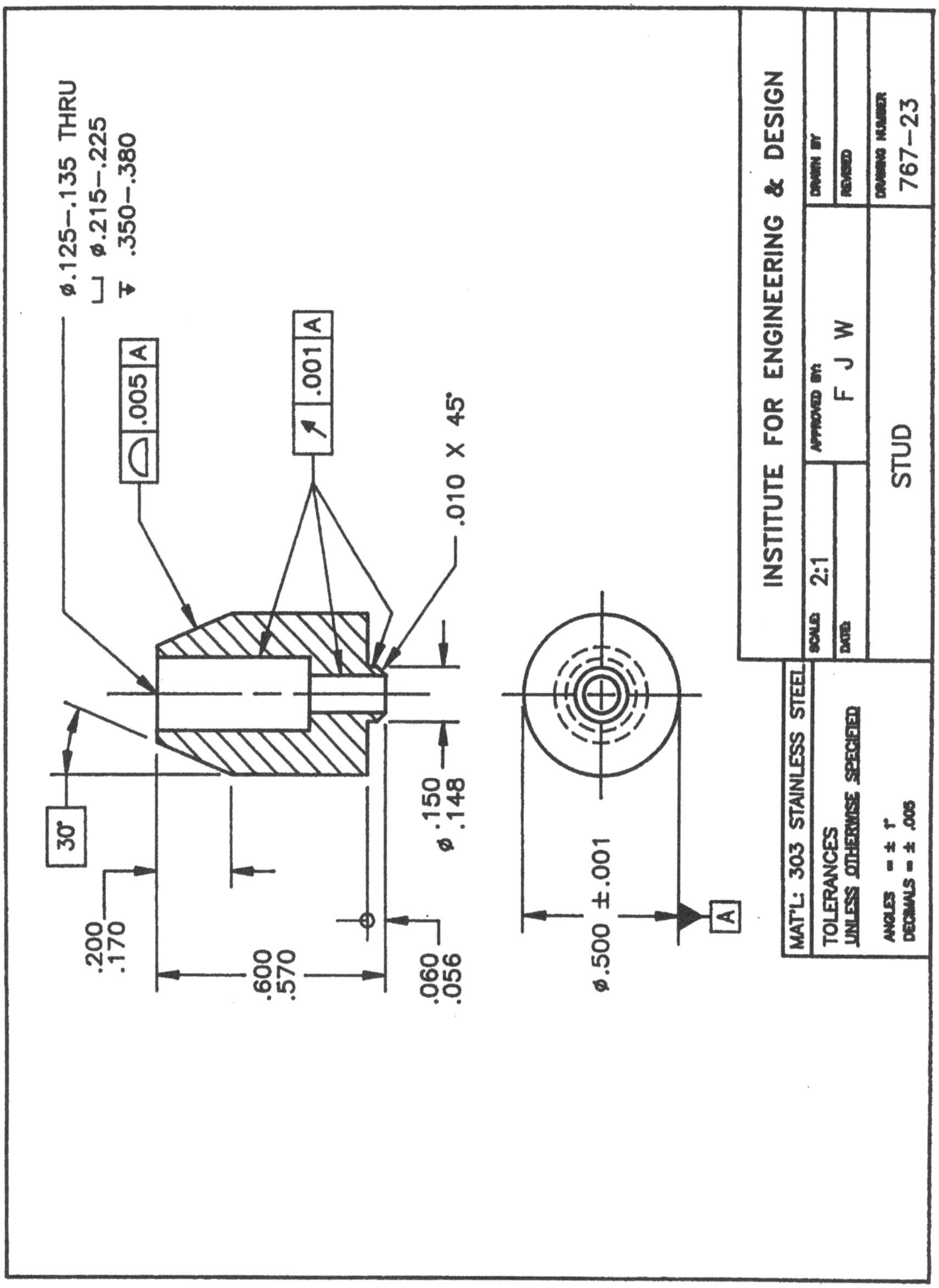

62. (continued) Using the following variables measurement data for Part 767-23-Stud on the previous page, assess the part and make observations regarding any conflicting measurements.

1. The outside diameter measures between .5002 and .5003.
   Okay or not okay?

2. The pilot hole sizes vary locally between .126 and .128.
   Okay or not okay?

3. The pilot hole has a maximum inscribed cylinder of .120.
   Okay or not okay?

4. The counterbore measures no more than .224 at all opposing points.
   Okay or not okay?

5. The counterbore will receive a cylindrical pin to its full depth that measures no more than .220 in diameter.
   Okay or not okay?

6. The surface from which the dimension origin symbol originates has a pit in it that is .007 deep. Okay or not okay?

7. All runout controlled surfaces have a full indicator movement that does not exceed .0006.
   Okay or not okay?

8. The conical feature has a pit in it that is .004 deep and the cone is off center of datum axis A a radius of .003. Okay or not okay?

TIP: If any diameter tapers by .002, the full indicator movement should be at least .001.

63. The following data has been collected with the set-up shown below. The slot size has been measured locally. Of all the two point opposing point measurements taken, the slot has uniformly measured 19mm wide. In position 1 shown, the height of the surface is 120.47 from the surface plate. In position 2 (a 180° rotation from position 1), the height of the opposing surface is 120.53 from the surface plate. Does this indicate that the slot is outside of its symmetry tolerance?

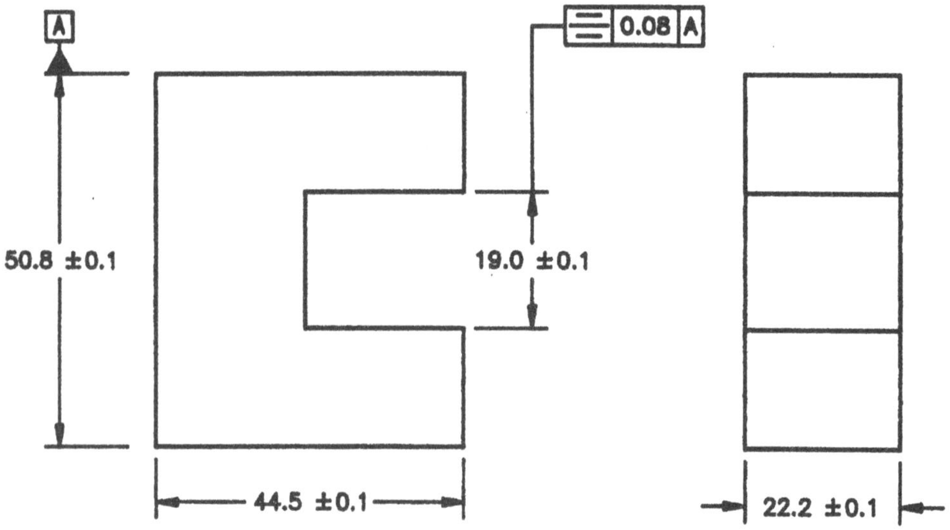

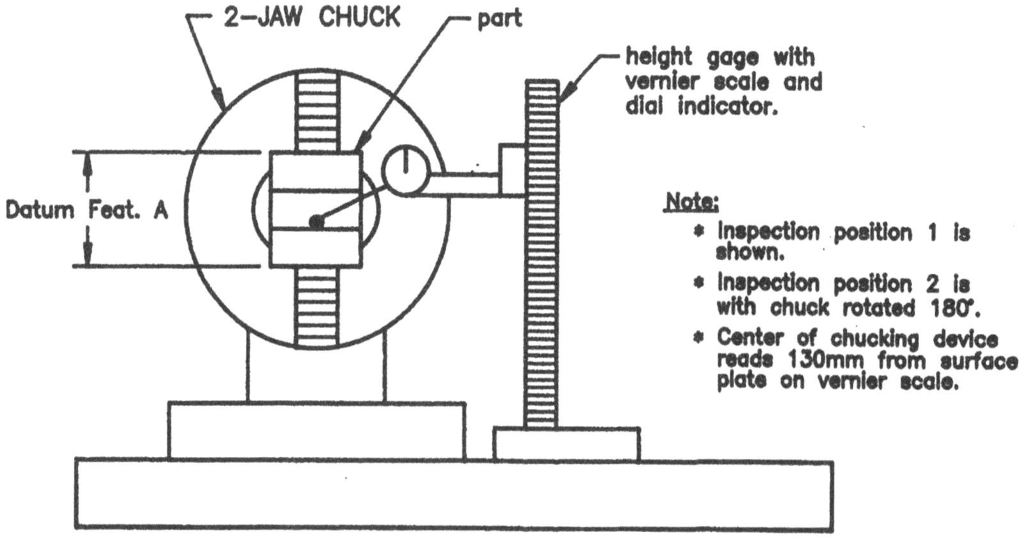

64. The following data has been collected with the set-up shown below. The width of the gage block (which is the largest that can be inserted into the slot) is 19mm. The top of the gage block measured in position 1 shown is 139.54. Does this indicate that the slot is outside of its positional tolerance?

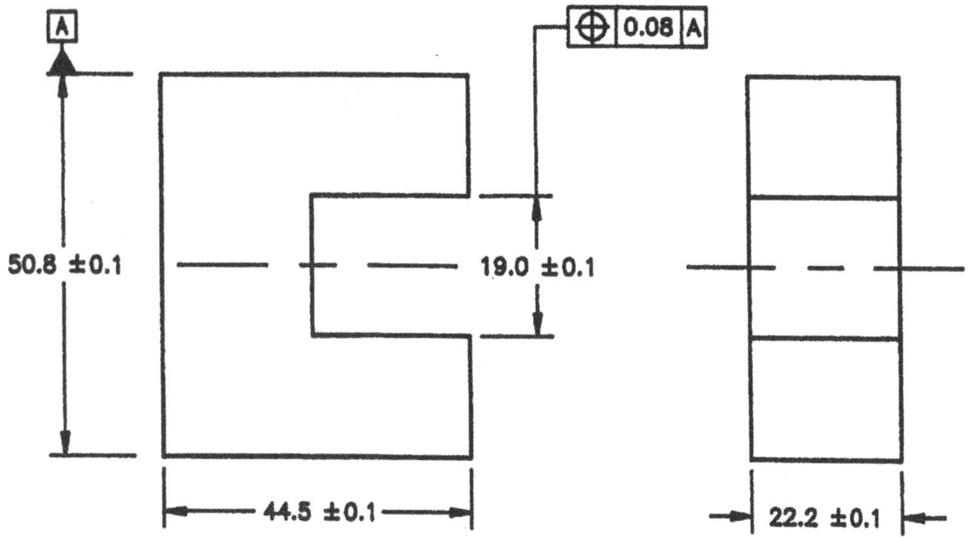

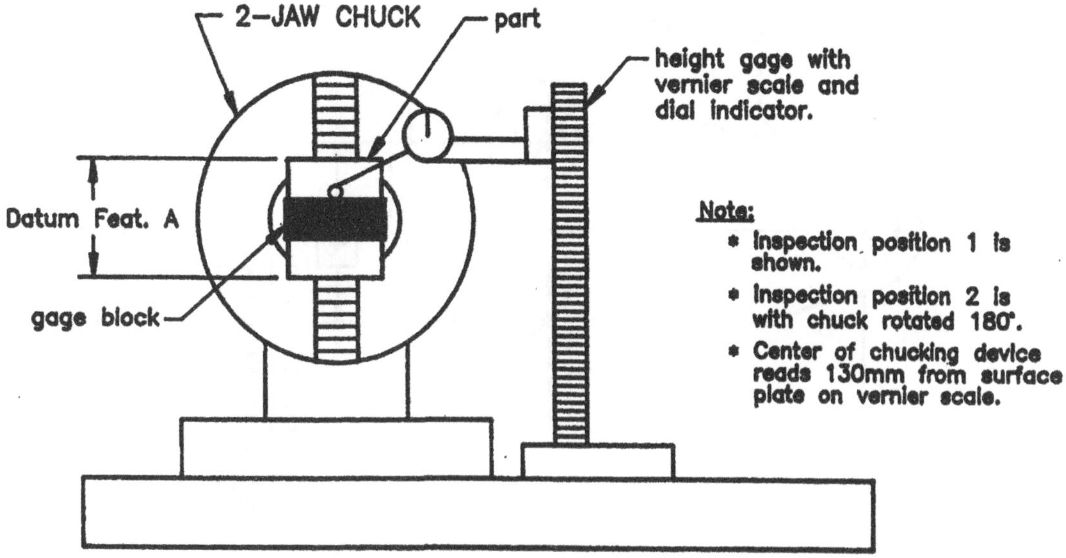

65.

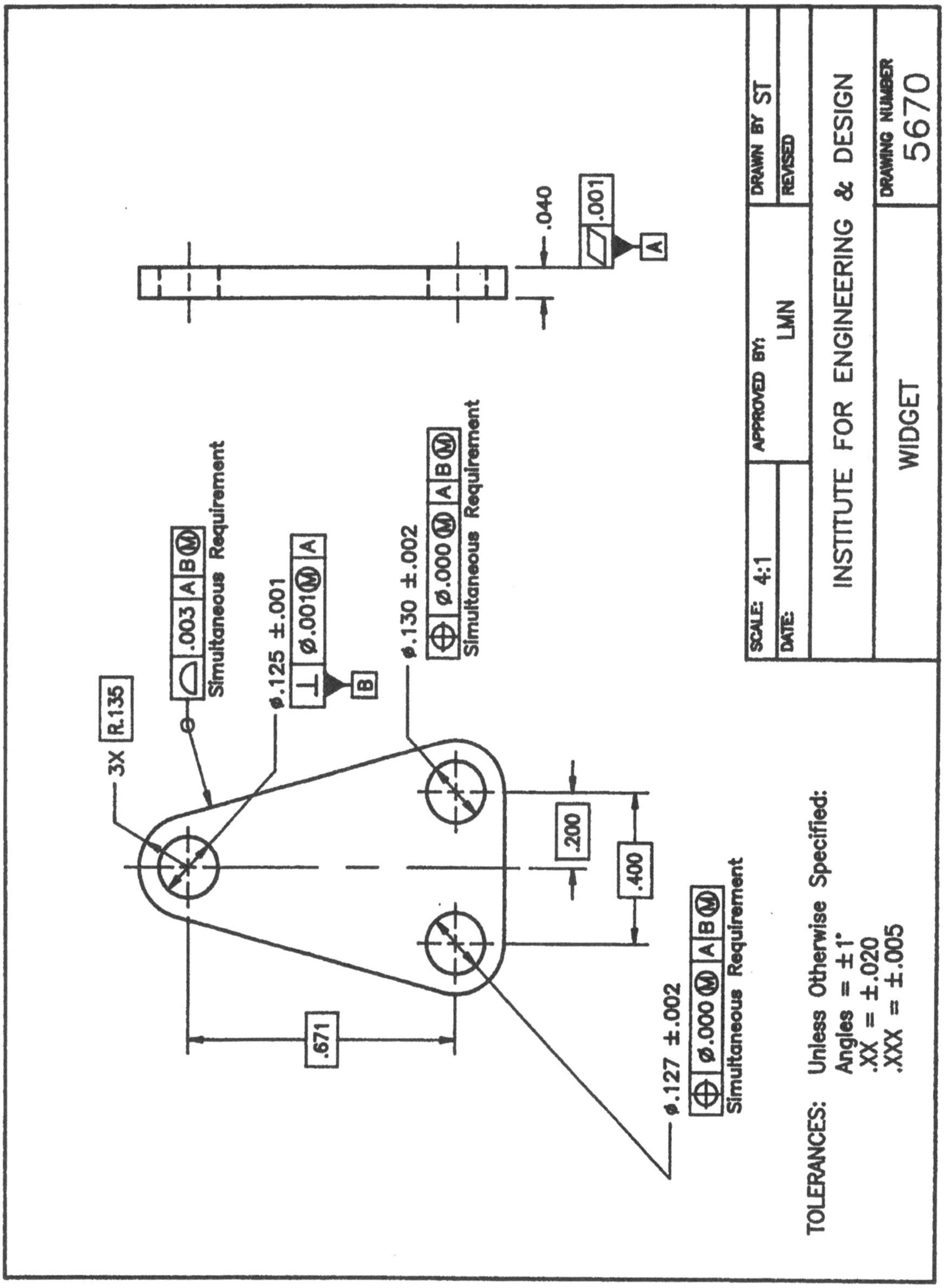
3X R.135
.003 A B Ⓜ
Simultaneous Requirement
⌀.125 ±.001
⊥ Ø.001Ⓜ A
B
⌀.130 ±.002
⌖ Ø.000 Ⓜ A B Ⓜ
Simultaneous Requirement
.200
.400
.671
⌀.127 ±.002
⌖ Ø.000 Ⓜ A B Ⓜ
Simultaneous Requirement
.040
.001
A
TOLERANCES: Unless Otherwise Specified:
Angles = ±1°
.XX = ±.020
.XXX = ±.005
SCALE: 4:1
DATE:
APPROVED BY:
LMN
DRAWN BY ST
REVISED
INSTITUTE FOR ENGINEERING & DESIGN
WIDGET
DRAWING NUMBER
5670

65. (continued) Please fill in the missing radii and diameters representing the allowed extremes of the part's profile and the virtual conditions of the holes.

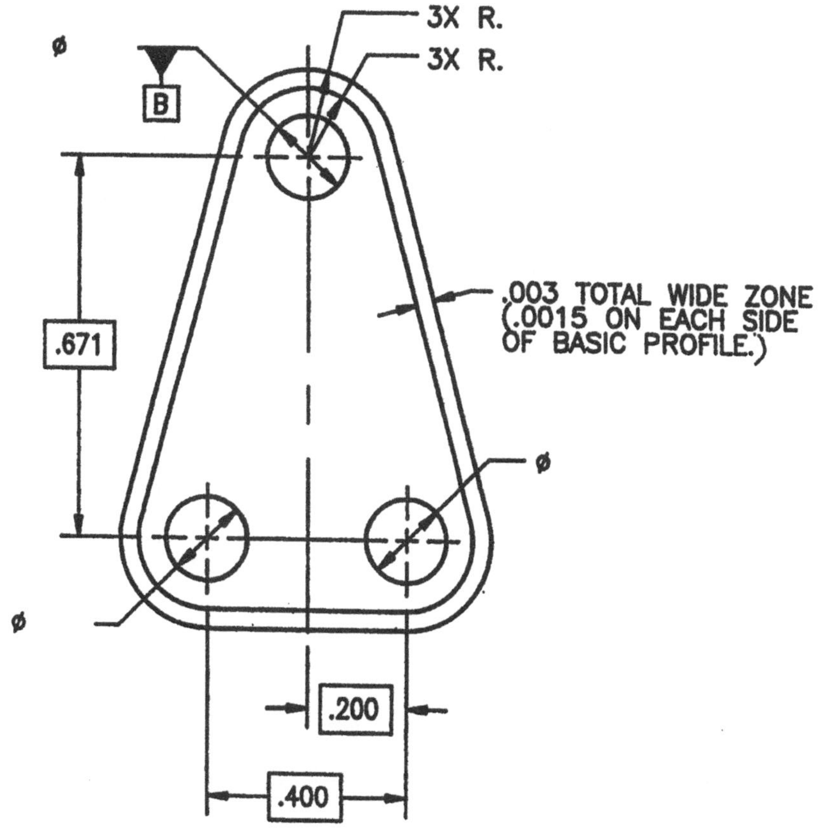

66.

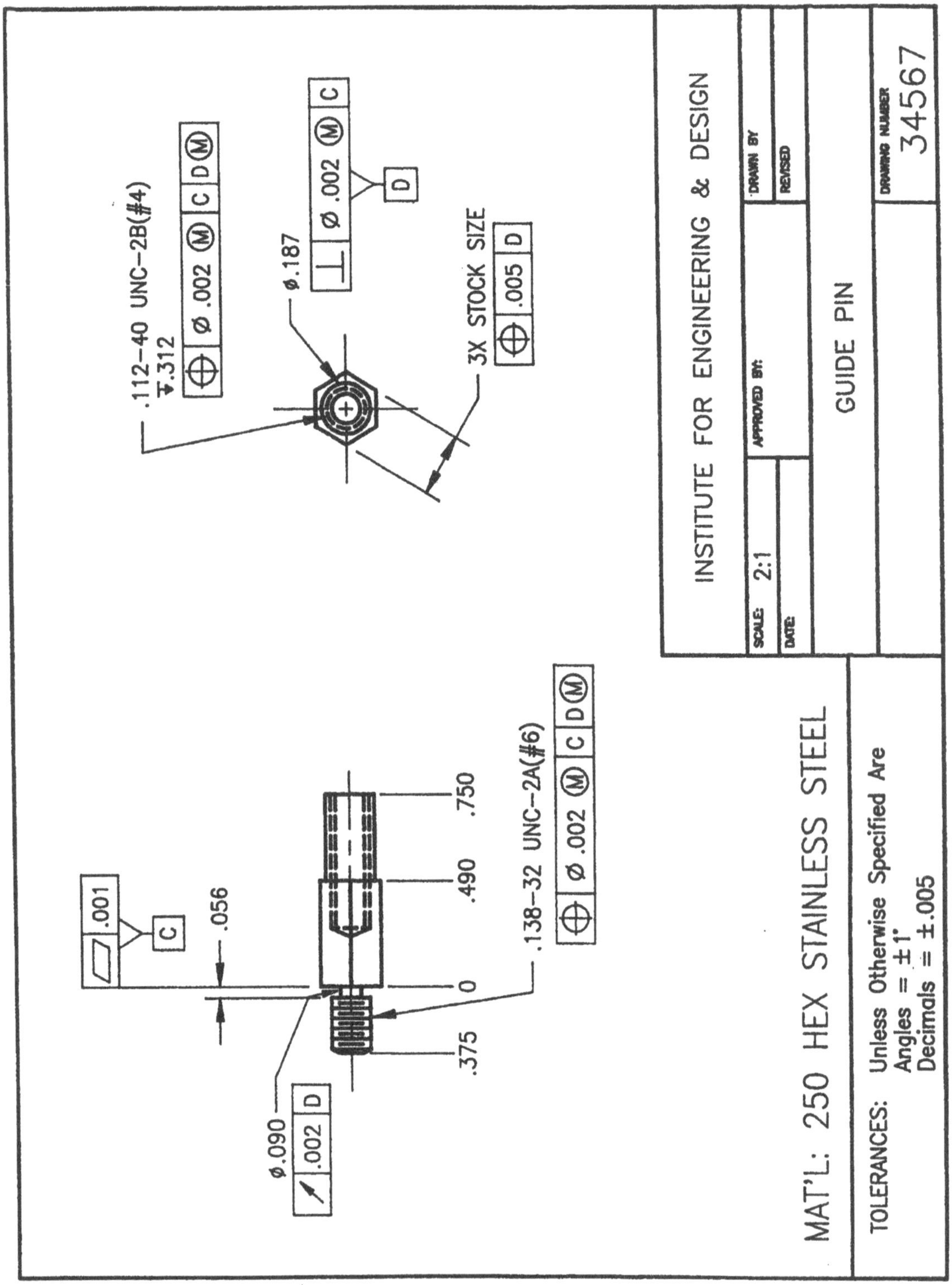
.112-40 UNC-2B(#4)
↧.312
⊕ | Ø.002 Ⓜ | C | D Ⓜ
Ø.187
⊥ | Ø.002 Ⓜ | C
D
3X STOCK SIZE
⊕ | .005 | D
.001 | C
C
.056
0
.490
.750
.375
.138-32 UNC-2A(#6)
⊕ | Ø.002 Ⓜ | C | D Ⓜ
Ø.090
↗ | .002 | D
INSTITUTE FOR ENGINEERING & DESIGN
SCALE: 2:1
DATE:
APPROVED BY:
DRAWN BY
REVISED
GUIDE PIN
DRAWING NUMBER
34567
MAT'L: 250 HEX STAINLESS STEEL
TOLERANCES: Unless Otherwise Specified Are
Angles = ±1°
Decimals = ±.005

66. (continued)

1) How deep a pit in datum feature C is acceptable?

2) How tall a bump is acceptable in the surface with the runout control?

3) How far off center radially may the axis of the diameter be that is controlled with circular runout?

4) How tapered may the diameter be that has a circular runout control on it?

5) What is the maximum bonus tolerance that could be allowed to the .112 threaded hole that is controlled with position due to the | ⌖ | ⌀.002Ⓜ | ?

6) What is the maximum datum shift that could be experienced due to the DⓂ for the hole discussed in question 5?

7) What is the maximum bonus tolerance that datum feature D could experience?

8) What is the shape and size of the position tolerance zone for the six-sided figure?

67. Please dimension and tolerance the sizes and positional controls for the gages depicted at the bottom of the page. Leave x's as are, but fill in all blanks. Make all positional tolerances zero at MMC. Use 10% of the difference between the LMC and the applicable virtual condition of the holes on the part as gage tolerance. Put it all into the plus tolerance on the gage pins.

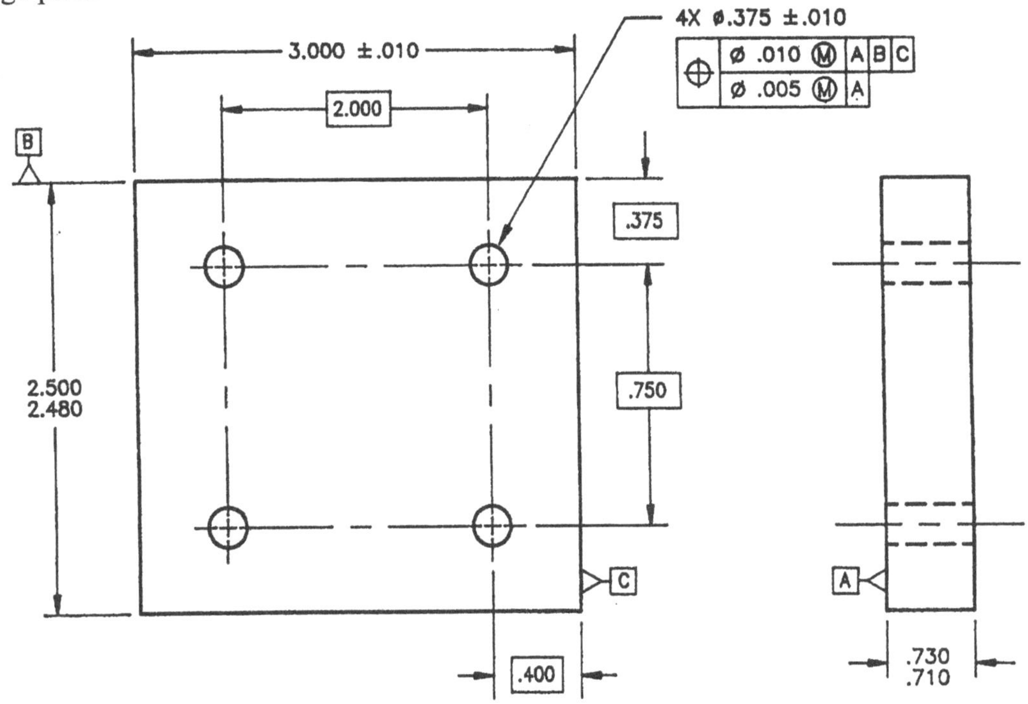

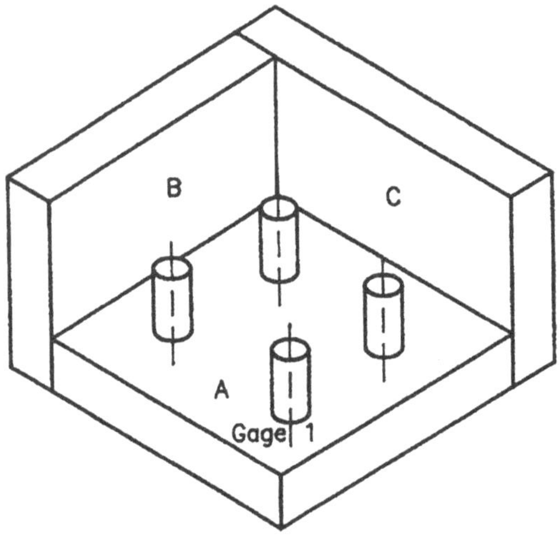

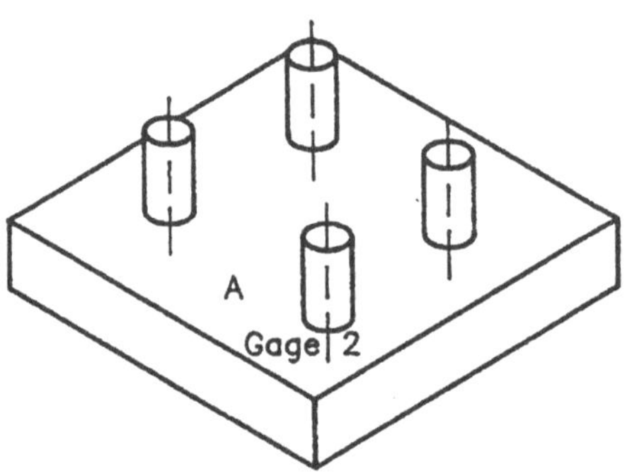

GAGE 1 – Datum to pattern
All four pins are located by the same basic dimensions (from the datum feature simulators) as used on the part drawing.

Pin sizes are ø ______ $^{+}_{-}$

with | ⌖ | ø ______ Ⓜ | A | B | C |

GAGE 2 – Feature to feature pin axis to pin axis basic dimensions are the same as gage 1 (and part drawing).

Pin sizes are ø ______ $^{+}_{-}$

with | ⌖ | ø .xxx Ⓜ | A | | |
|---|---|---|---|---|
| | ø ______ Ⓜ | A | | |

68. For the part shown at the top of this page, please answer the questions about each of the three gages shown (one on this page and two on the following page). These questions deal with the appropriate diameter of the gages' push pins and whether or not to project the tolerance zones on the holes that receive the push pins.

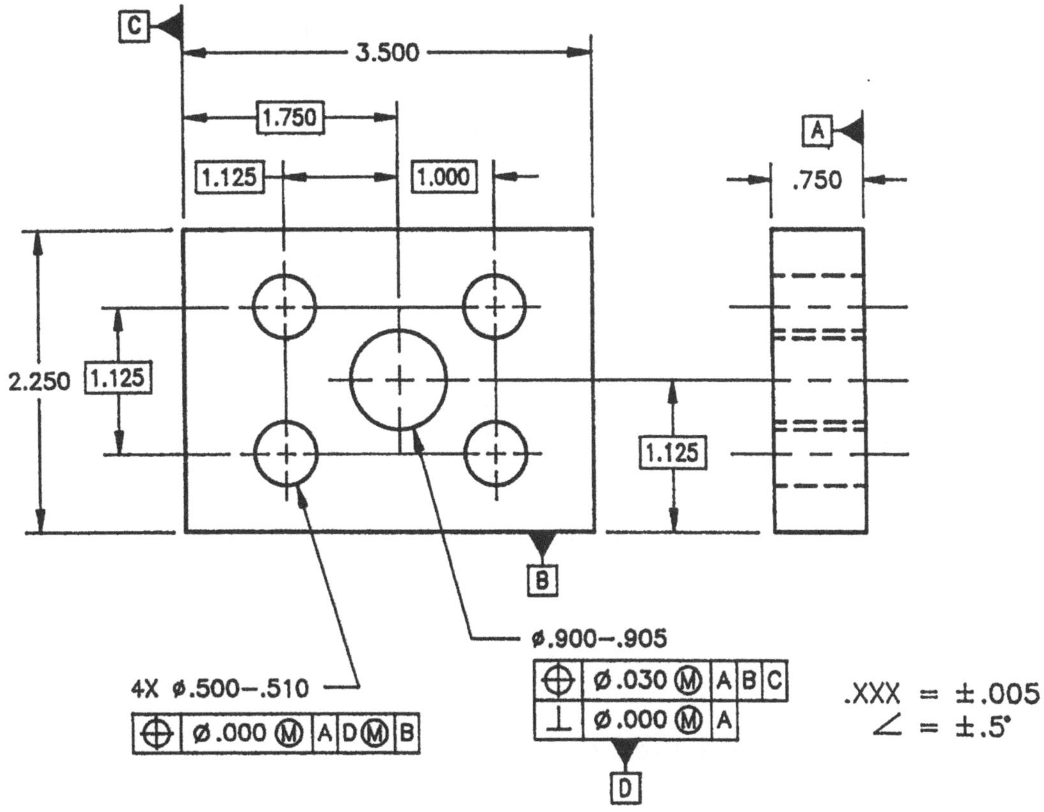

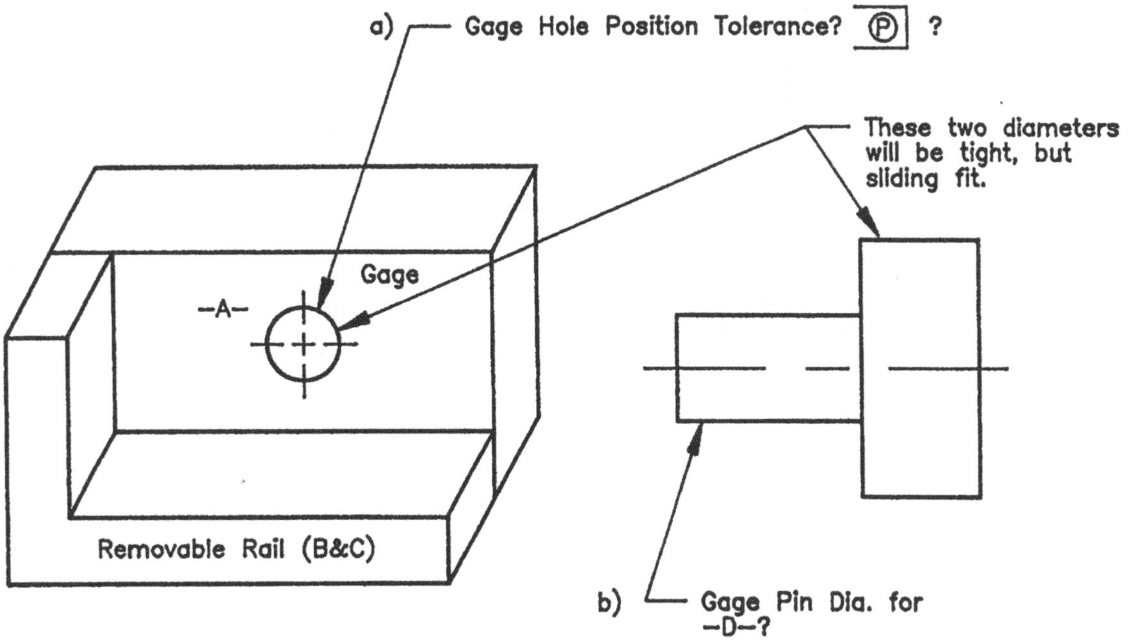

68. (continued)

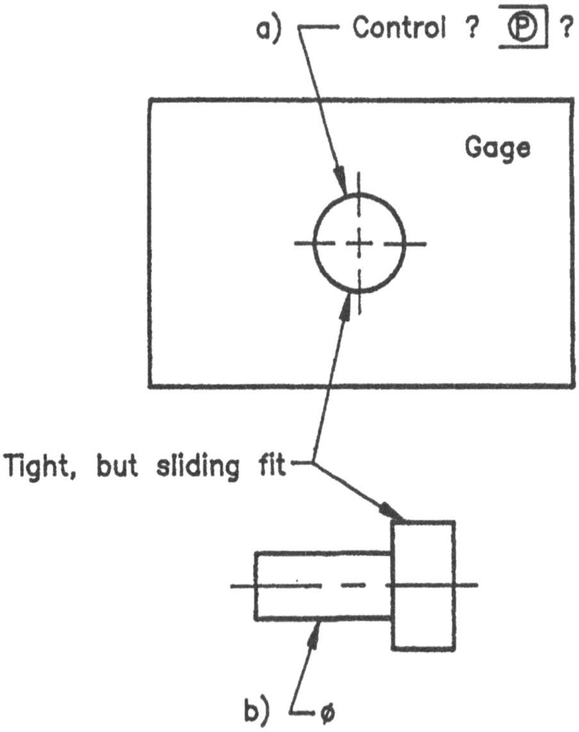

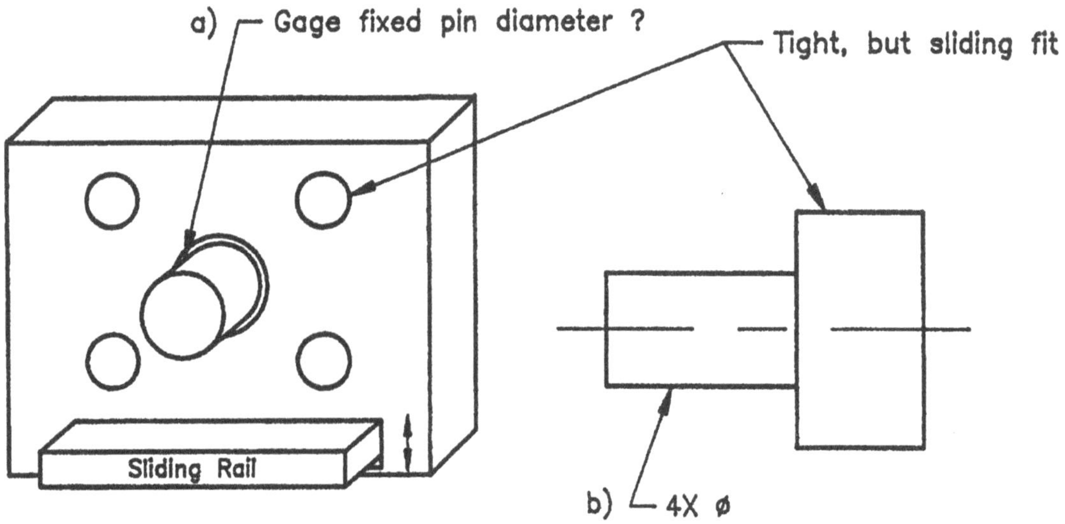

69. Please dimension and tolerance the sizes and positional controls for the gages depicted at the bottom of the page. Please fill in all blanks. Make all positional tolerances zero at MMC. Use 10% of the difference between the LMC and the applicable virtual condition of the holes on the part as gage tolerance. Put it all into the plus tolerance on the gage pins.

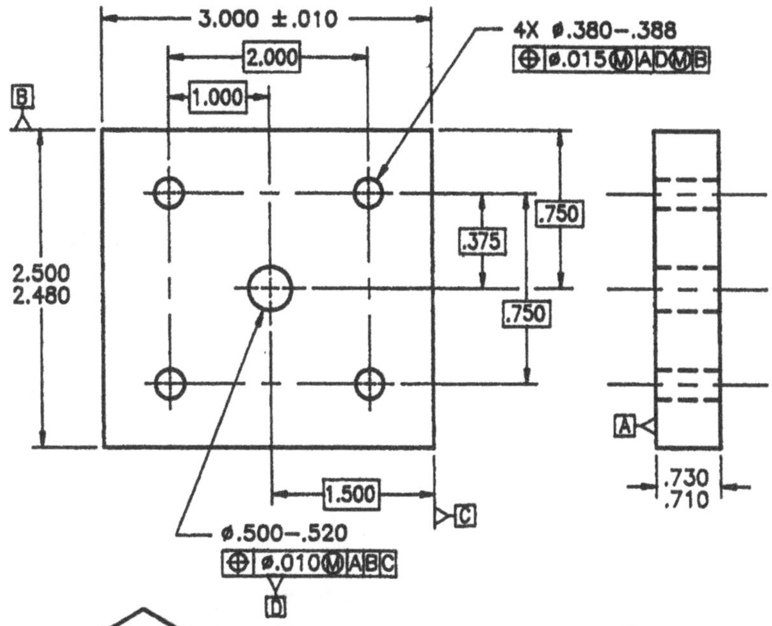

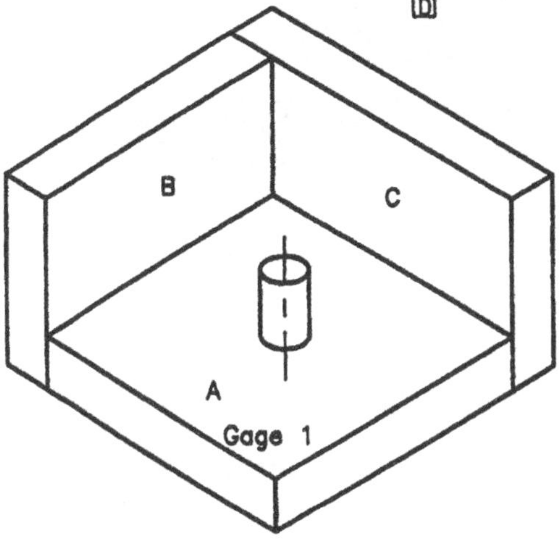

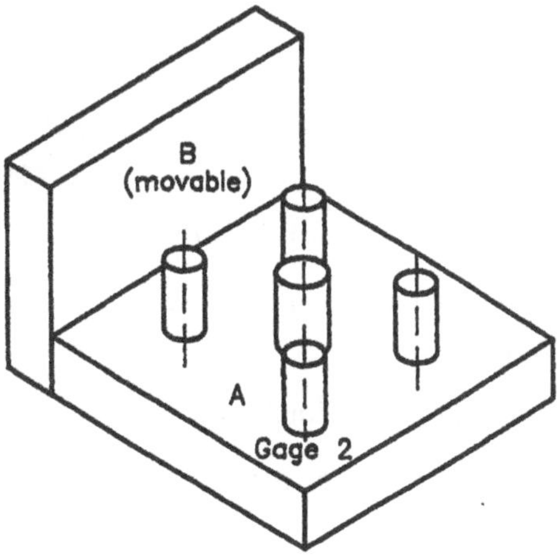

GAGE 1 – Gage inspects location of D. Pin axis is perpendicular to A, .750 basic from B and 1.500 from C.

Pin size is ø ______ + ______ − ______

with | ⊕ | ø ______ Ⓜ | A | B | C |

GAGE 2 – Gage inspects four hole pattern. Pin location is from axis of center pin. Each pin is perpendicular to A and .375 basic by 1.000 basic from center pin axis (representing D at its virtual condition of ø.490, see gage 1). B wall is for angular orientation (pattern rotation) only and therefore is movable in a parallel fashion. It requires a 2 high point min. contact from the part.

Pin size is 4X ø ______ + ______ − ______

with | ⊕ | ø ______ Ⓜ | A | D | B |

70.

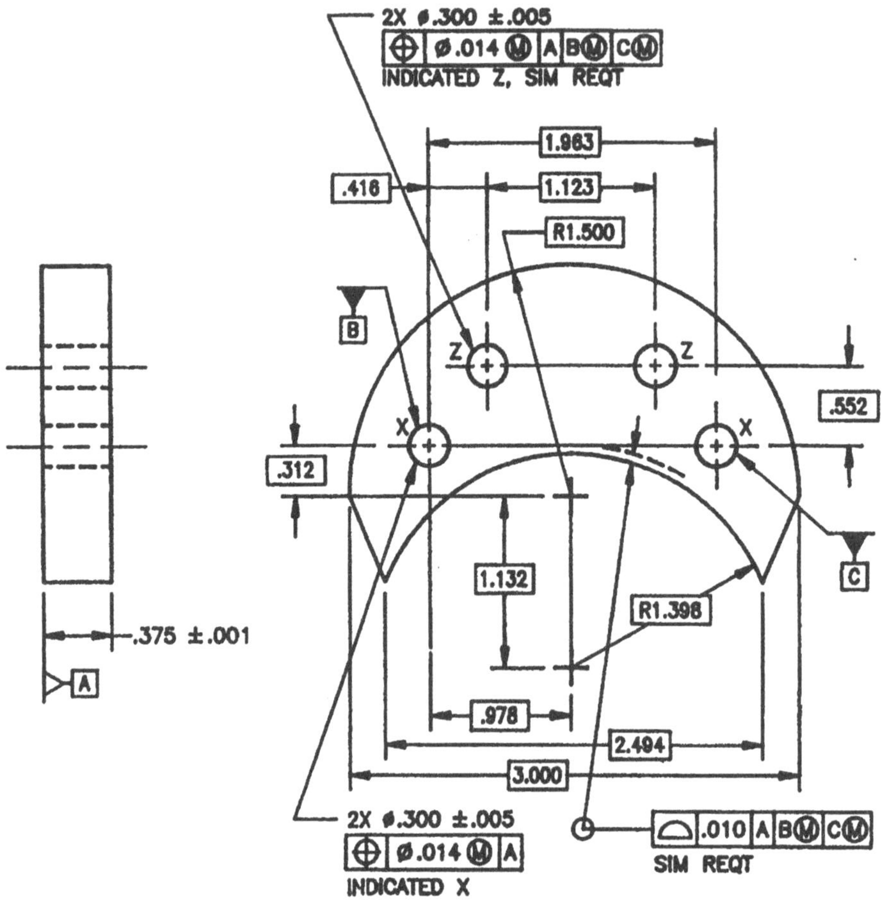

1) How flat must datum feature A be when produced?

2) For the two holes INDICATED X, what are they being positioned to besides datum feature A?

3) How many holes does the profile control tie the profile to?

4) What acts as the origin of measurement for distance for the two holes that are INDICATED Z?

5) What characteristic of geometry does datum A tolerance when used in each feature control frame?

6) Is the profile control a total wide tolerance zone of .010 or ± .010?

7) What is the virtual condition of the X holes? ...the Z holes?

8) What acts as the angular orientation datum?

9) If the holes are produced at ∅.300, how much positional tolerance are they entitled to? ...at ∅.295? ...at ∅.305?

10) Would it be more difficult for a coordinate measurement machine to establish the datums shown or a pattern datum consisting of all four holes? Why?

11) What measurement equipment would you prefer to inspect the profile control and why?

70. (continued)

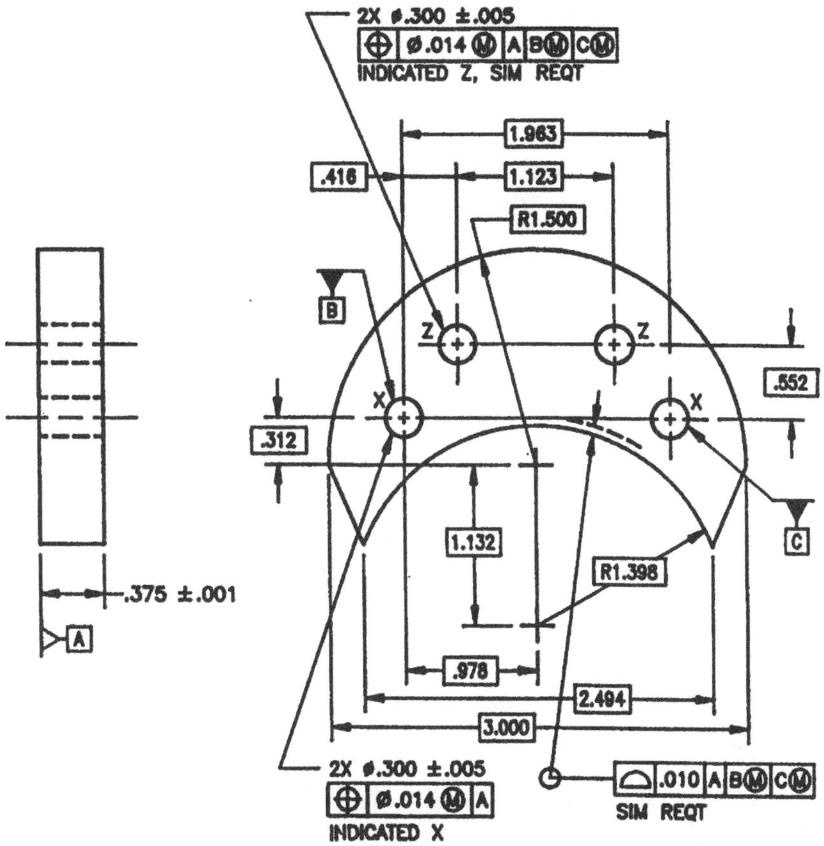

1) If the part thickness is measured with a micrometer at .374 and the median plane bows .003, is the part within its size tolerance?

2) If the two holes INDICATED X are produced at a size of ∅.300 and their axes a distance apart of 1.982, are they acceptable?

3) If the profile size shown as a basic 3.000 is produced at 2.980, does it exceed its tolerance?

4) If the two Z holes are produced at ∅.305 and their axes at a distance of 1.099 from each other, are they within tolerance?

5) If the holes are produced at ∅.300 and out-of-perpendicularity to datum A ∅.015, are they out of tolerance?

6) If the surface profile has a bump that exceeds the basic radius shown by .005, is the surface to be rejected as out of tolerance?

7) If the holes measure with a micrometer ∅.299 and can receive a gage pin (without regard to orientation or location) no larger than a ∅.292, do they exceed their size tolerance?

8) If the largest gage pin that can be inserted into datum feature B measures ∅.305, the largest gage pin that can be inserted into datum feature C measures a ∅.295 and a vernier caliper measurement from the outermost points on the gage pin diameters is a distance of 2.282, would that indicate that they exceed their positional tolerance?

71. Answer the following questions.
    1) What are the largest acceptable produced dimensions for the outside profile of this part?
    2) What is the maximum axial separation allowed between datum features B and C?
    3) What is the minimum axial separation allowed between datum features B and C?
    4) What is the maximum allowed distance between the axis of datum feature B and the edge of the part dimensioned .500 basic from it?
    5) What is the minimum allowed distance between the axis of datum feature B and the edge of the part dimensioned .500 basic from it?

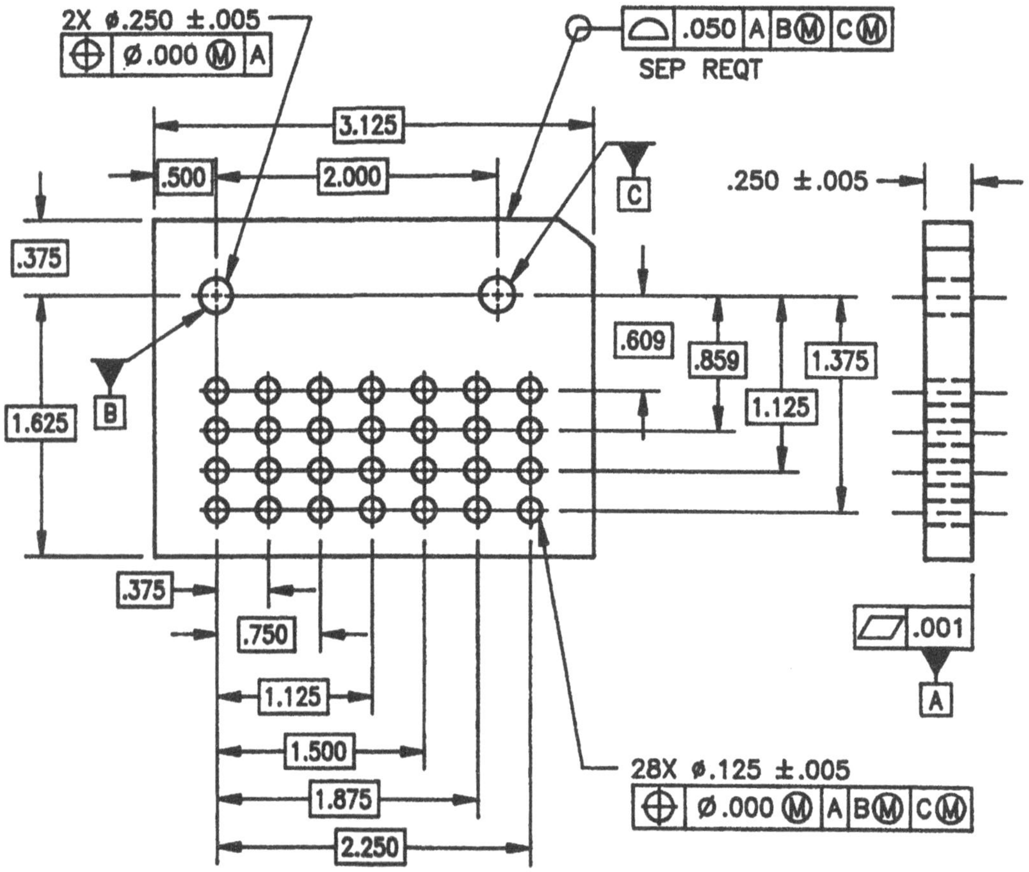

71. (continued)
   1) What are the largest acceptable produced dimensions for the outside profile of this part?
   2) What is the maximum axial separation allowed between datum features B and C?
   3) What is the minimum axial separation allowed between datum features B and C?
   4) What is the maximum allowed distance between the axis of datum feature B and datum plane D using the .500 basic dimension?
   5) What is the minimum allowed distance between the axis of datum feature B and datum plane D using the .500 basic dimension?

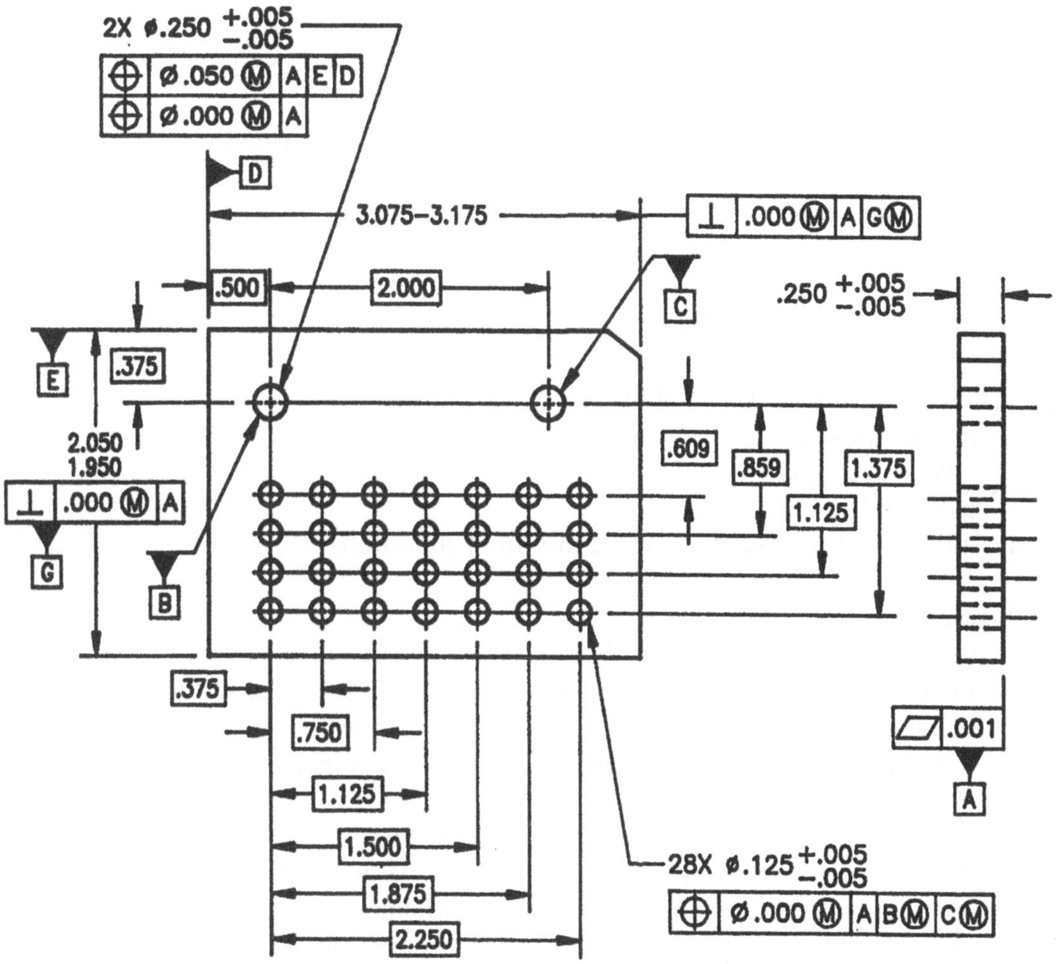

72. Complete the following questions by circling the appropriate answer.
    1) Unless otherwise specified, limits of size are measured in the following manner:
        a) LMC at cross sections (opposing points) and MMC for violations of a full form envelope.
        b) LMC and MMC are inspected for violations at every two diametrically opposed sets of points.
        c) LMC and MMC are both inspected with a full form envelope gage.
        d) LMC is a full form envelope check and MMC is a cross sectional check.

    2) When basic dimensions are given to locate a hole,
        a) no tolerance is allowed on the hole location.
        b) the location tolerance in the general title tolerance block on the drawing applies.
        c) the location tolerance is found in a position feature control frame.
        d) the tolerance will appear next to the basic dimension as a plus or minus tolerance.

    3) The correct formula for 'bonus tolerance' for a hole controlled with the MMC concept is
        a) LMC plus MMC.
        b) MMC minus virtual condition.
        c) actual size minus LMC.
        d) actual size minus MMC.

    4) Total runout can be inspected the best with
        a) a micrometer.
        b) a vernier caliper.
        c) a dial indicator.
        d) a fixed size receiver gage made at virtual condition.

    5) Position deviation from 'true position' can be calculated using the following formula:
        a) two times the square root of the sum of the squares of the x and y deviations.
        b) the difference between the sum of the square roots of the x and y deviations.
        c) the dividend of the squares of the x and y deviations.
        d) the square root of two times the difference in the x and y deviations.

    6) A coordinate measurement machine can most accurately simulate a primary datum by
        a) probing three arbitrary points on the primary datum feature.
        b) probing a hundred arbitrary points on the primary datum feature.
        c) probing the surface plate on which the highest points of the datum feature rest.
        d) probing the surface of the mating feature.

72. (continued)
    7) The datum axis generated by an external cylindrical datum feature is the axis of the
        a) minimum circumscribed cylinder.
        b) the least squares center.
        c) the best fit center of the probed points.
        d) the minimum radial separation center of the two cylinders that are able to contain all points of the cylinder's surface.

    8) An optical comparator is most often associated with the inspection of
        a) flatness.
        b) profile.
        c) perpendicularity.
        d) concentricity.

    9) Rocking datum features may
        a) be rejected.
        b) bow beyond their maximum material condition boundary.
        c) be restrained with enough torque to allow controlled features to be accepted.
        d) be adjusted to an optimum position.

    10) A cylindricity geometric characteristic symbol with a geometric tolerance of .003 allows
        a) pits as deep as .006.
        b) bumps as tall as .006.
        c) the geometric tolerance to be additive to the size tolerance.
        d) size variations for the cylinder's diameter as large as .006.

73. Complete the following questions by circling the appropriate answer.

1) Which of the following would be inappropriate to include in a feature control frame for a straightness control?
   a) a diameter symbol
   b) a maximum material condition symbol
   c) a datum reference
   d) a geometric tolerance

2) Runout is what type of control?
   a) surface to datum axis
   b) centerplane to datum axis
   c) an axis to datum axis
   d) derived median line to datum axis

3) Circular runout of a diameter does not control which of the following characteristics?
   a) coaxiality
   b) circularity
   c) straightness of the surface
   d) straightness of the axis

4) Position is a geometric control that is not capable of controlling
   a) perpendicularity
   b) distance
   c) coaxiality
   d) circularity

5) Of the following geometric characteristic symbols, which is the most restrictive in its control of surface characteristics?
   a) circular runout
   b) position
   c) total runout
   d) concentricity

6) Of the following geometric characteristic symbols, which is considered a 2-D line element control?
   a) circularity
   b) flatness
   c) concentricity
   d) cylindricity

7) Of the following geometric characteristic symbols, which is least capable of controlling flatness?
   a) perpendicularity
   b) angularity
   c) parallelism
   d) straightness

73. (continued)

8) Limits of size on a rectangular part are not a control of which of the following?
   a) flatness
   b) perpendicularity
   c) parallelism
   d) straightness

9) Virtual condition (MMC concept) is calculated for a hole by the following formula:
   a) MMC of the hole minus the geometric tolerance applicable at MMC.
   b) MMC of the hole plus the geometric tolerance applicable at MMC.
   c) MMC of the hole divided by the geometric tolerance applicable at MMC.
   d) MMC of the hole multiplied by the geometric tolerance applicable at MMC.

10) Positional tolerance for all holes in a floating fastener situation can be calculated in the following manner:
   a) MMC of the hole minus the LMC of the mating fastener.
   b) MMC of the hole minus the MMC of the mating fastener.
   c) LMC of the fastener minus the LMC of the hole.
   d) LMC of the hole minus the MMC of the hole.

11) What is the virtual condition of the following shaft?

   Ø8.3 – 8.5

⌖	Ø0.3 Ⓜ	A	B	C

   a) Ø8.0
   b) Ø2.49
   c) Ø8.8
   d) Ø27.67

74. Complete each of the following questions by circling the appropriate answer.

1) When a pattern of holes is positioned to a datum feature of size such as a hole and this datum feature is referenced at MMC, as the datum feature (hole) grows,
   a) each hole in the pattern receives an individual growth in its tolerance zone.
   b) the pattern of holes as a group may shift.
   c) the tolerance zone for each hole shrinks.
   d) the pattern of holes must maintain its relative location to the datum feature.

2) In the concept of composite positional tolerancing,
   a) the features within a pattern hold a tighter tolerance in relation to one another than to any datums that are listed only in the upper level control.
   b) the features within a pattern hold a looser tolerance in relation to one another than to any datums that are listed only in the upper level control.
   c) the features within a pattern hold an equally tight tolerance in relation to one another and to any datums listed only in the upper level control.
   d) the features within a pattern are held in place to one another by a positional tolerance, but are located to datums by a plus and minus tolerance.

3) When a pattern of holes is used to establish a datum reference frame from which to locate other part features,
   a) only one hole may be used to form the datum reference frame.
   b) only two holes may be used to form the datum reference frame.
   c) only three holes may be used to form the datum reference frame.
   d) an infinite number of holes may be used to form the datum reference frame.

4) When multiple patterns of features are located from the same datum reference frame, they are considered to be
   a) a simultaneous requirement.
   b) a separate requirement.
   c) a composite tolerance control.
   d) a combination tolerance control.

5) If a pattern of holes is positioned and no datum features are used in their positional feature control frame,
   a) the control is incorrect.
   b) the holes are considered positioned to one another.
   c) implied datum features are considered as the origin of measurement.
   d) the part must be accepted as produced.

6) Two hole patterns are positioned to the same primary planar datum feature, and this datum feature, due to imperfections of flatness, tends to rock. If the patterns of holes are considered as separate requirements,
   a) the part may be rocked one way to accept one pattern of holes and another way to accept the other pattern of holes during inspection.
   b) the part may not be rocked one way to accept one pattern of holes and another way to accept the other of holes during inspection.

74. (continued)
   c) the hole patterns must be inspected in two different (separate) orientations.
   d) the rocking datum must be brought under control by re-machining it (or another suitable re-manufacturing method) until it no longer rocks.

7) A four-hole pattern has size limits of ∅10.7 – 11.0 and a position tolerance to a datum reference frame of ∅0.3 at MMC. Its mating pin pattern has size limits of ∅9.7 – 10.0 and is positioned to a representative set of datums. How much position tolerance is available for the pins within the pin pattern at MMC?
   a) ∅0.7
   b) ∅1.0
   c) ∅0.5
   d) ∅0.4

8) For the circumstances given in Question 7, the position tolerance that would be allowed for a hole produced at ∅11 is
   a) ∅0.4
   b) ∅0.5
   c) ∅0.6
   d) ∅0.3

9) For the circumstances given in Question 7, what is the virtual condition (MMC concept) for the pins within the pin pattern?
   a) ∅10.7
   b) ∅11.0
   c) ∅10.5
   d. ∅10.4

75. The following example demonstrates factors to determine minimum wall thicknesses. It is a good beginning for learning how to calculate any wall thickness problem. I call it the "Drinking Straw" problem. The premise is that most parts can be turned in drinking straws (or gun barrels) with a little thought.

Please calculate the minimum wall thickness between the inside diameter and the outside diameter of this part.

Inner and Outer Boundaries for LMC

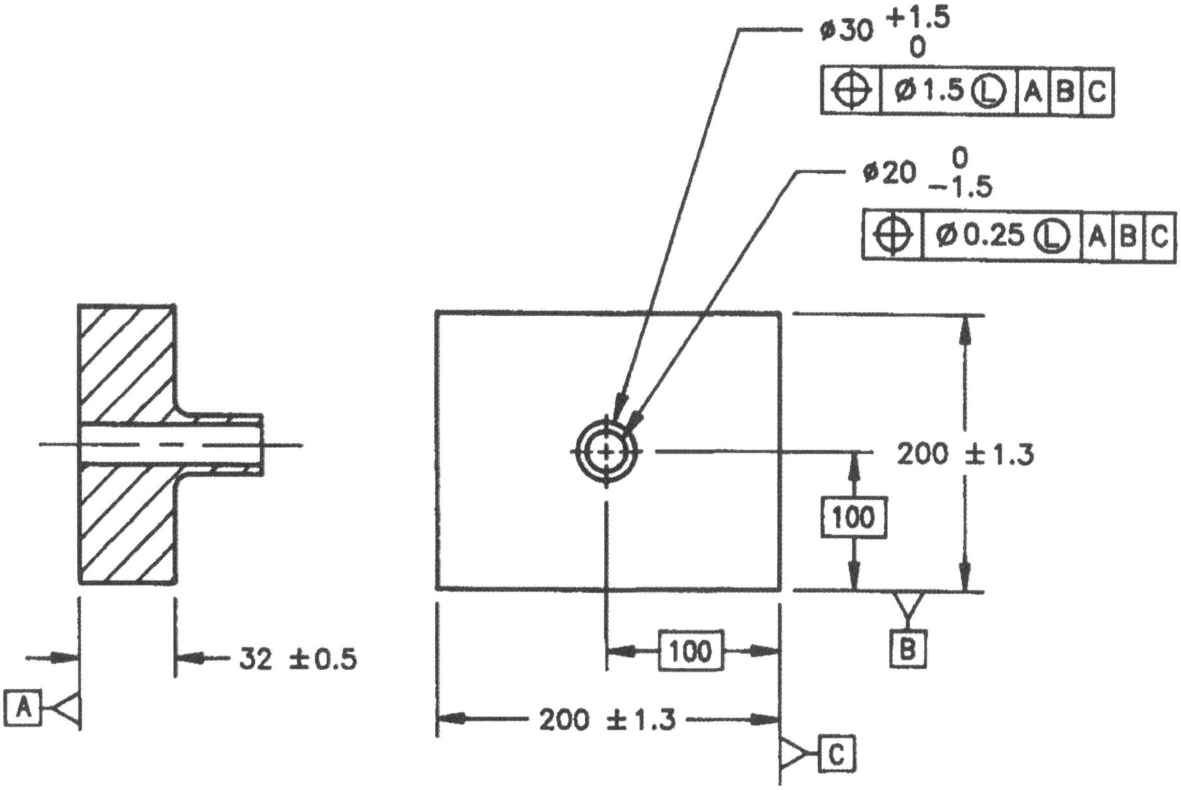

76. Displacement and a Simultaneous Requirement   Please calculate the minimum wall thickness between one of the four holes within the pattern and the outside diameter of the part.

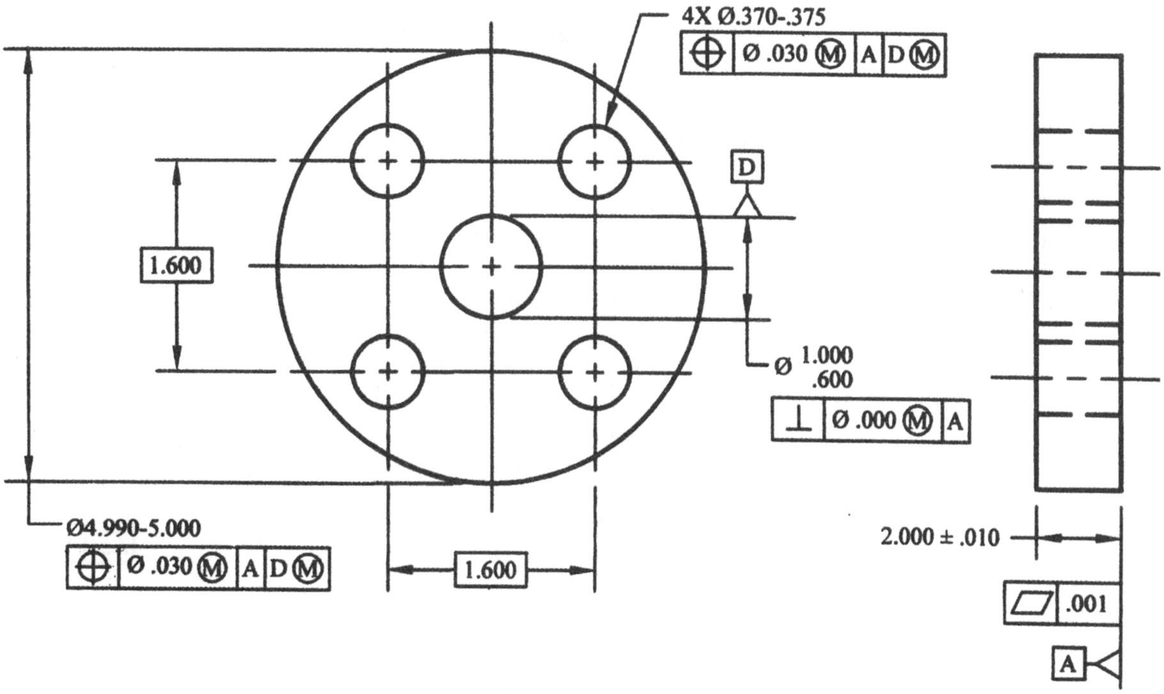

77. Displacement with a Separate Requirement   Please calculate the minimum wall thickness between one of the four holes within the pattern and the outside diameter of the part.

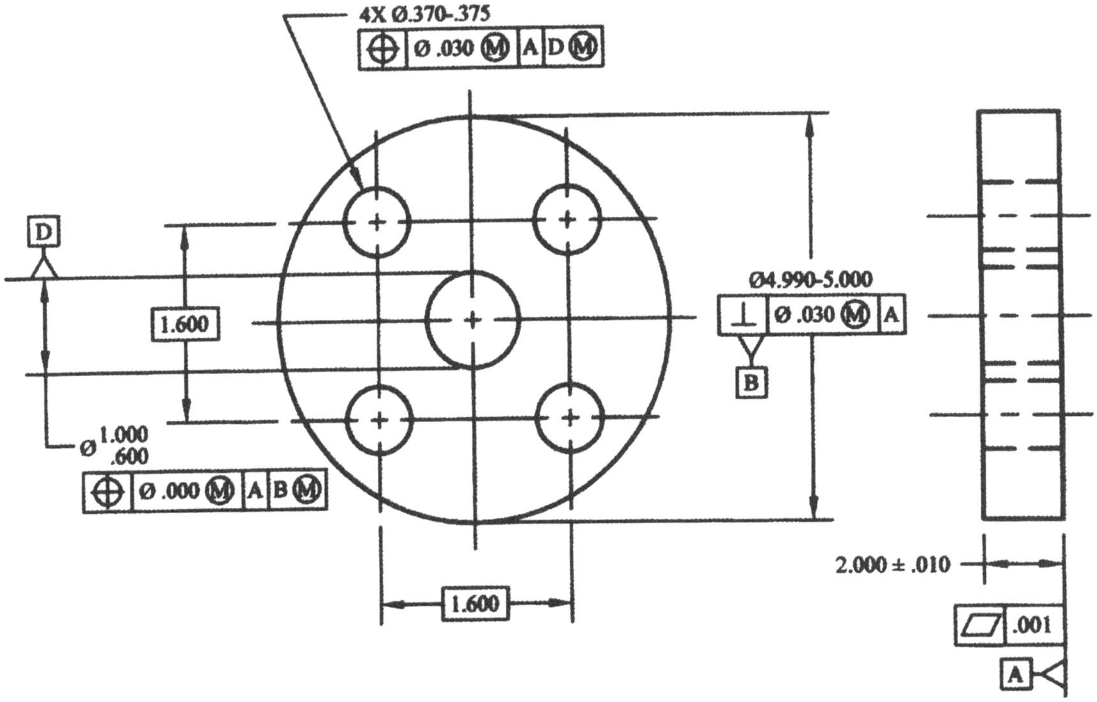

78. Dimension and tolerance Part 2 to mate with Part 1.

PART 1

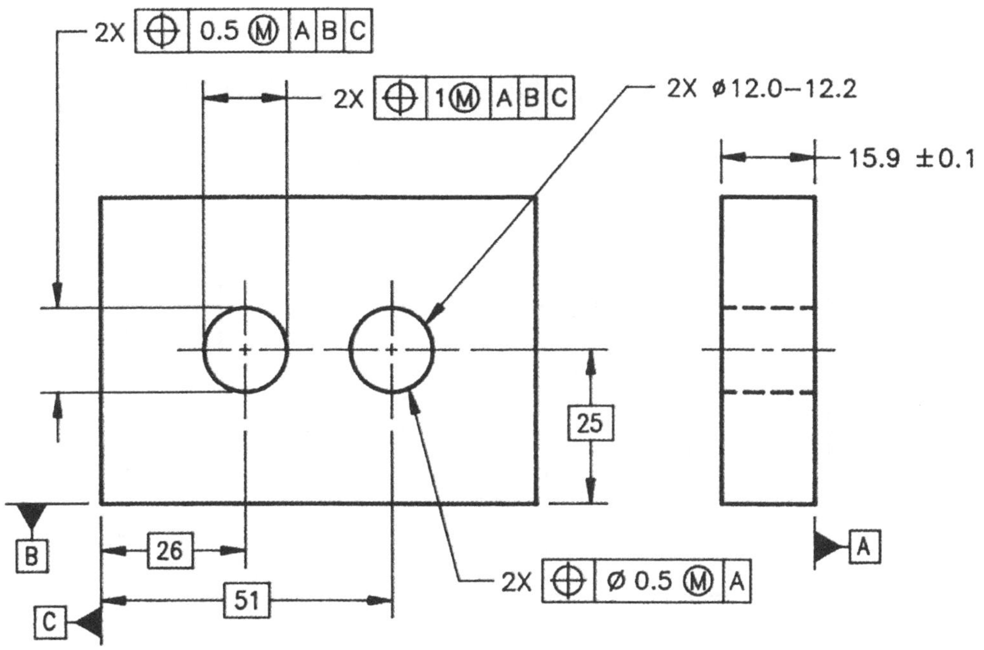

PART 2

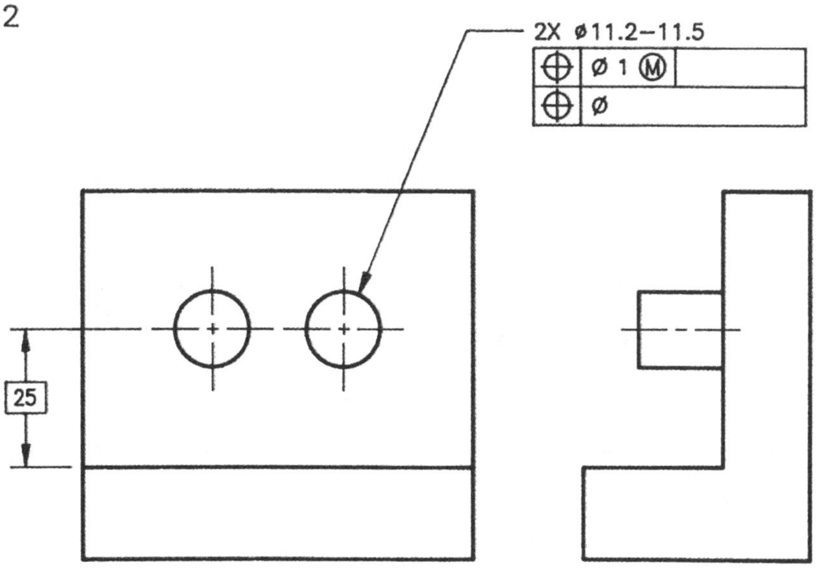

79. Dimension and tolerance Part 2 to mate with Part 1.

PART 1

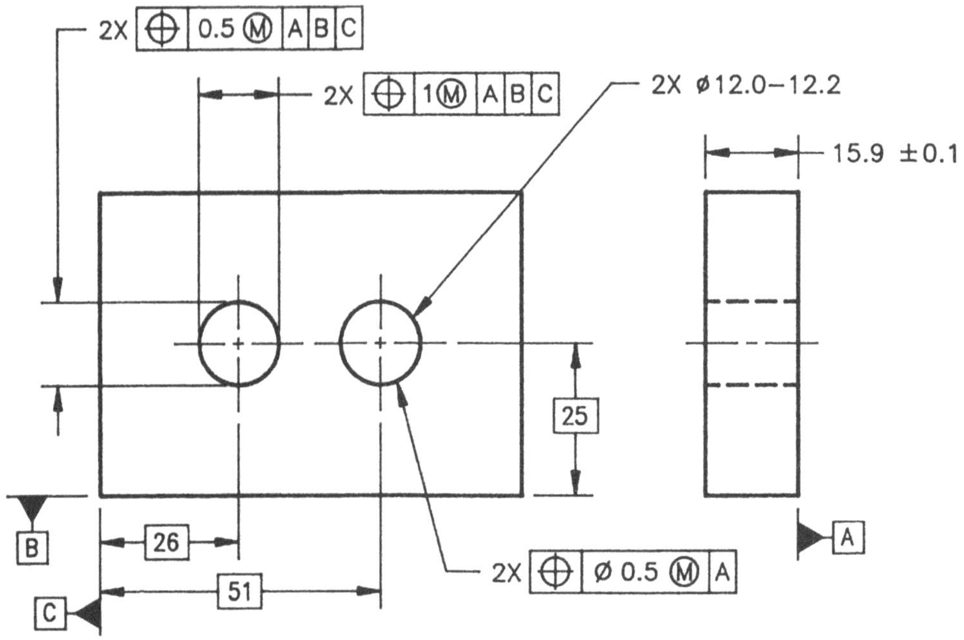

PART 2

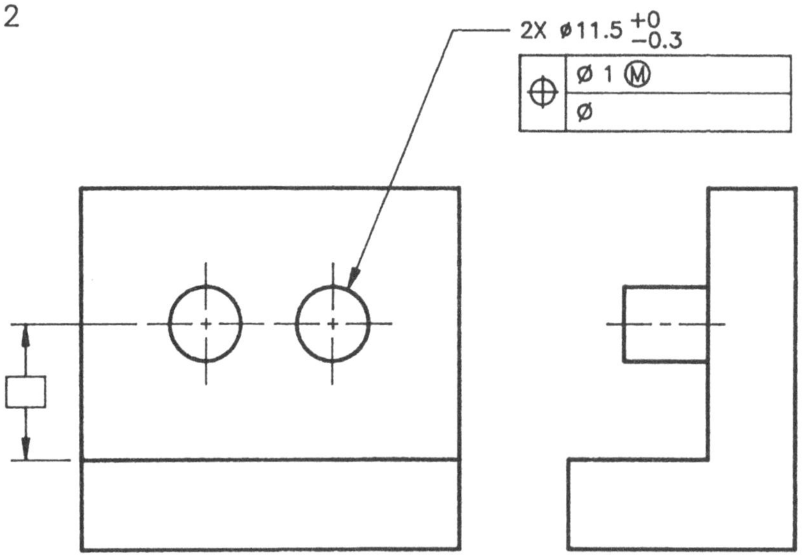

80. Convert this part to use composite positional tolerancing.
   a) Assign 3 datum features.
   b) Convert the general tolerance of ±.06 for the 1.62 and 1.50 dimensions by multiplying the square tolerance zone of .120 by 1.4 (the approximate square root of 2). This will circumscribe a cylindrical tolerance zone around the original square zone. This will serve as the Pattern Locating Tolerance Zone Framework.
   c) Calculate the Feature Relating Tolerance Zone Framework using the formula: MMC holes minus the virtual condition of the shafts equals the geometric tolerance for the holes. The virtual condition of the shafts is a diameter of .256.
   d) Assign the primary datum feature a flatness tolerance of .002.
   e) Assign the 4.0 and 5.0 sizes a perpendicularity tolerance of zero at MMC. Make one of these widths a fourth datum feature to orient the other width to.

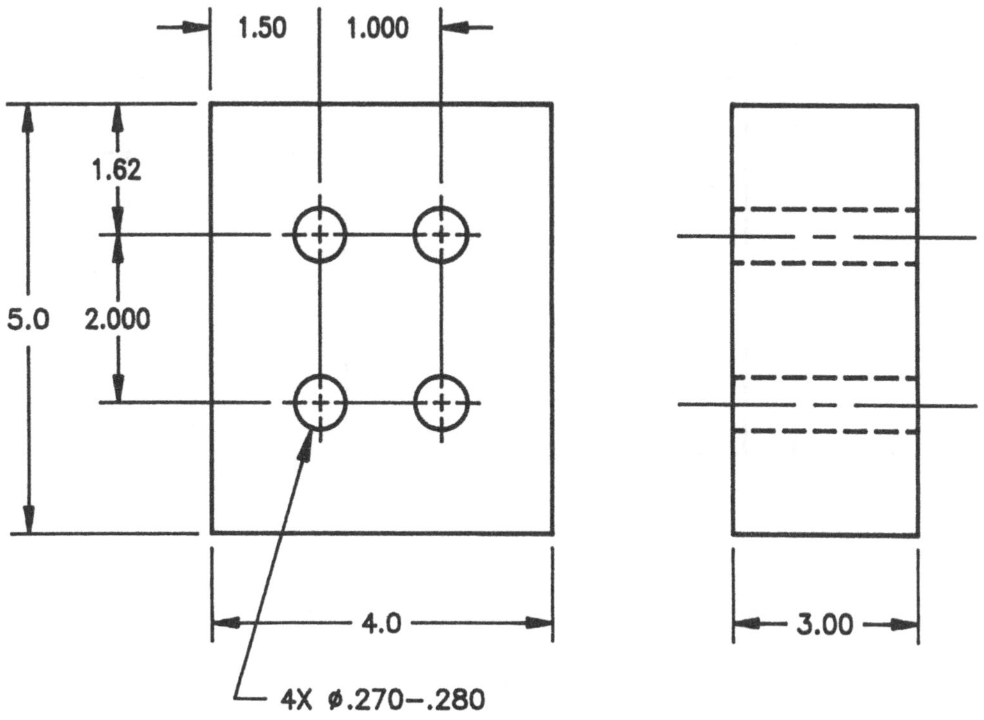

.X = ±.130
.XX = ±.06
.XXX = ±.010
∠ = ±1°

81. This is the mating part for the previous problem. The last part must fit down inside of this cavity and the holes must fit over the pins in the cavity.

a) Assign all dimensions and tolerances.
b) Square the part up as you did on the previous part. Assign the pins in the cavity only a single level of position tolerance.
c) Assign the primary datum feature a flatness tolerance of .002.
d) Assign the pin sizes to be a diameter of .240-.256.
e) Assign the cavity sizes the same size tolerances used on the mating part.
f) Make the cavity big enough to receive the mating part from the previous page.

[TIP: Make certain the cavity's MMC is increased by the difference between the mating holes' virtual condition boundaries.]

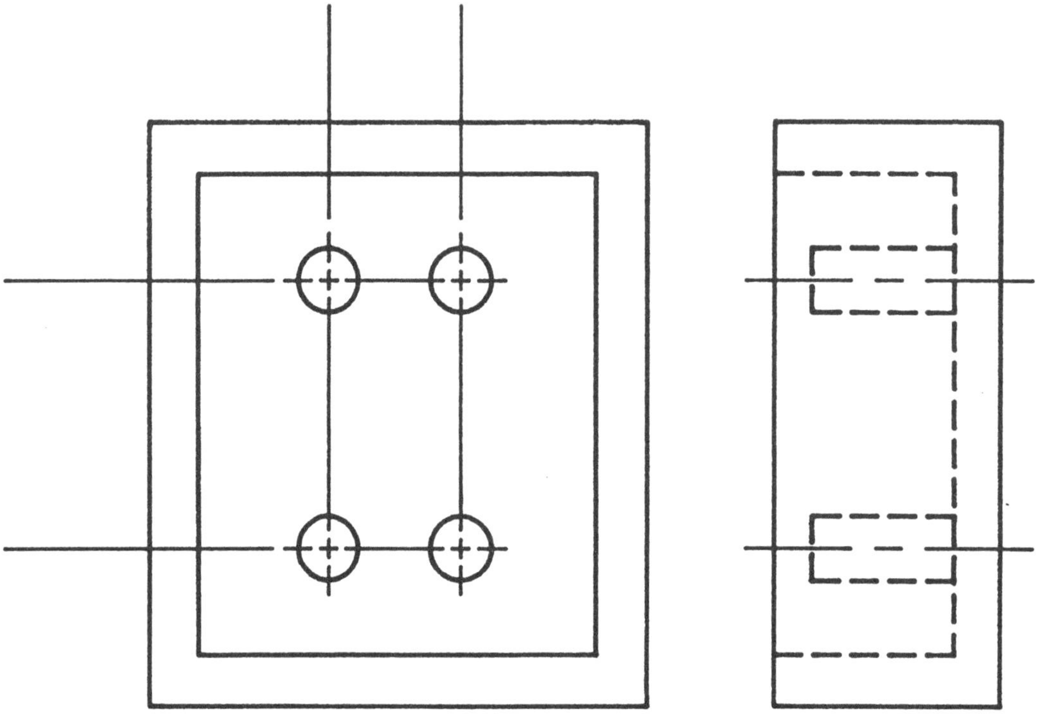

82. Please follow the directions on this page using the following two pages.

Core Board and Receiver Problem

Using the illustrations on the next two pages, complete the following questions. The .056 - .058 holes in the core board will each have the body of a nail with a .070 head (LMC) pressed into them. Nail is shown below. All nail heads must contact pogo pins on the receiver board.

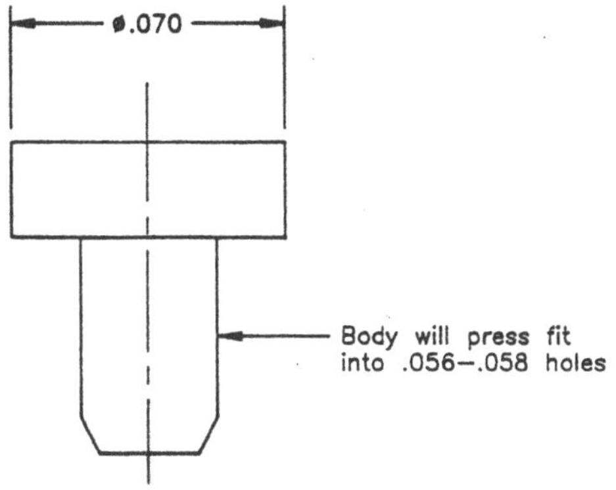

a) Calculate possible movement of heads to determine reduction of .070 contact area.

b) Determine the possible shift from the datum features by subtracting the LMC of the datum pins from the LMC of the datum holes.

c) Each pogo pin will be considered a point contact. Using the shift calculated in b) and the positional tolerance allowed each pogo pin, determine the amount of movement in terms of a diameter away from true position allowed each pogo pin.

d) Will the pogo pin points contact the nail head in all cases? If so, by how much in terms of a diameter? If not, by how much could it miss (in terms of a diameter)?

82. (continued)

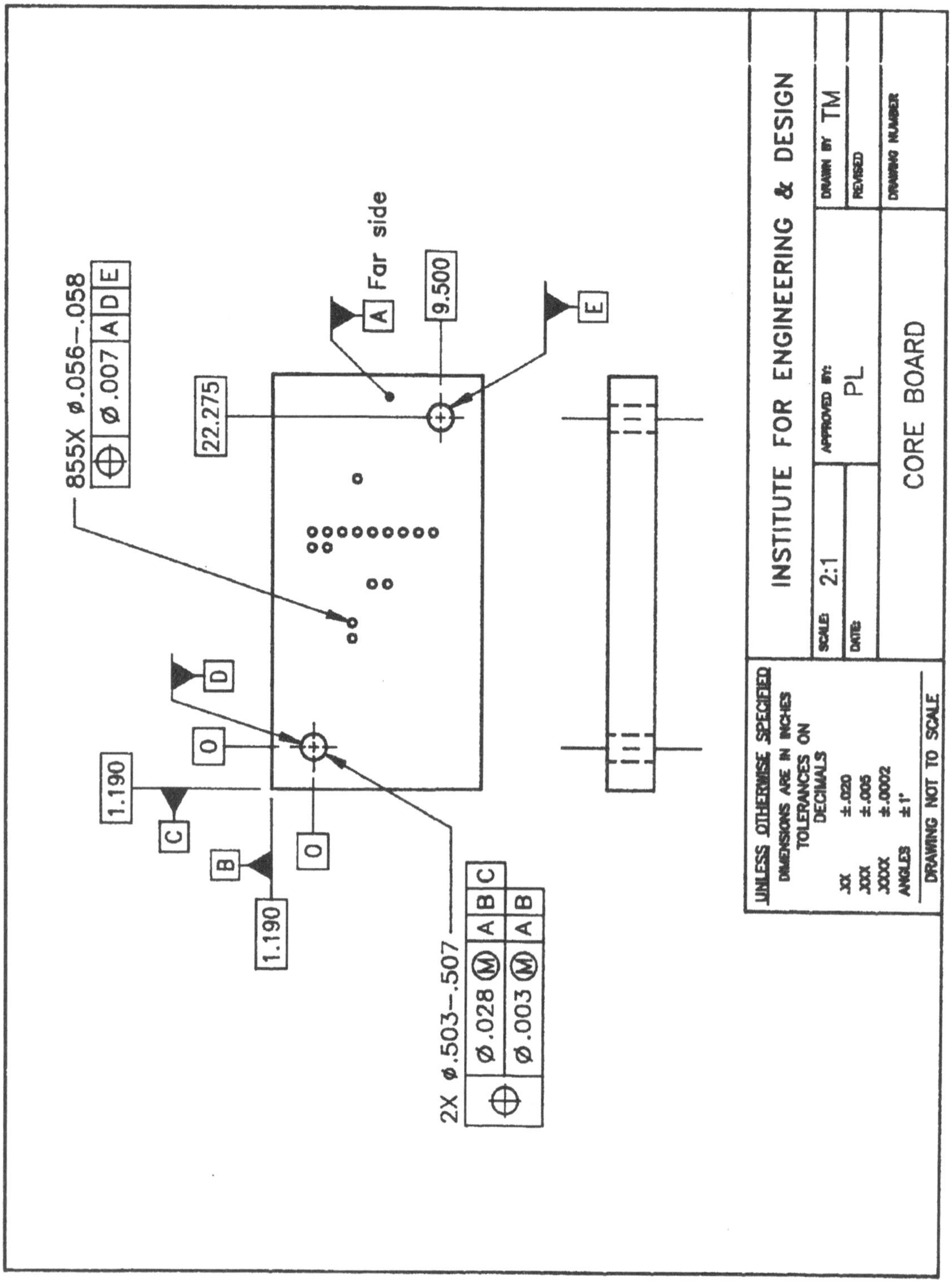

82. (continued)

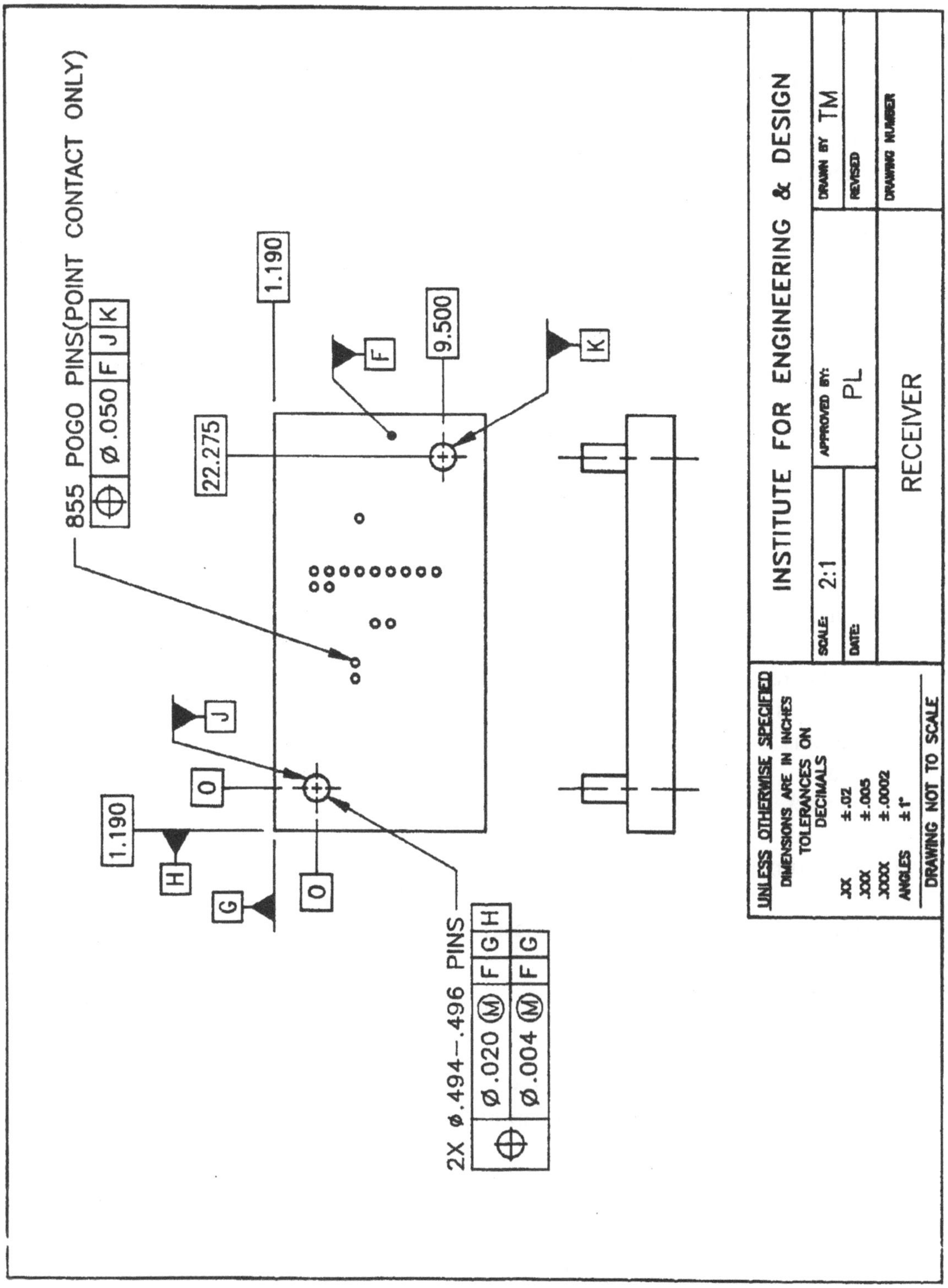

83. Terms and Definition pertaining to Datums and Datum Features

1) Datums are:
   a. real
   b. simulated
   c. perfect
   d. irregular and flawed
   e. derived from datum features

2) Features are:
   a. real
   b. surfaces
   c. perfect
   d. always measurable for size
   e. sometimes measurable for size

3) Datum features are:
   a. never measurable
   b. real
   c. perfect
   d. sometimes size features
   e. surfaces

4) Surfaces are:
   a. features
   b. datums
   c. real
   d. perfect
   e. flawed
   f. sometimes chosen to serve as datum features

5) Datum features should be chosen based on:
   a. function
   b. smoothness as produced
   c. repeatability
   d. accessibility
   e. representation of mating features
   f. the part itself with no consideration for mating features

6) List five common physical datum feature simulators.

7) A dimension with a box around it, like [100], is called?
   a. size dimension
   b. basic dimension
   c. plus and minus tolerance
   d. limit dimension

83. (continued)

8) What is the [A] symbol called?
   a. datum feature symbol
   b. tolerance
   c. allowance
   d. geometric characteristic symbol

9) What kind of tolerance is expressed by 50 ± 1?
   a. limit dimensioning
   b. location tolerance
   c. bilateral plus and minus tolerancing
   d. unilateral plus and minus tolerancing

10) What is the Ⓜ called when used after the geometric tolerance?
   a. datum identification symbol
   b. Zone descriptor
   c. basic size
   d. maximum material condition symbol

11) What is third datum feature in the feature control frame called?
   a. primary
   b. tertiary
   c. geometric characteristic
   d. modifier

12) What is the Ⓜ called when used after a datum feature of size?
   a. a geometric characteristic symbol
   b. datum feature symbol
   c. tolerance
   d. maximum material boundary symbol

13) If a Ø2 appears in a feature control frame, it is:
   a. the hole size.
   b. the size tolerance.
   c. basic size.
   d. the geometric tolerance

14) List at least two components (other than the perpendicularity symbol) that you feel are needed in a perpendicularity control.

15) A clearance fit implies the mating features, when produced within size limits,
   a. will always encounter material interference.
   b. will sometimes encounter material interference.
   c. will never encounter material interference.

83. (continued)

16) An interference fit implies the mating features, when produced within size limits,
    a. will always encounter material interference.
    b. will sometimes encounter material interference.
    c. will never encounter material interference.

17) In a fixed fastener assembly condition, the geometric tolerance to be divided between the two parts containing the mating features is equal to,
    a. the difference between the Least Material Conditions of mating features.
    b. the difference between the Maximum Material Conditions of mating features.
    c. the difference between the shaft MMC and the hole LMC.
    d. the difference between the shaft LMC and the hole MMC.

84. MMC and LMC Please identify the Ⓜ and Ⓛ of these.

HOLES:

1) 62.5 ± 0.1 Ⓜ ______________________

Ⓛ ______________________

2) 59-62 Ⓜ ______________________

Ⓛ ______________________

3) 83.5 ± 0.05 Ⓜ ______________________

Ⓛ ______________________

4) 69.5-70.5 Ⓜ ______________________

Ⓛ ______________________

5) 1.245-1.265 Ⓜ ______________________

Ⓛ ______________________

SHAFTS:

6) 7.5-8.1 Ⓜ ______________________

Ⓛ ______________________

7) 71 ± 0.2 Ⓜ ______________________

Ⓛ ______________________

8) 12.45-12.65 Ⓜ ______________________

Ⓛ ______________________

9) 21.885-22.875 Ⓜ ______________________

Ⓛ ______________________

10) 0.85-0.90 Ⓜ ______________________

Ⓛ ______________________

85. Size Controls Form

1) Unless otherwise specified, individual features of size must:
   a. not violate size limits of MMC at any cross section.
   b. not violate an envelope of perfect form at LMC.
   c. not violate an envelope of perfect form at MMC.

2) The concept discussed in Question 1 deals with the:
   a. Conrad Principle
   b. Taylor Principle
   c. Johnson Principle

3) Least material condition on most parts is checked
   a. at cross sections.
   b. for an envelope.
   c. never checked because it serves no purpose.

4) When checking for maximum material condition violations in a hole with size limits of Ø50 - 51, what size and type of gage would you use?
   a. a full form plug gage with a diameter of 50
   b. a full form plug gage with a diameter of 51
   c. measurement tool capable of checking size at any cross section

5) What effect on size measurements would adding the note "Perfect form at MMC not required" have?

6) Ø25 ± 0.1 AVG. This is a hole in metal. It is the I.D. of a thin walled tube.
   What is the AVG. MMC?
   What is the AVG. LMC?

7) Ø.500 ± .005 This is a flexible rubber shaft.
   What is the MMC?
   What is the LMC?

8) .625 - .635 This is a slot milled into a thick rigid plate.
   What is the MMC?
   What is the LMC?

9) .010 ± .002 This is a steel tab that is non-rigid.
   What is the MMC?
   What is the LMC?

85. (continued)

10) Ø100 ± 1

⌖	Ø0Ⓛ	A	B	C

This control refers to a glass shaft outside diameter. A hole runs through its center making it look like a drinking straw. It is mounted on a plate.
What is the MMC?
What is the LMC?
How does the use of the LMC symbol change the way the size is measured?

11) Ø37.5 ± 0.2

—	Ø0.5Ⓜ

This is a metal rod (shaft) that is twenty-five feet long.
What is the MMC?
What is the LMC?
How does the use of straightness of the derived median line change the way size is measured?

12) 10.0 - 10.1 This forms two sides of a rectangular part. It is considered an external feature of size. It is rigid.
What is the MMC?
What is the LMC?

86. Rules, Concepts, Characteristics and Untoleranced Dimensions

1) What purpose does referencing a datum feature in a feature control frame serve?

2) Please categorize all the geometric characteristics as: INDIVIDUAL (not related to datums); RELATED (related to datums); INDIVIDUAL or RELATED:

3) When we check an individual feature for form, what are we checking to?
   a. a datum
   b. a perfect geometric form of itself
   c. a point, line or plane

4) Once on a drawing, how can we tell if a feature is individual or related?

5) Name three ways to eliminate the requirement for an envelope of perfect form at MMC for a feature of size.

6) If you do one of the three things stated in Question 5, can you exceed MMC at any cross sectional check?

7) If you assign the LMC concept and symbology to the geometric tolerance in the feature control frame, where is the envelope of perfect form size requirement?
   a. at MMC
   b. it does not exist
   c. at LMC

8) Which material condition symbol (modifier) provides the most tolerance to draw from when used?

9) Which parts are considered to be most economical to manufacture?
   a. those with features controlled at MMC
   b. those with features controlled RFS

10) What material condition symbol provides for the easiest use of functional (receiver-type) gaging?

11) Is it ever possible to use the Ⓜ symbol both for a feature of size and a datum feature of size in one Feature Control Frame? Yes No

For example: | ⌖ | ⌀0.5Ⓜ | A | BⓂ | C |

12) When features are controlled RFS, they are given an original tolerance of form, orientation, profile, runout or location. Where can we draw extra geometric tolerance from?
   a. the size limits
   b. the title block
   c. we can't

86. (continued)

13) As a minimum, the feature control frame must contain:
    a. a geometric characteristic symbol and datum feature reference.
    b. a geometric characteristic symbol and a geometric tolerance.
    c. a datum feature reference, tolerance and a material condition symbol.

14) Draw the symbols for the following geometric characteristics and categorize them as form, orientation, profile, runout or position.
    a. Flatness
    b. Straightness
    c. Roundness
    d. Cylindricity
    e. Profile of a Line
    f. Profile of a Surface
    g. Parallelism
    h. Perpendicularity
    i. Angularity
    J. Feature Control Frame
    k. Position
    l. Runout (Circular)
    m. Concentricity
    n. Runout (Total)

15) Identify the following symbols:
    a. Ⓜ
    b. ↧
    c. ∅
    d. Ⓟ
    e. ⟨ST⟩
    f. Ⓛ
    g. ⊔

87. Datums

1) What geometrically perfect elements are representative of datums?

2) How are datum features simulated in manufacturing or inspection?

3) What criteria is used to choose a datum feature when tolerancing?

4) What is a datum reference frame and what is it used for?

5) What six degrees of spatial freedom does a part have before it is restrained?

6) How many high points of contact are required to simulate a primary datum plane from a planar datum feature?

   A secondary datum plane constructed from a planar surface after the primary is constructed as described above?

   A tertiary datum plane constructed from a planar surface after the primary and secondary are constructed as described above?

7) Are three datums always necessary for part orientation? If yes, why? If no, why not?

8) Is a datum feature perfect?

9) Is a datum feature ever given form control of its own? If not, why is it not necessary? If so, why it is necessary?

10) How is the order of precedence of datum features shown in a feature control frame?

11) Why does the order of precedence of datum features make such a difference in feature orientation?

12a) If a shaft is chosen as a datum feature (RFS), how is the datum feature simulated?

12b) If a hole is chosen as a datum feature (RFS), how is the datum feature simulated?

13) Of what use is an angular orientation datum in a datum reference frame that includes a cylindrical datum feature?

14) How many datum planes are derived from a cylindrical datum feature?

87. (continued)

15) Why are datum targets necessary?

16) How are datum features controlled RMB simulated?

17) At what size or condition are secondary datum features of size that have their own tolerance of orientation or position simulated as datums?

18) At what size is a primary datum feature of size represented for the features controlled to it?

19) What is a compound datum feature and what is it used for?

20) When a pattern of holes is controlled to a datum feature hole, as the datum feature hole departs from MMB within its limits, what happens to the controlled hole pattern and how is it calculated?

21) Do screw threads make good datum features?

22) If more than one datum reference frame is specified for a part, what effect will this have on manufacturing and inspection?

23) Why would a designer choose to use only a portion of a surface as a datum feature instead of the entire surface?

24) In what forms are datum targets specified?

25) If a designer wants to show a datum target as originating from the opposite side of a part than the view it has been shown in, how may it be done without another view?

26) What kind of contact is required from the physical datum feature simulation equipment, in a datum target area, on a non-cylindrical surface?

27) How is the shape and location of a datum target area defined?

28) What does the dimension origin symbol look like and of what use is it?

88. Virtual Condition and Resultant Condition Boundaries

1) When a feature has errors of orientation, location and certain tolerances of form, its effective mating boundary is virtually altered at that orientation or at that location or by its form.
T F

2) In virtual condition (MMC concept), the effective mating boundary of a hole is virtually increased and the shaft size virtually decreased, even though they still measure within their limits of size.
T F

3) A virtual condition boundary (MMC concept) of the feature represents the worst condition of assembly of the mating components.
T F

4) If a feature is allowed a virtual condition, it may exceed its limits of size as an individual feature when measured.
T F

5) How does calculating the virtual condition of an external feature of size differ from calculating the virtual condition (MMC concept) of an internal feature of size?
T F

6) Functional gages are made to the virtual condition (MMC concept) of features they inspect.
T F

7) Functional gages are usually only made to check features controlled with MMC. Otherwise, for checking features controlled RFS, a functional gage would have to be made to take into account every size the feature could be produced at.
T F

89. Form Controls

FLATNESS

1) Describe the configuration of a flatness tolerance zone?

2) Is flatness considered individual or related to datums?

3) Is a datum feature reference allowable in the feature control frame?

4) May a diameter sign be included in the feature control frame?

5) Is flatness a surface or an axis control?

6) Is flatness a two-dimensional or a three-dimensional control?

7) To obtain a high level of certainty (for example: 90% certainty) of how flat a surface is, when measuring flatness will a few points from the surface suffice, or does this require hundreds of points per square inch?

8) If a surface is controlled with flatness, may the controlled surface in combination with its opposing side on the part violate an envelope of perfect form at MMC?

9) Must the flatness tolerance refine the size tolerance or may it be larger? Why?

10) May a MMC symbol be included in the feature control frame if flatness is applied to s single surface?

11) How is the flatness control directed to the controlled surface?

12) May flatness be applied on a unit or a rate basis?

13) May flatness be applied to control a derived median plane?

STRAIGHTNESS

General:

1) Are datum feature references allowable/required in the feature control frame?

2) What view should the straightness tolerance be applied in?

Surface Straightness:

3) In straightness of a surface, what is the shape of the tolerance zone, and what must lie within it if the feature is to be accepted?

4) What material condition symbols may be included in the feature control frame?

5) May a diameter sign appear in the feature control frame?

89. Form Controls (continued)

6) Without a straightness feature control frame, what controls the straightness of a part surface?

7) Is perfect form of the controlled surface and its opposing surface required at MMC?

8) Is bonus tolerance ever possible?

If so, when?

If not, why not?

Straightness of the derived median line:

9) In straightness of a derived median line, what is the tolerance zone shape and what must be contained within it for the feature to be acceptable?

10) Is perfect form of the controlled feature required at MMC?

11) Is this control considered individual or related to datums?

12) How might straightness of a derived median line that is controlled with a MMC symbol be gaged?

13) How might straightness of a derived median line that is controlled with the RFS concept be gaged?

14) Is a virtual condition allowed for both controls described in Questions 12 and 13?

15) Must the straightness tolerance be smaller than the feature's size tolerance?

CIRCULARITY

1) Is circularity a two- or a three-dimensional control?

2) What is the tolerance zone shape?

3) What material condition symbols are allowed in the feature control frame?

4) Are datum feature references allowable?

5) Is a feature controlled with circularity allowed to exceed its size limits?

6) Is any extra tolerance derived from the controlled feature's size limits as the feature departs from MMC?

7) How is circularity checked?

89. Form Controls (continued)

8) Does the tolerance in the feature control frame have to be smaller than the feature's size limits?

9) Is circularity considered individual or related to datums?

10) Can circularity controls be assigned to cones?

11) May a diameter sign be included in the circularity feature control frame?

12) How is the feature control frame connected to the controlled surface?

13) Circularity controls the surface
   a. between two concentric cylinders.
   b. at every cross section.
   c. but is mainly an axis control.

14) Without a circularity feature control frame, what would normally control a feature's roundness?

15) A circularity control is assumed to apply
   a. at LMC.
   b. RFS.
   c. at MMC.

CYLINDRICITY

1) Name at least two differences between circularity and cylindricity.

2) List three different ways cylindricity can be checked. Order them with the best way first and the least desirable method last.

3) What is the tolerance zone configuration a cylindricity control generates?

4) Is cylindricity a surface or an axis control?

5) May a least material condition symbol be used in the feature control frame?

6) Can cylindricity and circularity be applied to the same surface? If so, which tolerance must be smaller and why?

7) Does a cylindricity tolerance ever reference datum features inside the feature control frame?

8) Must the controlled feature, if produced at MMC, have perfect form?

9) Is the cylindricity tolerance additive to the controlled feature's MMC size limit?

10) Is bonus tolerance ever allowed?

89. Form Controls (continued)

11) A cylindricity tolerance is considered
    a. form.
    b. orientation.
    c. location.
    d. profile.
    e. runout.

12) A produced feature's cylindricity is compared to a perfect geometric counterpart of
    a. a datum.
    b. an axis.
    c. itself.

13) List all items allowed in the cylindricity feature control frame.

14) A cylindricity tolerance controls roundness and
    a. perpendicularity.
    b. flatness.
    c. straightness and taper.
    d. angularity.

15) The tolerance applies simultaneously to circular and ____ elements of the controlled surface.
    a. longitudinal
    b. perpendicular
    c. flat
    d. positional

90. Orientation Controls

1) What is the configuration of the tolerance zone defined by a parallelism of a planar surface control?

2) What does that tolerance zone take its orientation from?

3) If we wanted the following orientation tolerance zone | ∠ | 0.5 | for a planar surface to be exactly 35° from datum plane R as well as perfectly perpendicular to datum plane S, what could we do to the given feature control frame to reflect that?

4) If a part has a maximum height of 1.000. Is a .002 parallelism control additive to that height?

5) Is at least one datum feature reference always required in the feature control frame for angularity?

6) How does the tolerance configuration for the angularity in the control shown in Question 3 differ from the tolerance zone that would be generated if the 35° was ± 2° ?

7) What material condition symbols would be appropriate in feature control frames for orientation controls if they were used on a single planar surface?

8) What is usual configuration and orientation of a controlled feature's perpendicularity tolerance zone if it is used on a planar surface?

9) What does the use of a datum feature reference add to the orientation geometric control?

10) ∅.250±.005 Hole

∠	∅.005Ⓜ	A

If the controlled hole was produced at a diameter of .250, what angularity tolerance would apply?

11) Does the controlled hole in Question 10 have a virtual condition? If so, what is it? If not, why not?

12) Would it be appropriate to include a maximum material boundary symbol in the feature control frame for datum features that are planar surfaces? Please explain.

13) How many contact points are required to establish datum A from a planar surface if A is primary?

14. Is an orientation geometric control related to datums or individual?

90. Orientation Controls (continued)

15) For parallelism used on a planar surface to a datum plane generated by an opposing planar surface, must the tolerance in the feature control frame be smaller than the size tolerance of the part's opposing sides that are included in the control?

16) ∅25±0.5 Hole

⊥	∅0.5Ⓜ	AⓂ	B

If the controlled hole departs from MMC to LMC, how much extra perpendicularity tolerance can be drawn from the size?

17) List the MMC and LMC of the datum feature in Question 16.

18) If the datum feature A referenced in Question 16 is a hole and has a size tolerance of ∅50±0.1, as it departs from MMB to LMC, how much additional shift of the controlled hole's tolerance zone is gained?

19) If both the controlled hole and datum feature A were made at LMC, how much could their axes be out of perpendicularity to each other?

20) What is the shape of the tolerance zone generated by this geometric control?

91. Profile

1) In a profile of a curved surface control, the profile must be dimensioned with
   a. plus and minus dimensions.
   b. limit dimensions.
   c. radial dimensions.
   d. basic dimensions.

2) What is the difference between profile of a line and profile of a surface?

3) For profile controls, datum feature references are
   a. allowable but not required.
   b. not allowed.
   c. required.

4) If a datum (or datums) was used for a profile control, what purpose would be served?

5) Are material condition symbols allowed for the feature being controlled? Why or why not?

6) Is profile of a surface more often individual or related to datums?

7) Is profile of a line more often individual or related to datums?

8) If both profile of a surface and profile of a line are used on the same feature, which must have the smaller profile tolerance?

9) How is profile checked?

10) What is the tolerance zone shape of profile of a line?

11) What is the tolerance zone shape of profile of a surface?

12) How is profile tolerance designated as bilateral or unilateral?

13) How is the extent of where the profile control applies on a surface designated?

14) Suppose the entire outer periphery of the part profile must be controlled. How may it be done?

92. Runout

1) Circular runout checks are taken
   a. at cross sections.
   b. over the entire surface without lifting the indicator.

2) Runout checks rely on
   a. fine indicator movement.
   b. full indicator movement.
   c. full indicator motivation.

3) Runout checks are taken
   a. while the feature is stationary.
   b. while the feature rotates 180°.
   c. while the feature rotates 90°.
   d. while the feature rotates 360°.

4) Runout checks are taken on the controlled feature as the feature rotates about
   a. the datum axis.
   b. its own axis.
   c. any axis on the part.

5) Total runout includes a variety of geometric controls. List them.

6) What does circular runout control?

7) Give an example of when circular runout would be a suitable control.

8) Runout is a
   a. datum axis to feature axis control.
   b. datum surface to feature axis control.
   c. surface to surface control.
   d. feature surface to datum axis control.

9) Is runout a control related or individual?

10) Are datums allowable or required?

11) Must the tolerance in the feature control frame be smaller than the controlled feature's size limits?

12) In its 360° rotation relative to the datum axis, must the controlled feature remain within an envelope of perfect location at MMC?

92. Runout (continued)

13) While stationary, must the controlled feature fit within a boundary of perfect form at MMC?

14) May a maximum material condition symbol be included in a runout feature control frame?

15) If circular runout and total runout are used on the same feature, which control must have the smaller runout tolerance?

93. Location

CONCENTRICITY

1) Concentricity is a _____ control.
   a. datum axis to entire feature surface
   b. datum axis to cross sectional feature surface
   c. feature median points to datum axis

2) Concentricity is always assumed to apply at
   a. RFS.
   b. MMC.
   c. LMC.

3) Is a concentricity control a refinement of controls already existing in the size limits?

4) Must a tolerance zone descriptor be included in all concentricity feature control frames?

5) What is the tolerance zone shape in a concentricity control?

6.) Is concentricity a related or individual control?

7) Concentricity is a ______ control.
   a. orientation
   b. form
   c. location
   d. runout
   e. profile

8) If a feature is given a concentricity control,
   a. it probably rotates in assembly (or use).
   b. it probably does not rotate in assembly (or use).

9) Under what circumstances would a designer want to apply a concentricity control to a feature?

10) A feature is given a concentricity control. While that part is in a stationary state, is the feature allowed to violate an envelope of perfect form at MMC?

93. Location (continued)

11) If a cylindrical feature is controlled with concentricity, when the feature rotates about the datum axis, may its sweep cover an area that violates the stationary feature's envelope of perform at MMC?

SYMMETRY

12) Symmetry is a _______ control.
    a. centerplane to datum
    b. cloud of median points to datum plane
    c. datum centerplane to centerplane

13) Symmetry is always assumed to apply at
    a. RFS.
    b. MMC.
    c. LMC.

14) Is a symmetry control a refinement of controls already existing in the size limits?

15) May a tolerance zone descriptor, such as a diameter sign, appear in the feature control frame?

16) Describe the tolerance zone for a symmetry control.

17) Is symmetry required to use a datum?

93. Location (continued)

18) Symmetry is a __________ control.
   a. form
   b. orientation
   c. profile
   d. runout
   e. location

19) If a feature is given a symmetry control,
   a. it probably mates in the assembly.
   b. it probably does not mate in the assembly; otherwise, position at MMC would be used.

20) If a slot is controlled with symmetry, must the feature have perfect form (flatness, straightness and parallelism) if produced at MMC?

POSITION

21) What key controls are generally needed to define the position of a hole pattern?

22) What are the components of a datum reference frame?

23) From what elements on the part are datum planes established?

24) What might cause a rocking effect in a datum plane?

93. Location (continued)

25) Why is tolerance accumulation not a problem when chain dimensioning is use to locate the true position of the feature?

26) What is the purpose of the datum reference frame?

27) What information is contained within the feature control frame?

28) In addition to the original tolerance of position, when a feature is controlled at MMC, extra tolerance is sometimes available. Where is the extra positional tolerance drawn from?

FLOATING FASTENERS

29) Define what a floating fastener situation is.

30) What are the floating fastener formulae?

31) Calculate floating fastener tolerance for the following situations:

a. MMC fastener = .250
MMC hole = .265

b. MMC hole = .520
MMC fastener = .500

32) Calculate fastener MMC size from the following:

a. MMC hole = .775
Positional tolerance = .025

b. Positional tolerance = .030
MMC hole = 1.030

93. Location (continued)

33) Calculate hole MMC size for the following situations.

a. Fastener MMC - .3125
Tolerance = .010

b. Tolerance = .020
Fastener MMC = 6.25

PROJECTED TOLERANCE ZONES

34) What determines the distance a tolerance zone is projected?

35) Why is the use of projected tolerance zones necessary?

36) On what types of features are projected tolerance zones used?

37) How is positional tolerance calculated for a fixed fastener situation?

38) Show two different ways a projected tolerance zone can be depicted on a drawing.

39) How can features controlled with projected tolerance zones be inspected for their positional or perpendicularity tolerance?

'New Rules' Workbook Section

Workbook Section

1. Please complete the following feature control frame to indicate that the datum feature simulator for datum feature C translates (moves).

21X ⌀4.0-4.3

⌖	⌀0 Ⓜ	A	B Ⓜ	C Ⓜ

2. Complete the following feature control frame to indicate a unilateral profile of a surface control that is 0.8 all inside of the basic profile.

⌓	0.8	A	B Ⓜ	C Ⓜ

3. Eliminate the envelope of perfect form at maximum material condition by using the appropriate symbol after the size limits of the following hole.

⌀10.5-10.7

4. Use the correct symbology to make the following two diameters treated as though they are one diameter.

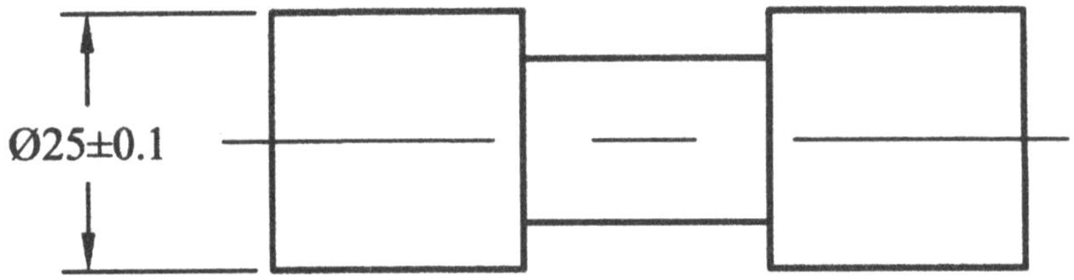

5. When the symbol Ⓜ is used after a datum feature of size referenced in a feature control frame, it stands for: (Circle correct answer)
   a. Minimum Material Condition
   b. Maximum Material Condition
   c. Maximum Material Boundary
   d. Minimum Material Boundary

6. When the symbol Ⓛ is used after a geometric tolerance in a feature control frame, it stands for: (Circle correct answer)
   e. Least Material Boundary
   f. Least Material Condition
   g. Least Squares Center
   h. Least Boundary Condition

7. Put a symbol in the blank area on the elbow of the leader line from the feature control frame that means ALL OVER, APPLIES IN ALL VIEWS SIMULTANEOUSLY.

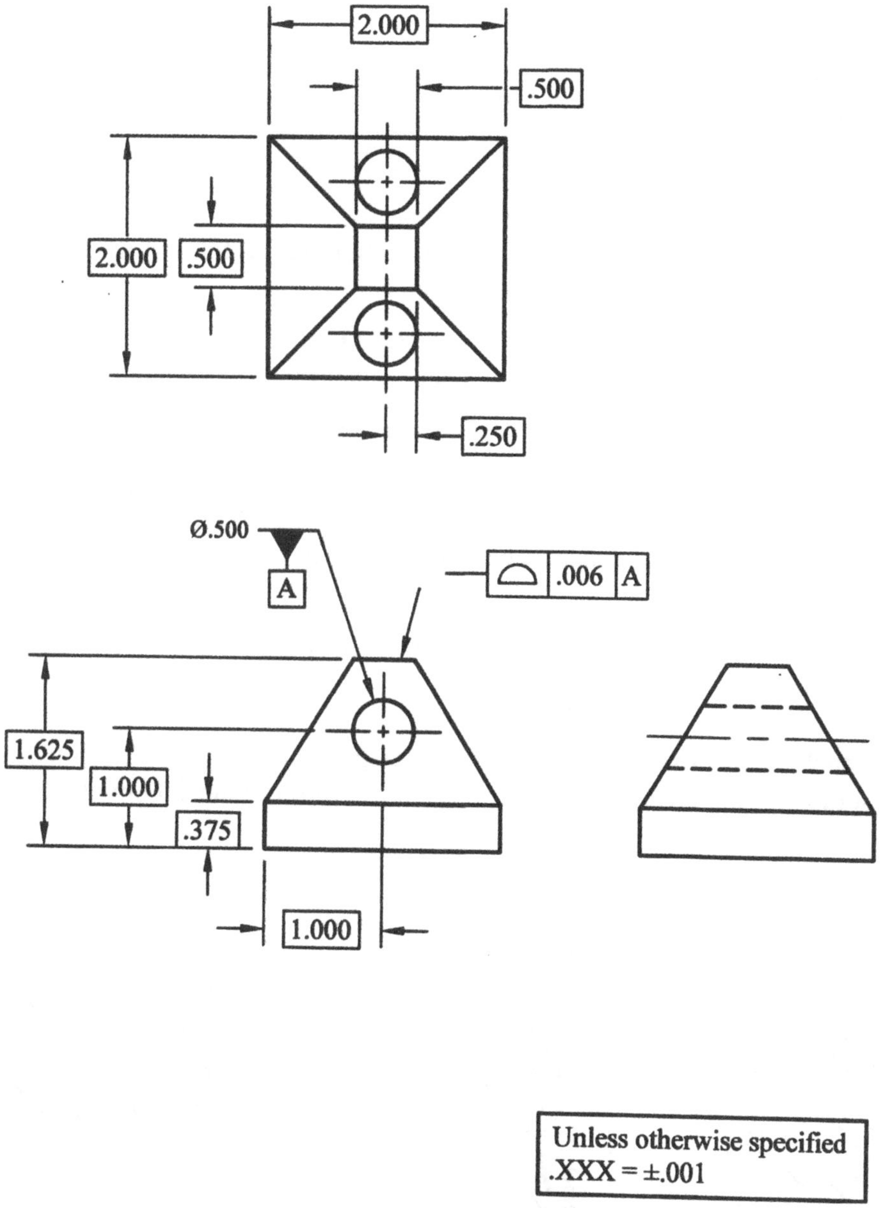

8. When no material condition symbol is used inside a feature control frame after any datum features of size referenced, what is the abbreviation for the symbol that is implied? (Circle correct answer)
   a. MMC
   b. RMB
   c. LMC
   d. MMB
   e. RFS
   f. LMB

9. What does each of the abbreviations in question 8 stand for?
   a. MMC stands for ____.
   b. RMB stands for ____.
   c. LMC stands for ____.
   d. MMB stands for ____.
   e. RFS stands for ____.
   f. LMB stands for ____.

10. What does the symbol that includes the SF letters in the following drawing stand for?

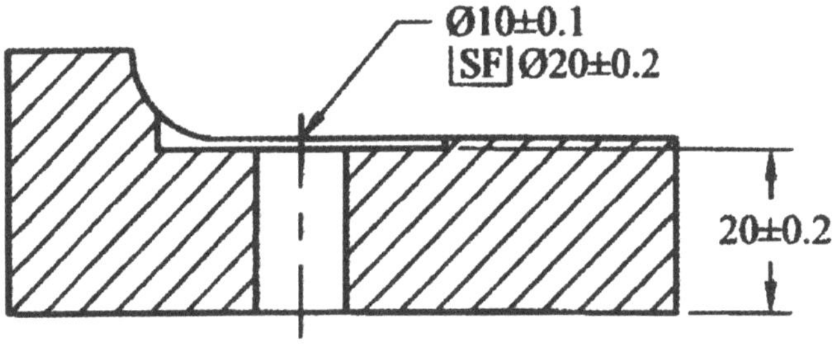

11. According to the rules in ASME Y14.5-2009, may the Ⓢ symbol be used in a feature control frame?

12. Label each of the envelopes to which arrows point in the following illustrations as either the "actual mating envelope" or the "actual minimum material envelope".

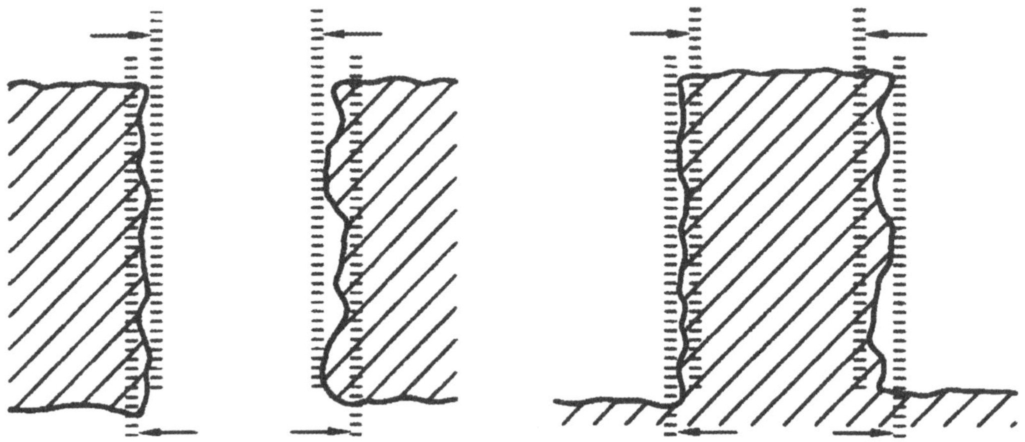

13. What geometric characteristic symbol (if placed below the 25±0.5) would be capable of creating the following tolerance zone for the part shown? (Circle the correct answer below the illustrations)

Part Specification

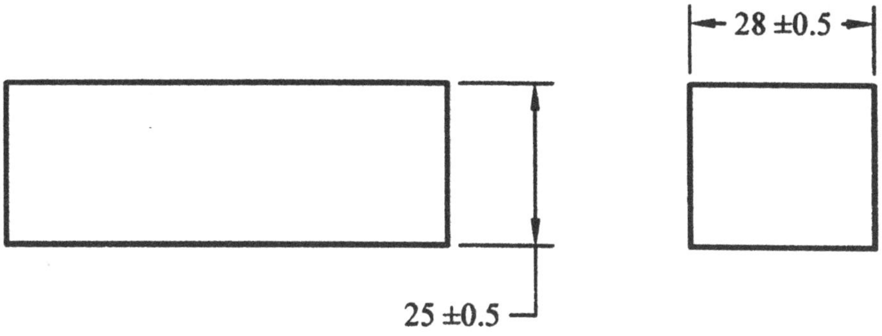

Tolerance Zone

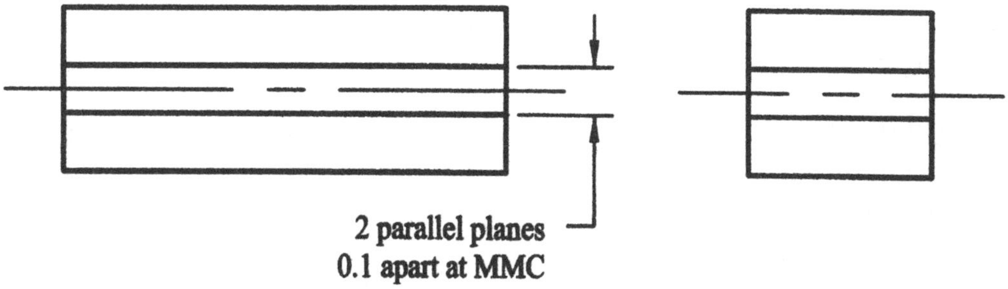

a. | ⏤ | 0.1 Ⓜ |
b. | ○ | 0.1 |
c. | ⌒ | 0.1 |
d. | ⏥ | 0.1 Ⓜ |

14. An "Actual Mating Envelope" is a similar perfect feature(s) counterpart of smallest size that can be contracted about an external feature(s) or of largest size that can be expanded within an internal feature(s) so that it coincides with the surface(s) at the highest points. What is the difference between an "Unrelated Actual Mating Envelope" and a "Related Actual Mating Envelope"? (Circle one)
   a. Unrelated is not tied to the feature control frame, but related is.
   b. Related is oriented or located to a datum(s), while unrelated is not.
   c. Unrelated is not a "Simultaneous Requirement" with other feature patterns, but Related is related to other feature patterns.
   d. Related is a combined control using one feature control frame stacked on another.

15. Which geometric characteristic symbol that is categorized as "Orientation" may be used to tolerance any angle no matter how many degrees it is specified as?
    a. Angularity
    b. Perpendicularity
    c. Parallelism
    d. Profile of a Surface

16. Where axes, center planes or surfaces are shown coincident on a drawing and geometric tolerances establish the relationship between the features, what applies?
    a. a plus and minus tolerance
    b. a zero basic dimension
    c. a general tolerance note
    d. a reference dimension

17. Where a coordinate system is shown on the drawing, unless otherwise specified, it shall be:
    a. left handed
    b. labeled in the negative direction
    c. neither labeled in the positive or negative directions
    d. right handed

18. Per ASME Y14.5-2009, unless otherwise specified, a datum feature simulator is:
    a. a true geometric counterpart
    b. physical
    c. theoretical
    d. rails, machine table, surface planes, collet chucks and parallels used to fixture parts.

19. Match the following symbols with what they represent by listing the correct letter of the definition after each symbol.
    a. Continuous Feature of Size (treats multiple features of size as one)
    b. Translation symbol (datum feature simulator translates/moves)
    c. All Over (control applies in all views-commonly used with profile of a surface)
    d. Unequally Disposed or Unilateral Tolerance
    e. Independency (size tolerance does not control form)

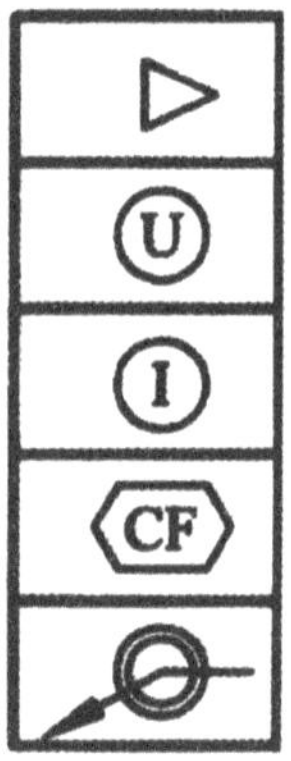

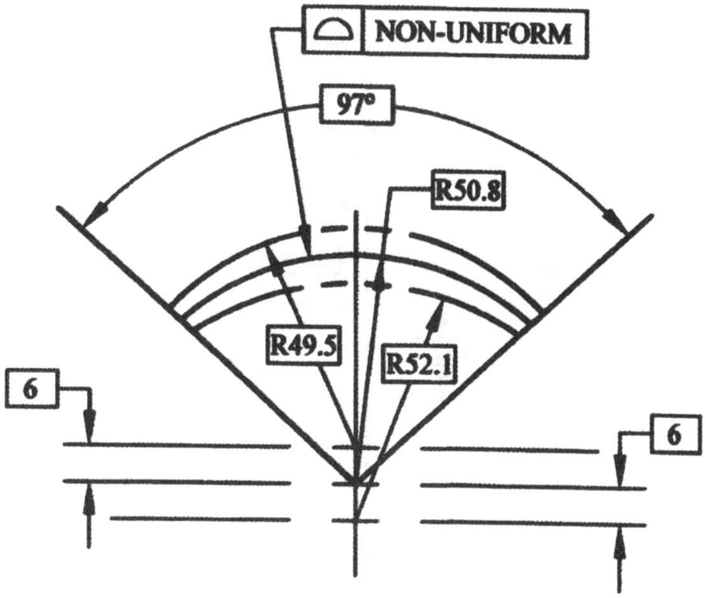

20. The illustration above is of a NON-UNIFORM profile of a surface control. The surface being toleranced by this control must reside within the tolerance zone depicted by the phantom lines. The purpose of such a tolerance zone can be to: (Circle all correct answers)
    a. smooth the transition between profile segments
    b. allow greater variations in the surface form at the ends of the zone
    c. allow abrupt surface variations at the ends of the profile zone
    d. allow smaller surface variations at the ends of the profile zone
    e. allow larger surface variations in form in the middle of the zone

21. The following illustration compares Composite Profile Tolerancing (one profile symbol) with Two Single Segment Profile Tolerancing (two profile symbols).

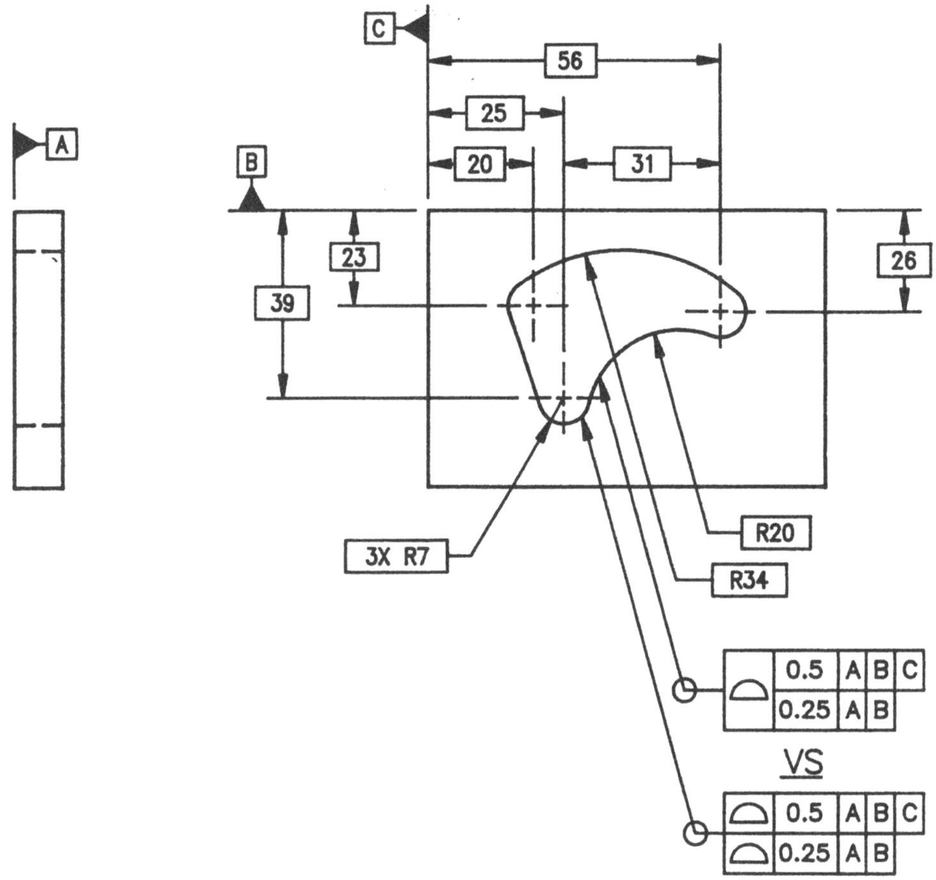

In the composite profile control:

a. What is the tolerance on the size and form of the hole?
b. What is the tolerance on the perpendicularity of the hole to datum plane A?
c. What is the tolerance on the angular orientation of the hole to datum plane B?
d. What is the tolerance on the location of the hole to datum plane B?

In the two single segment profile control:

e. What is the tolerance on the size and form of the hole?
f. What is the tolerance on the perpendicularity of the hole to datum plane A?
g. What is the tolerance on the angular orientation of the hole to datum plane B?
h. What is the tolerance on the location of the hole to datum plane B?

22. The following illustration a composite profile control with another profile control that applies INDIVIDUALLY to each hole.

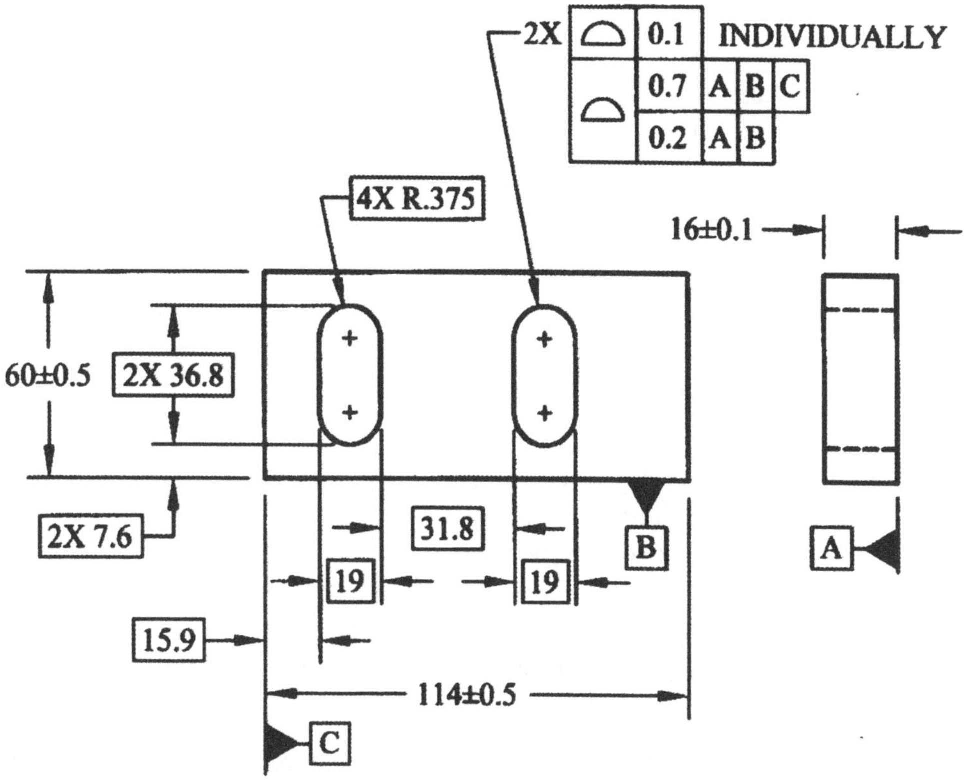

a. What is the tolerance on the size and form of the holes?
b. What is the tolerance on the location of the holes to datum B?
c. What is the tolerance on the location of the holes to datum C?
d. What is the tolerance on the location of the holes to each other?
e. What is the tolerance on the perpendicularity of the holes to datum plane A?
f. What is the tolerance on the angular orientation (rotation) of the 2-hole pattern to secondary datum plane B?
g. What is the tolerance on the angular orientation (rotation) of the 2-hole pattern to tertiary datum plane C?

23. The following illustration shows a conical feature used as a datum feature.

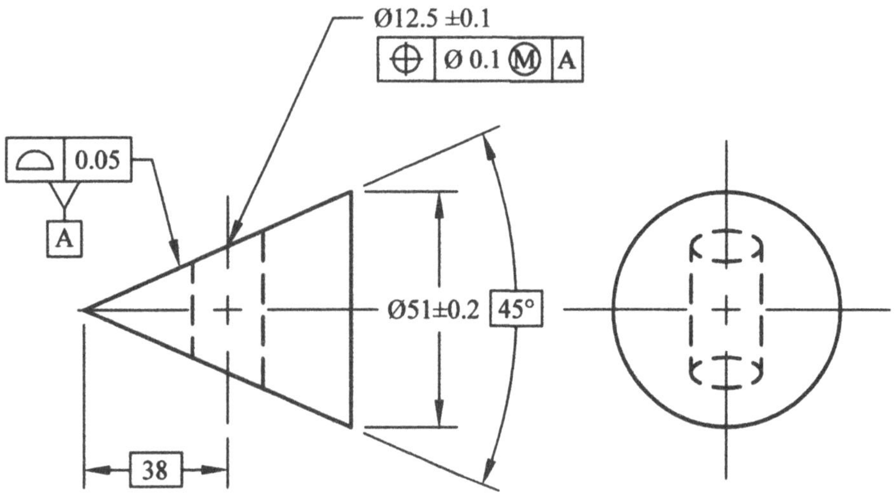

a. How many spatial degrees of freedom are controlled by referencing datum feature A in the position control on the hole?
b. Does datum feature A generate:
   1. an axis only
   2. an axis and a plane (from the point at the apex of the cone)
   3. an axis and a plane from the right edge in the view on the left
   4. a plane only

24. The following illustration shows a conical datum feature. In the position feature control frame the spatial degrees of freedom that each datum reference controls are listed in brackets after the datum reference.

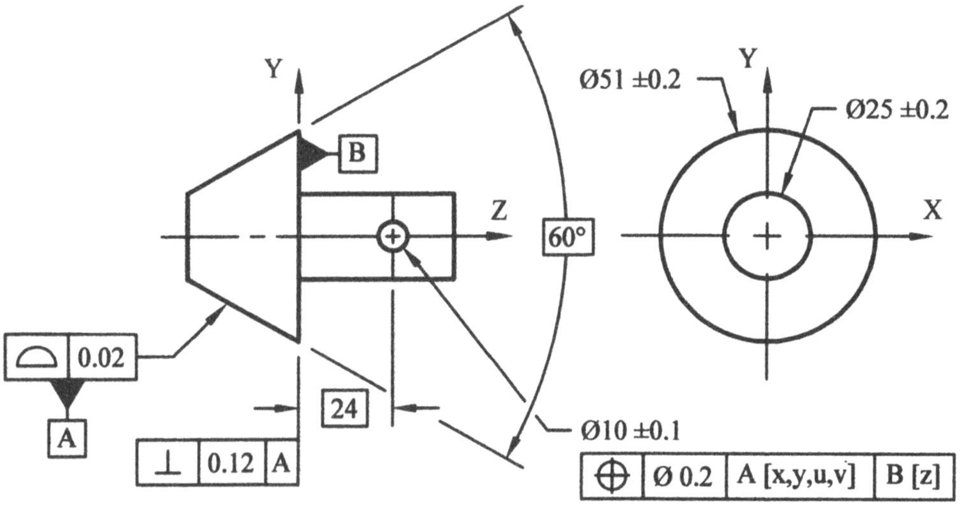

a. What controls the rotational spatial degree of freedom known as w?
b. If datum feature B was referenced as primary instead of secondary, would it be necessary to specify the bracketed z after B in the feature control frame?
c. If datum feature B was not referenced in the position control, how would the dimensions on the part drawing change?

25. **Detail Drawing**: Datum Feature Pattern Referenced at Regardless of Material Boundary (RMB)

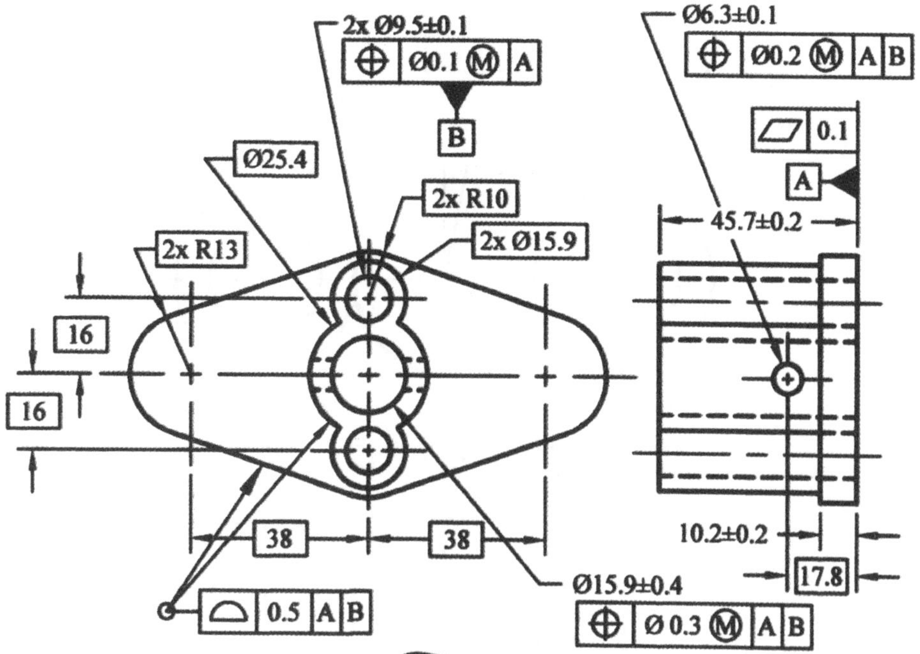

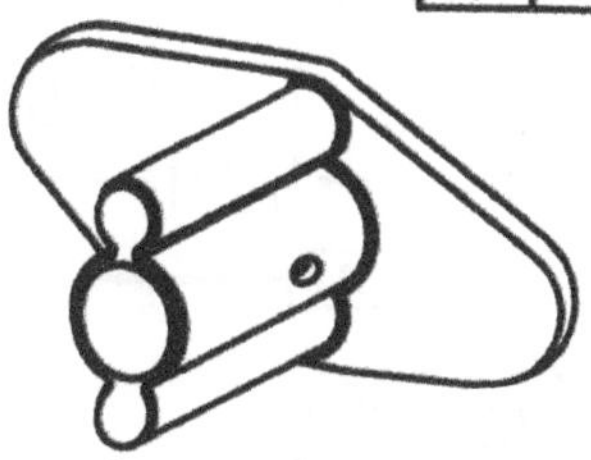

Fixture: The fixture shown below is to be used to mount the part on for inspection of the profile and position controls. The leader line points at one of the two fixturing pins. Describe how the pins would be sized to represent the two-hole pattern at RMB.

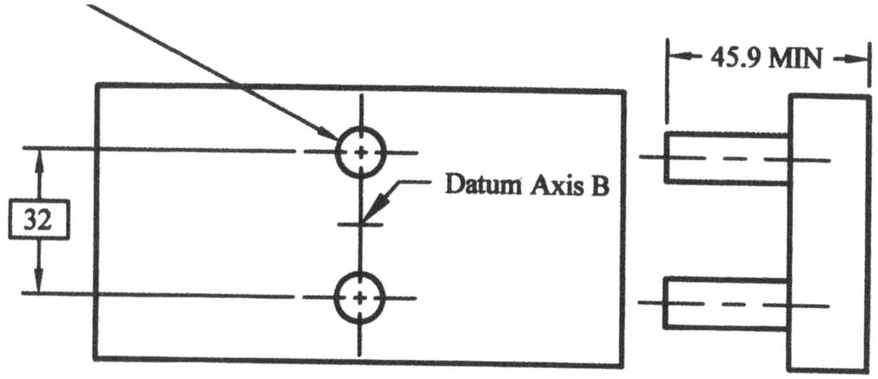

26. Detail Drawing of Datum Feature Pattern Referenced at Maximum Material Boundary (MMB).

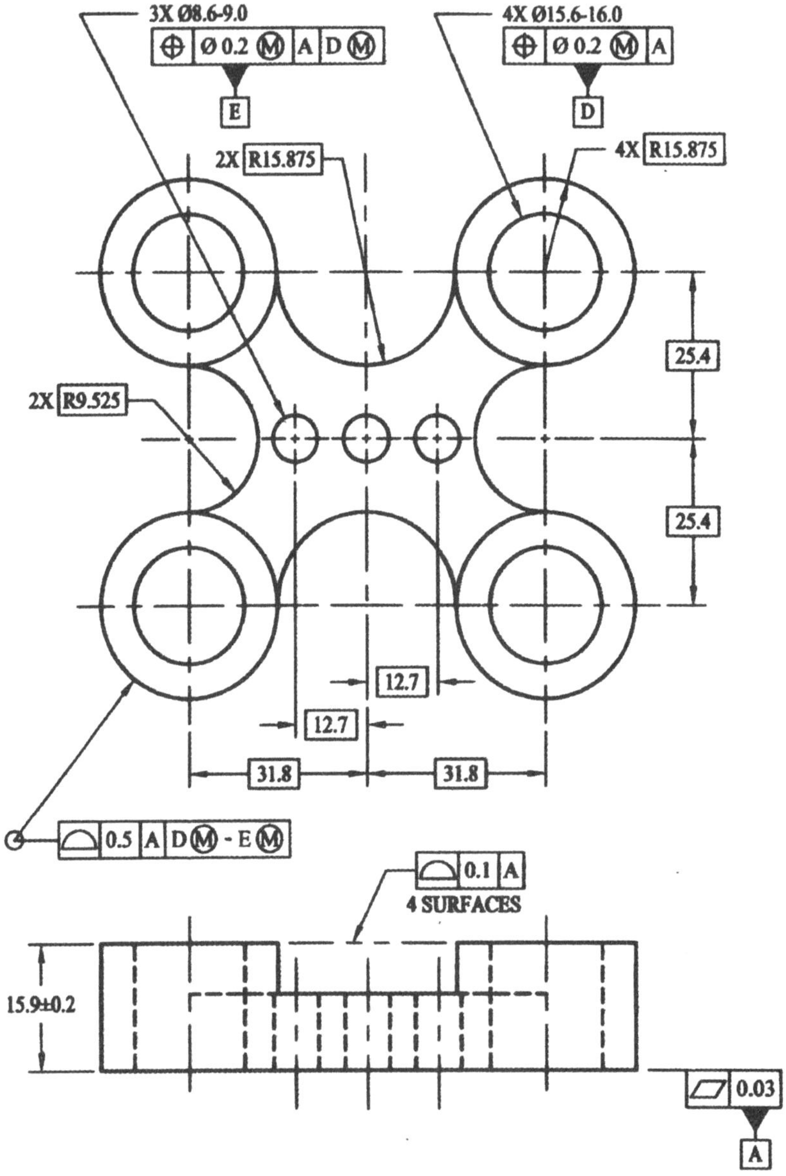

Describe or draw the fixture that could be used to mount the part shown above to inspect its profile of a surface control.

27. Detail Drawing with Degrees of Spatial Freedom Indicated by Showing X, Y and Z Axes on the Drawing.

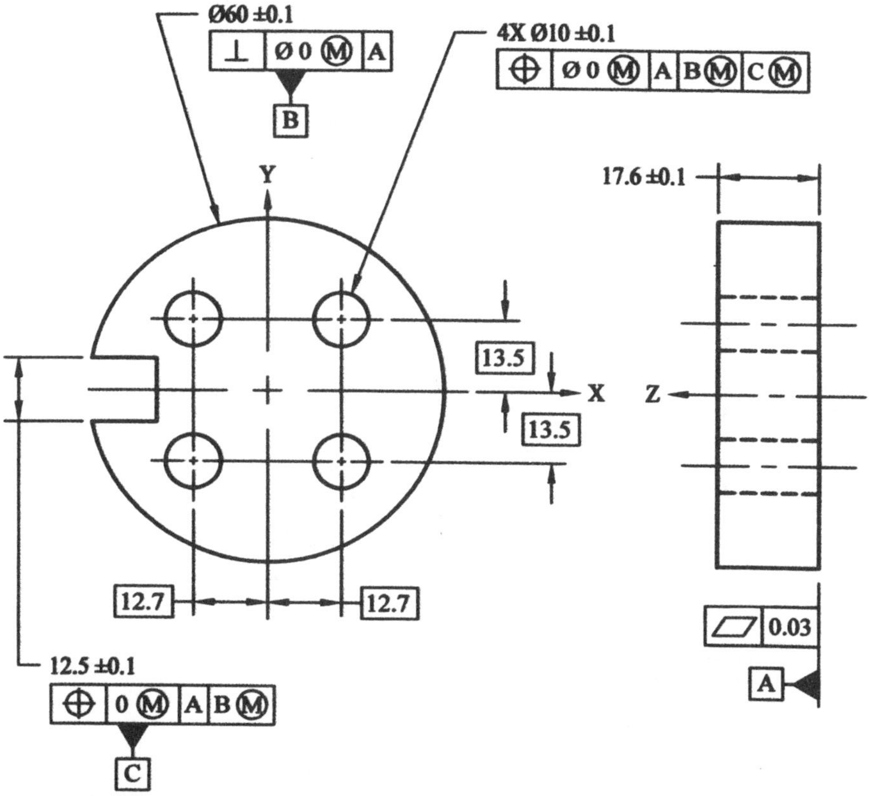

There are 6 spatial degrees of freedom. 3 are linear (x, y and z) and 3 are rotational (u, v and w). Which datum features control each of these? Specify the datum feature that controls each degree of spatial freedom listed below.

a. Movement in the x direction is controlled by datum________.

b. Movement in the y direction is controlled by datum________.

c. Movement in the z direction is controlled by datum________.

d. Rotation in the u direction is controlled by datum________.

e. Rotation in the v direction is controlled by datum________.

f. Rotation in the w direction is controlled by datum________.

28. Overriding the Degrees of Spatial Freedom Controlled by Datums.

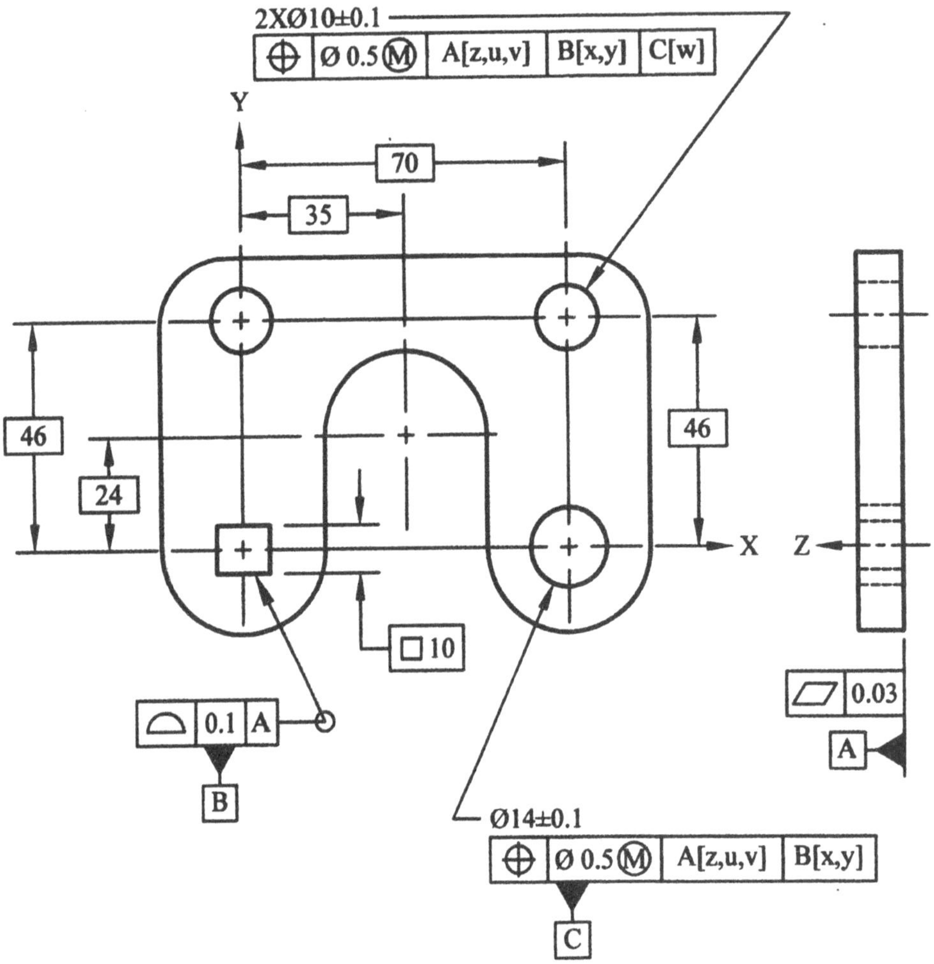

The positional controls shown in the illustration above designate which spatial degrees of freedom are controlled by the datum references. In the 4-hole positional control, what degrees of spatial freedom would each datum reference have controlled if not overridden by the bracketed designations?

a. Datum A would have controlled the following degrees of spatial freedom: ________________________________.

b. Datum B would have controlled the following degrees of spatial freedom: ________________________________.

c. Datum C would have controlled the following degrees of spatial freedom: ________________________________.

29. Multiple Datum Reference Frame Identification. Sometimes it is desirable to use more than one datum reference frame on a single part. When this is done, the spatial degrees of freedom may be delineated on the field of the drawing. This can be done by showing the X, Y and Z axes created for each datum reference frame and stating the datum reference frame each applies to in brackets after each X, Y and Z shown.

For Example:

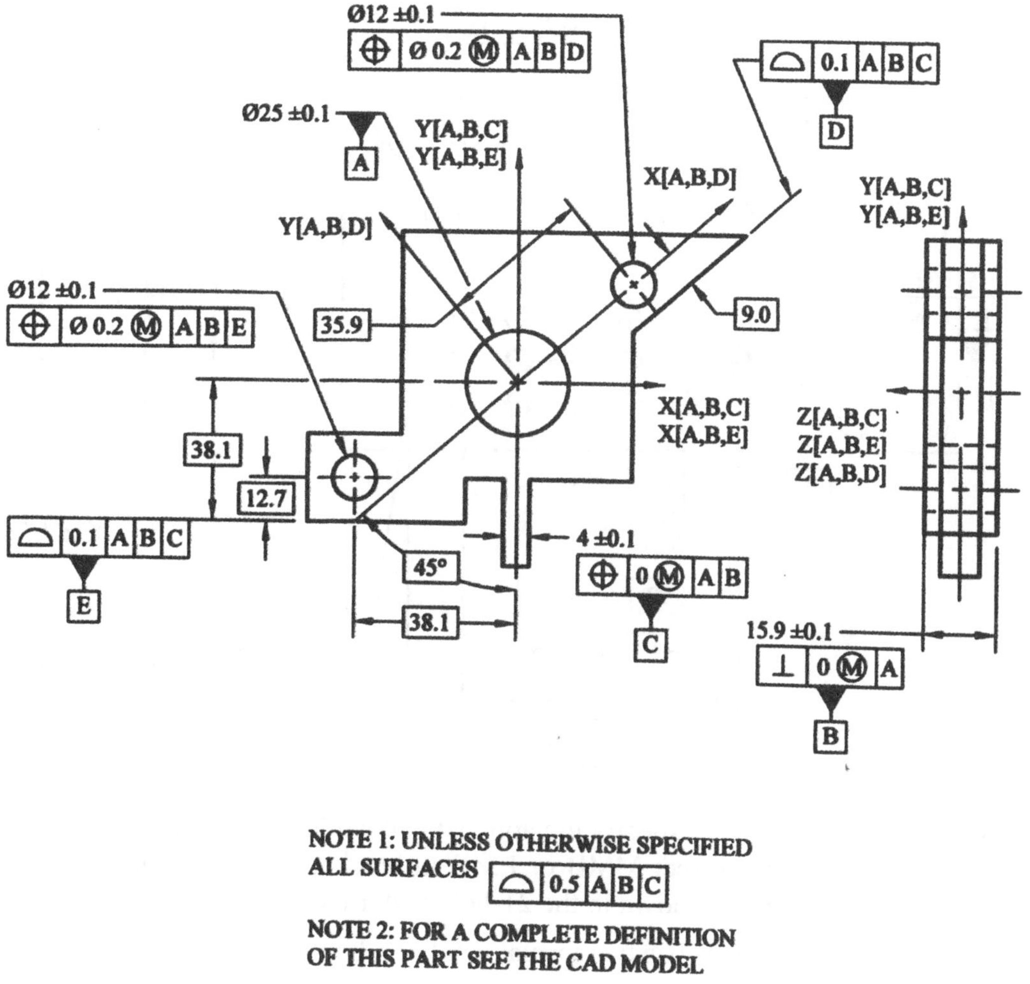

In this example:

a. Could an angularity geometric characteristic symbol be used in place of the perpendicularity symbol in the control on datum feature B?______________.

b. Which datum planes do the datum reference frames use to control angular orientation?

1. Datum reference frame A,B,C uses__________________.
2. Datum reference frame A,B,D uses__________________.
3. Datum reference frame A,B,E uses__________________.

c. What is the virtual condition of the Ø12±0.1 holes?__________.

d. What is the flatness tolerance on datum feature E?___________.

e. What is the location tolerance on datum feature D?___________.

30. Specifying the correct Maximum Material Boundary.

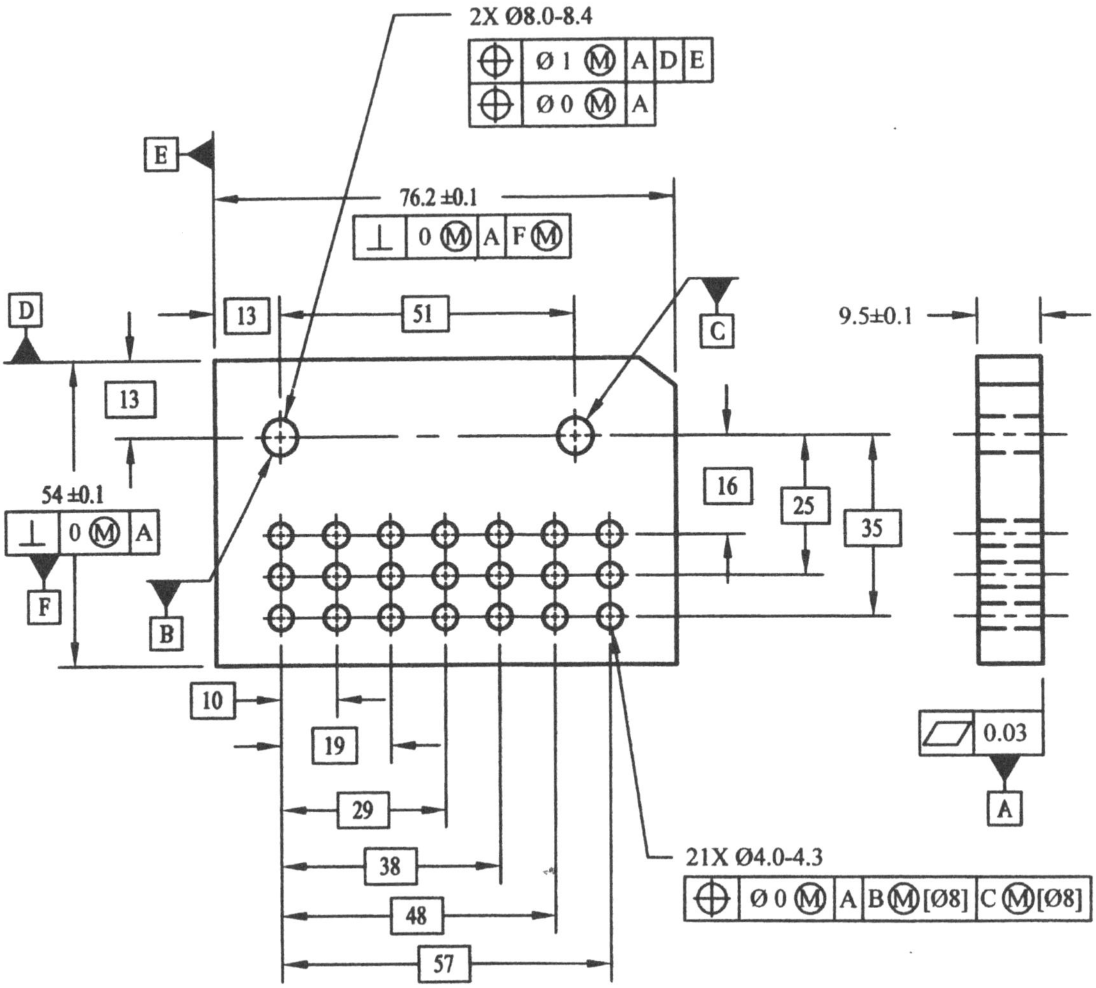

Sometimes it may desirable to specify the applicable boundary generated by a datum feature. This done by putting the correct MMB or LMB in brackets after the datum feature and its material boundary symbol as shown in the 21-hole pattern's feature control frame above.

a. The two hole pattern has two virtual condition boundaries. What are they?

b. If the ∅8 was not specified in brackets after datum features B and C, how would one know what the applicable Maximum Material Boundary for each is?

c. What are the inner boundaries for the 54 and 76.2 widths?

d. What size would the gage pins be for a gage that inspects thc position of the 2-hole pattern to datums A, D and E?

e. If a gage was designed for the 21-hole pattern to inspect its position tolerance, how many gage pins on a plate would be needed and what would their sizes be?

f. What would the size and shape be of a gage meant to inspect the perpendicularity of the 76.2 millimeter with?

31. Please fill in the correct MMB or LMB in the blank spaces on the following illustrations.

Correct Maximum Material Boundary Calculations

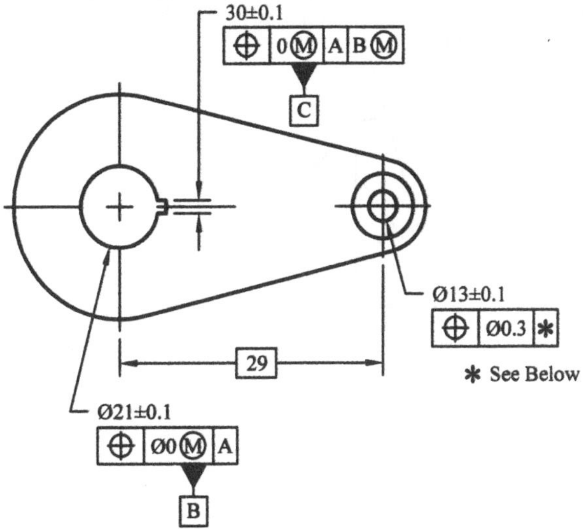

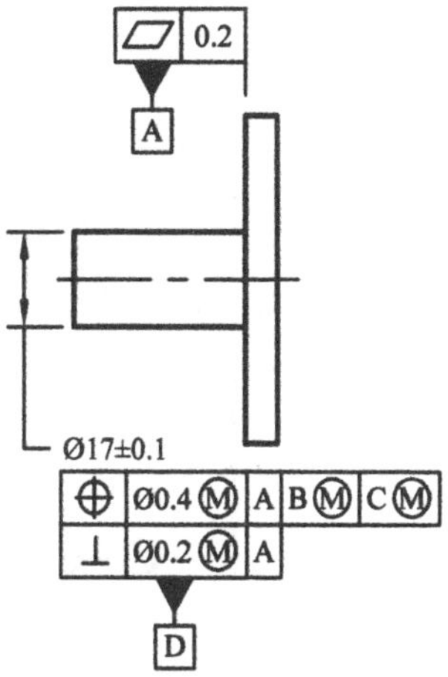

Option	Geometric Control	Maximum Material Boundary of datum feature D
(a)	⊕ | Ø 0.3 | DⓂ	
(b)	⊕ | Ø 0.3 | A | DⓂ	
(c)	⊕ | Ø 0.3 | A | B | DⓂ	

Correct Least Material Boundary Calculations

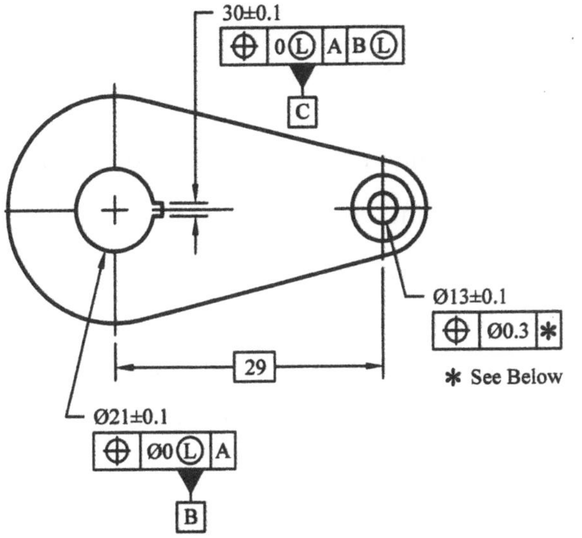

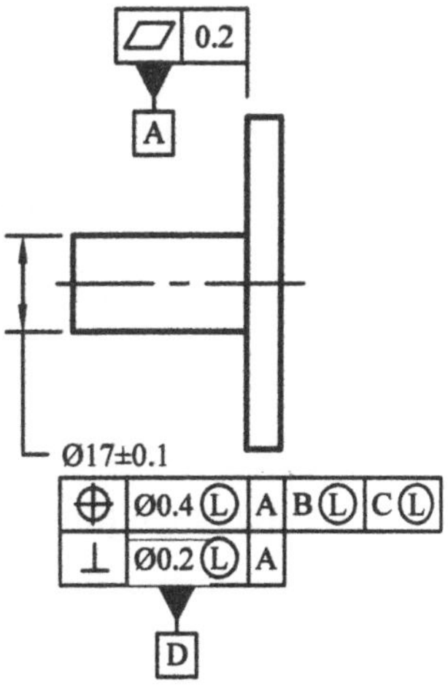

Option	Geometric Control	Least Material Boundary of datum feature D
(a)	⊕ | Ø 0.3 | DⓁ	
(b)	⊕ | Ø 0.3 | A | DⓁ	
(c)	⊕ | Ø 0.3 | A | B | DⓁ	

32. Answer the questions for the following illustration.

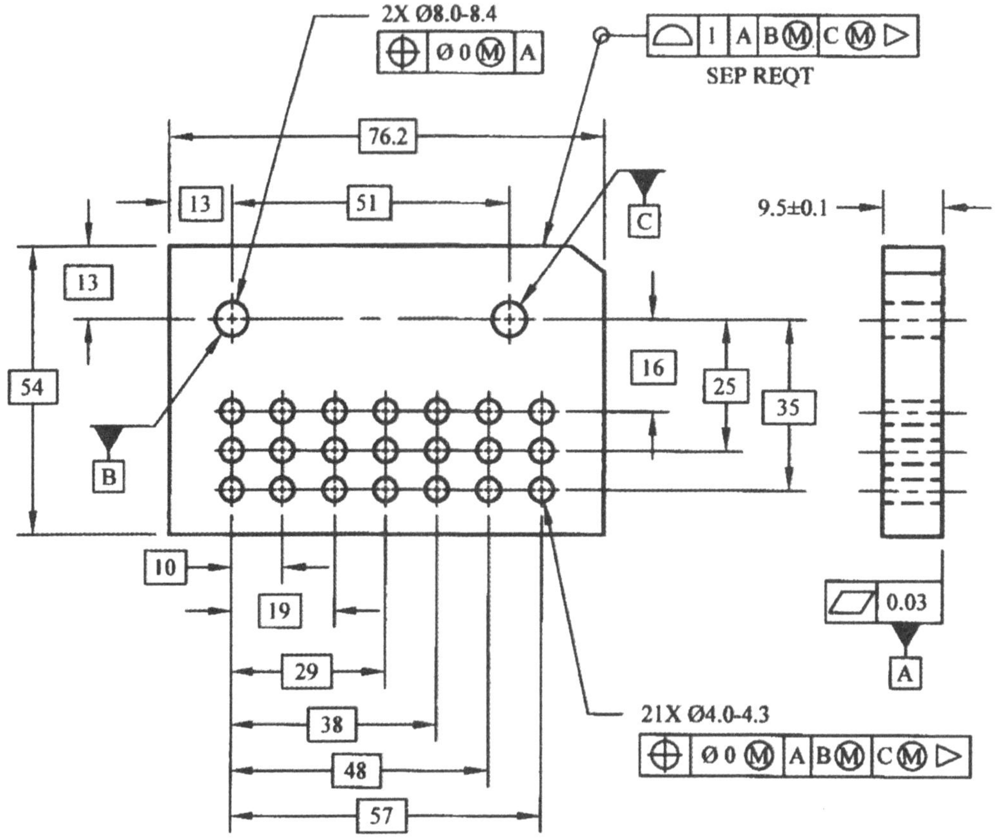

For the 21-hole pattern:

a. What does the symbol Ⓜ used after the zero tolerance in the feature control frame (FCF) stand for?

b. Would the datum feature simulator for datum feature B be stationary or moveable?

c. What does the symbol Ⓜ used after datum feature B in the FCF stand for?

d. What does the symbol used after the Ⓜafter datum feature C stand for?

e. What size gage pin would be used to represent datum feature B?

f.. What size gage pin would be used to represent datum feature C?

g. How would the datum feature simulator used to represent datum feature C differ from the one used for datum feature B?

For the profile of a surface control:

h. What does the SEP REQT note stand for?

i. How would datum feature shift due to datum feature B's departure from its virtual condition differ if the SEP REQT note was not on the drawing?

33. Using BASIC or BSC after a datum feature reference in a feature control frame.

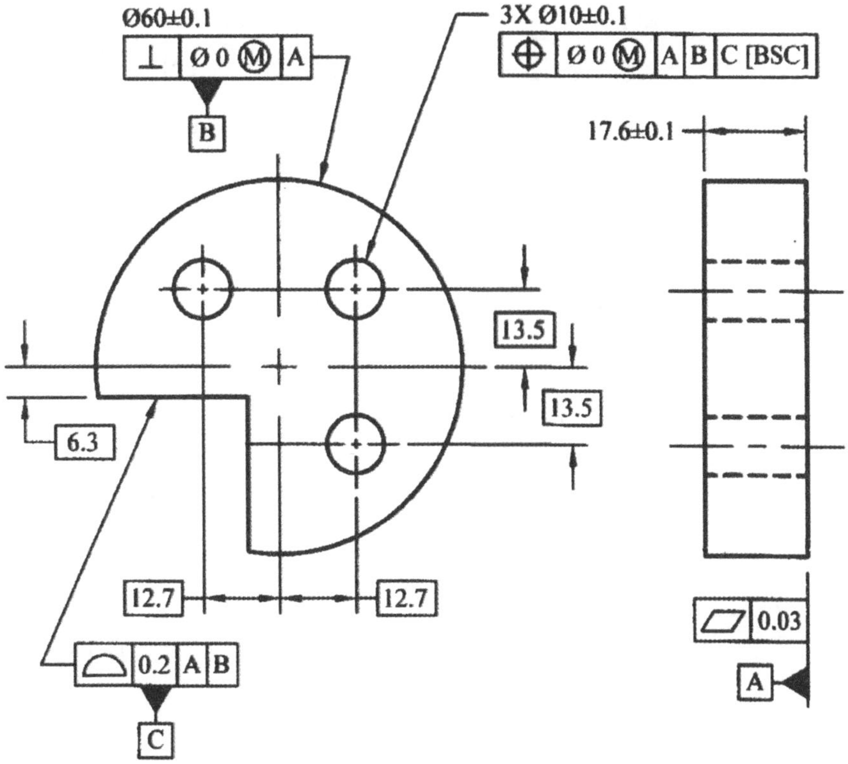

1. The datum feature simulator for datum feature C:
   a. Is represented at 6.4.
   b. Is represented at 6.3.
   c. Progresses normal from the MMB until it makes maximum contract with datum feature C.
   d. Is represented at 6.2.

2. What is the inner boundary of datum feature B?

3. What is the outer boundary of datum feature B?

4. What is the inner boundary of each of the 3 holes?

5. What is the outer boundary of each of the 3 holes?

6. What is the minimum wall thickness between one of the three holes and the outside diameter of the part?

34. Using MMB after a planar angular orientation datum feature reference in a feature control frame.

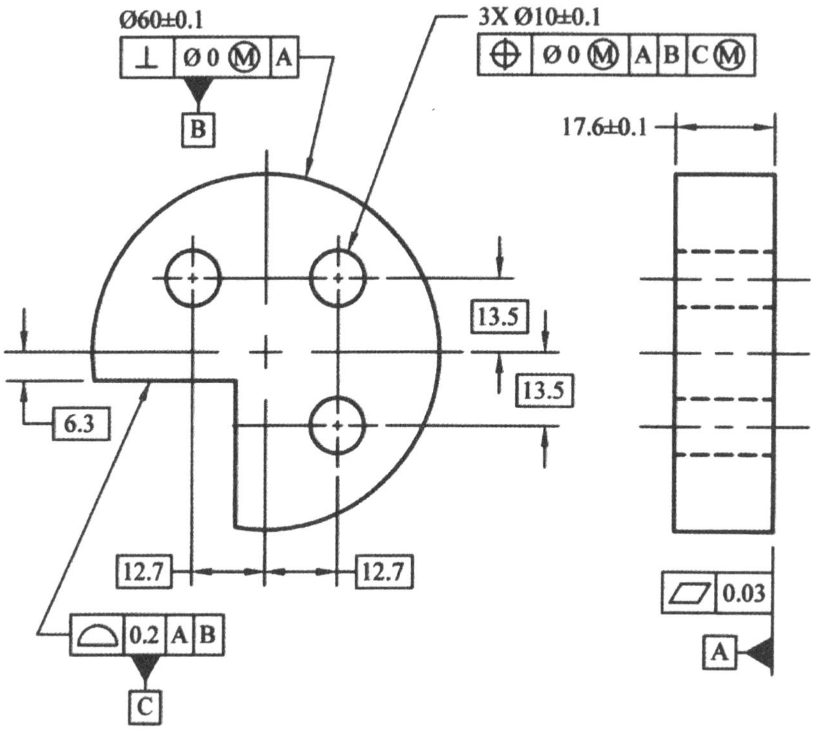

1. The datum feature simulator for datum feature C:

   a. Is represented at 6.4.

   b. Is represented at 6.3.

   c. Progresses normal from the MMB until it makes maximum contract with datum feature C.

   d. Is represented at 6.2.

35. Implying RMB after a planar angular orientation datum feature reference in a feature control frame.

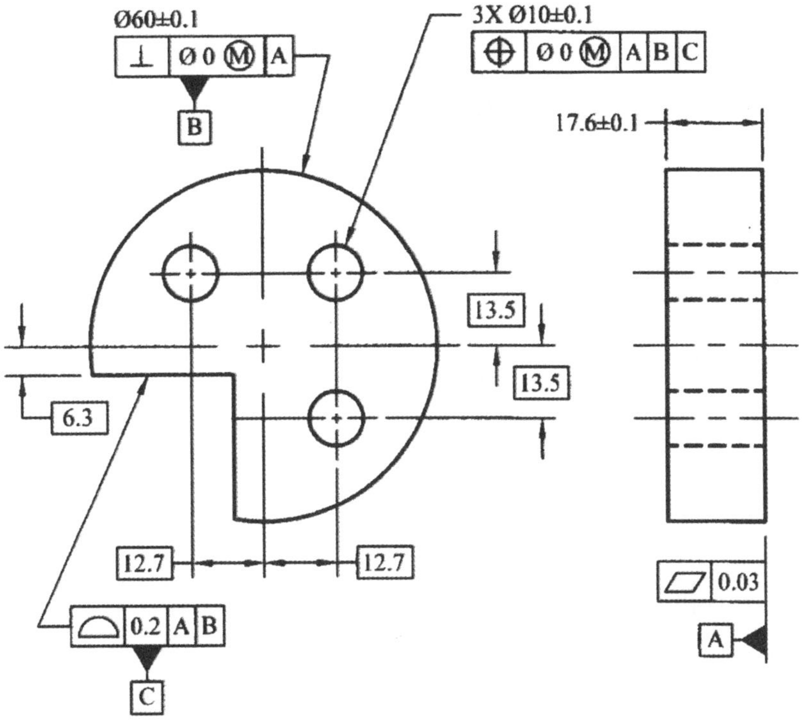

1. The datum feature simulator for datum feature C:
   a. Is represented at 6.4.
   b. Is represented at 6.3.
   c. Progresses normal from the MMB until it makes maximum contract with datum feature C.
   d. Is represented at 6.2.

36. Using a Translation Modifier after an offset angular orientation datum feature width.

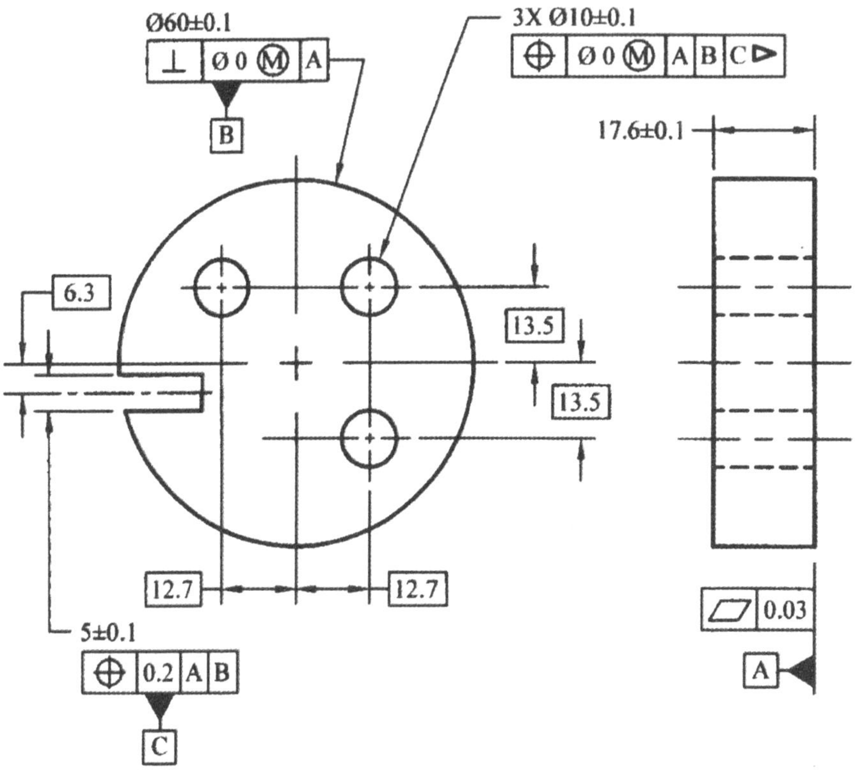

1. The datum feature simulator for datum feature C:
   a. Is represented at its maximum separation, perpendicular to datum plane A and its center plane is located 6.3 from datum axis B.
   b. Is represented at its maximum separation, perpendicular to datum plane A and its center plane is free to move within the position tolerance zone from datum axis B.
   c. Is represented at its maximum material boundary of 4.7, perpendicular to datum plane A and its center plane located 6.3 from datum axis B.
   d. Is represented at its least material boundary of 5.3, perpendicular to datum plane A and it center plane located 6.3 from datum axis B.

37. Using RMB after an offset angular orientation datum feature width.

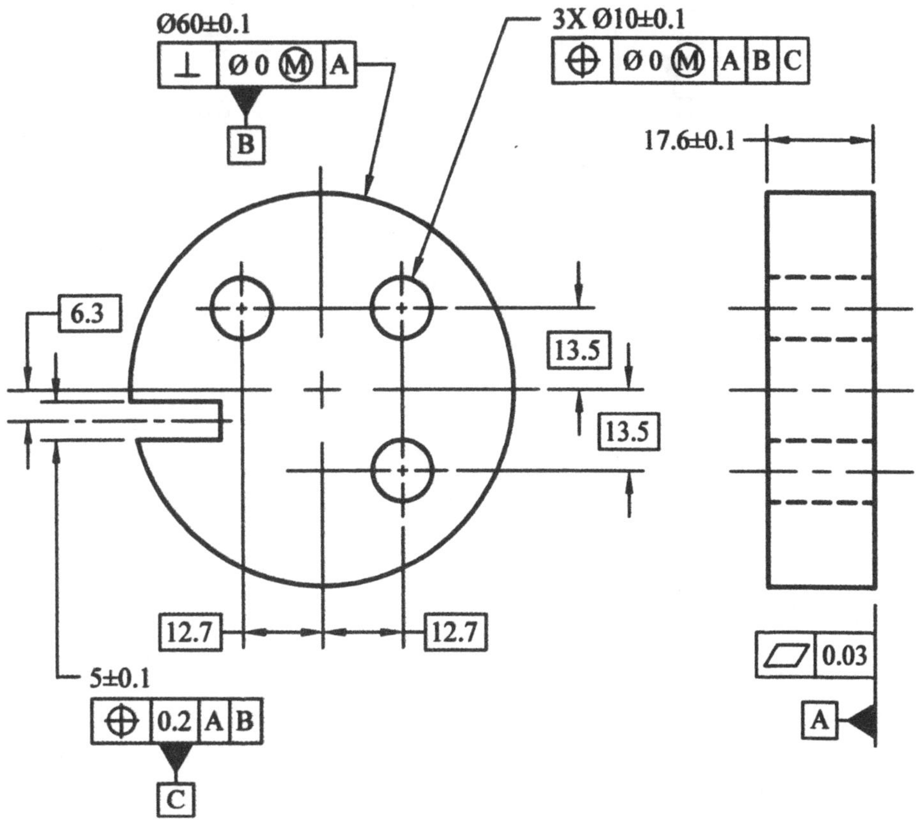

1. The datum feature simulator for datum feature C:
   a. Is represented at its maximum separation, perpendicular to datum plane A and its center plane is located 6.3 from datum axis B.
   b. Is represented at its maximum separation, perpendicular to datum plane A and its center plane is free to move within the position tolerance zone from datum axis B.
   c. Is represented at its maximum material boundary of 4.7, perpendicular to datum plane A and its center plane located 6.3 from datum axis B.
   d. Is represented at its least material boundary of 5.3, perpendicular to datum plane A and it center plane located 6.3 from datum axis B.

38. Using MMB after a centered semi-cylindrical angular orientation datum feature.

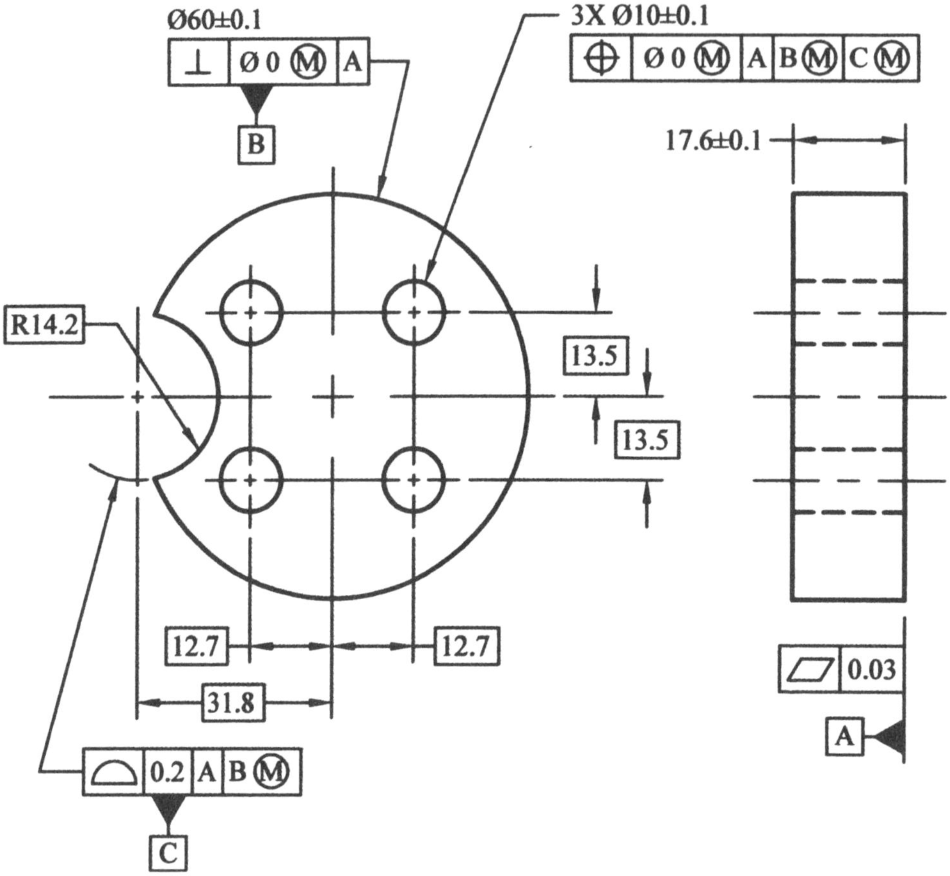

1. The datum feature simulator for datum feature C:
   a. Is represented at a radius of 14.1, perpendicular to datum plane A and its center plane is located 31.8 from datum axis B.
   b. Is represented at a diameter of 14.1, perpendicular to datum plane A and its center is free to move within the profile tolerance zone from datum axis B.
   c. Is represented at its maximum inscribed cylinder size, perpendicular to datum plane A and its center plane located 31.8 from datum axis B.
   d. Is represented at its least material boundary of 14.3, perpendicular to datum plane A and it center plane located 31.8 from datum axis B.

39. Using RMB after a centered semi-cylindrical angular orientation datum feature.

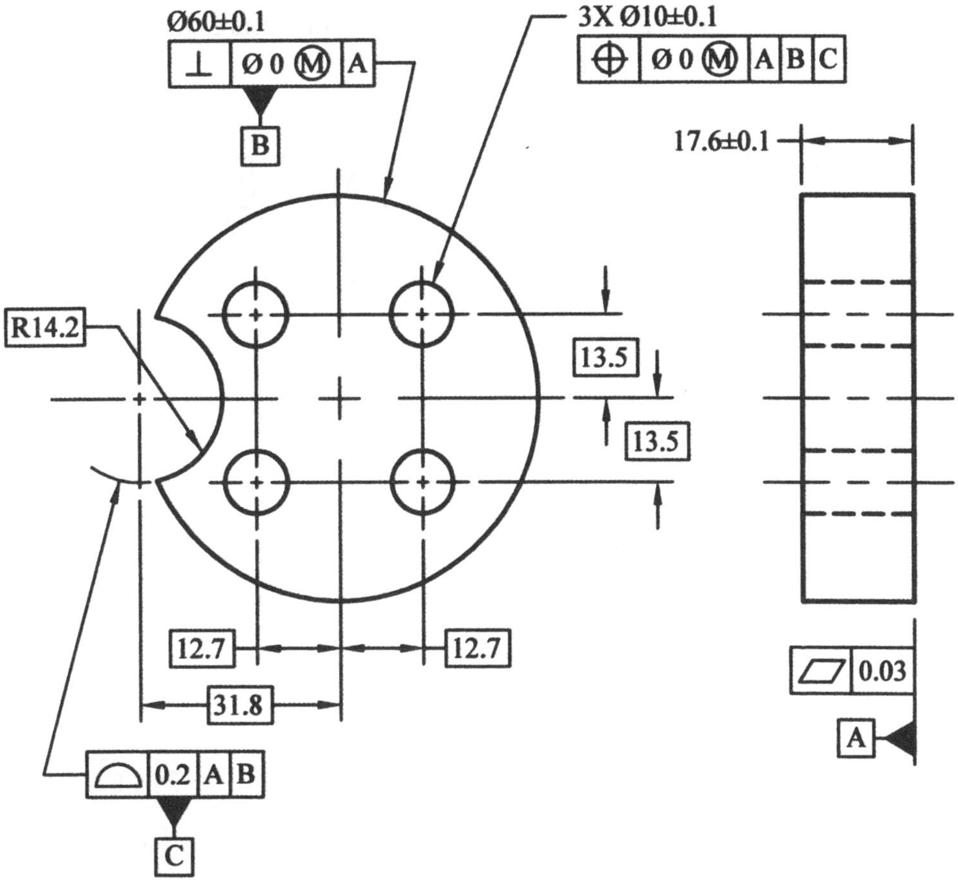

1. The datum feature simulator for datum feature C:
   a. Is represented at a radius of 14.1, perpendicular to datum plane A and its center plane is located 31.8 from datum axis B.
   b. Is represented at a diameter of 14.1, perpendicular to datum plane A and its center is free to move within the profile tolerance zone from datum axis B.
   c. Is represented at its maximum inscribed cylinder size, perpendicular to datum plane A and its center plane located 31.8 from datum axis B.
   d. Is represented at its least material boundary of 14.3, perpendicular to datum plane A and it center plane located 31.8 from datum axis B.

40. Using the illustration of the part detail shown at the top of the page, please dimension (but don't tolerance) the diameter of the gage pins, the gage inside diameter and the distance from the center of the gage (datum feature simulator B) to datum feature simulator C.

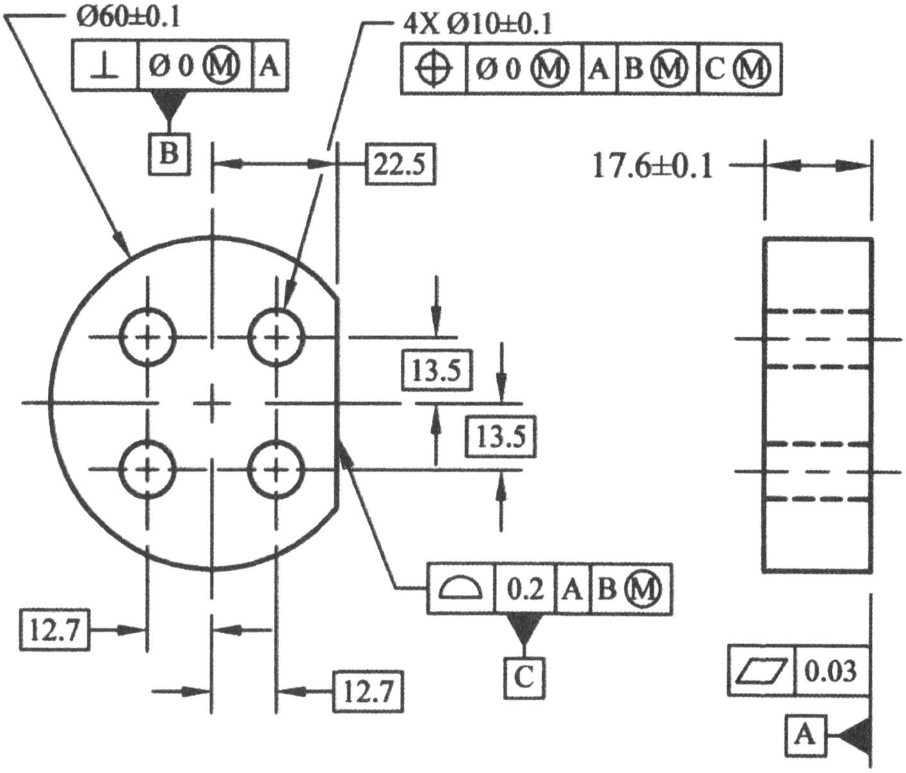

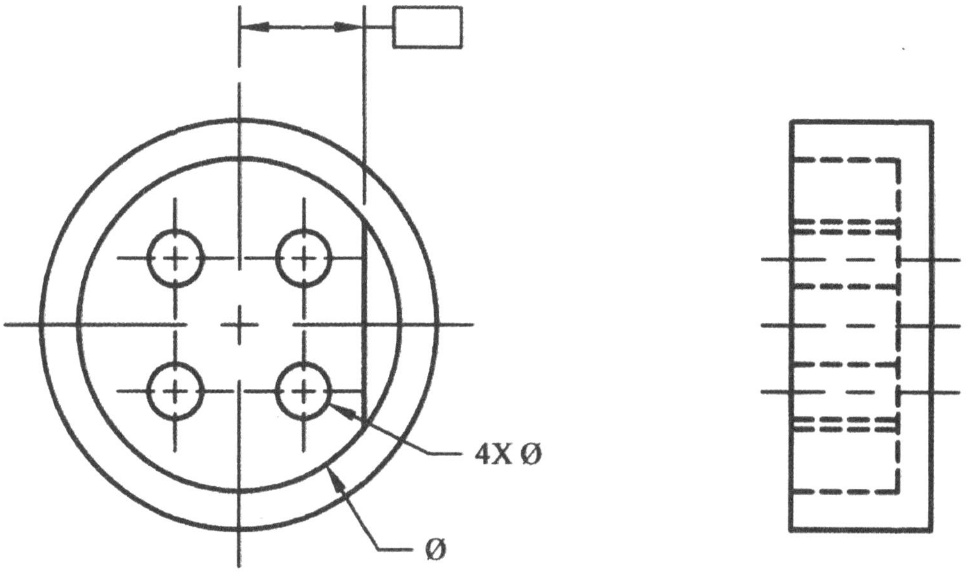

Functional Gage

41. For the detail drawing shown in item (a) please dimension (not tolerance) the partial gage shown in item (b) to represent the datum feature simulator and also the gage element to represent the outside diameter of the part.

(a) Outer Boundary of a Pattern of Pins Referenced at MMB.

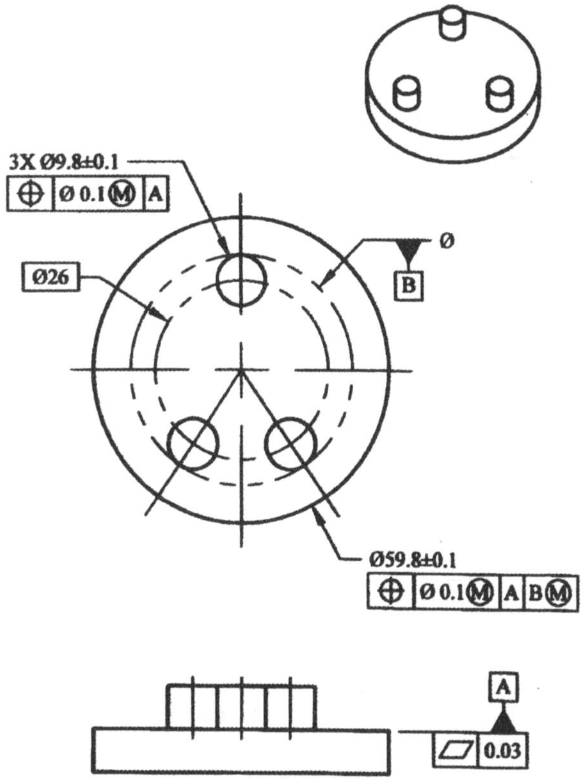

(b) A Partial Gage Design for the part in Illustration a.

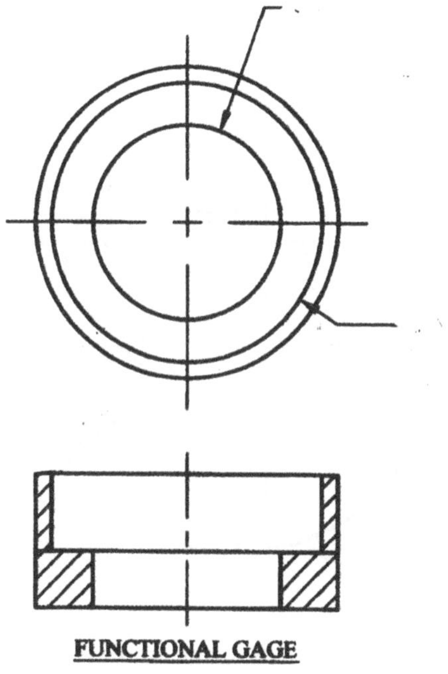

42. For the detail drawing shown in item (a) please dimension (not tolerance) the gage element shown in item (b) to represent datum feature simulator B.

(a) [Inner boundary of a pattern of pins referenced at MMB]

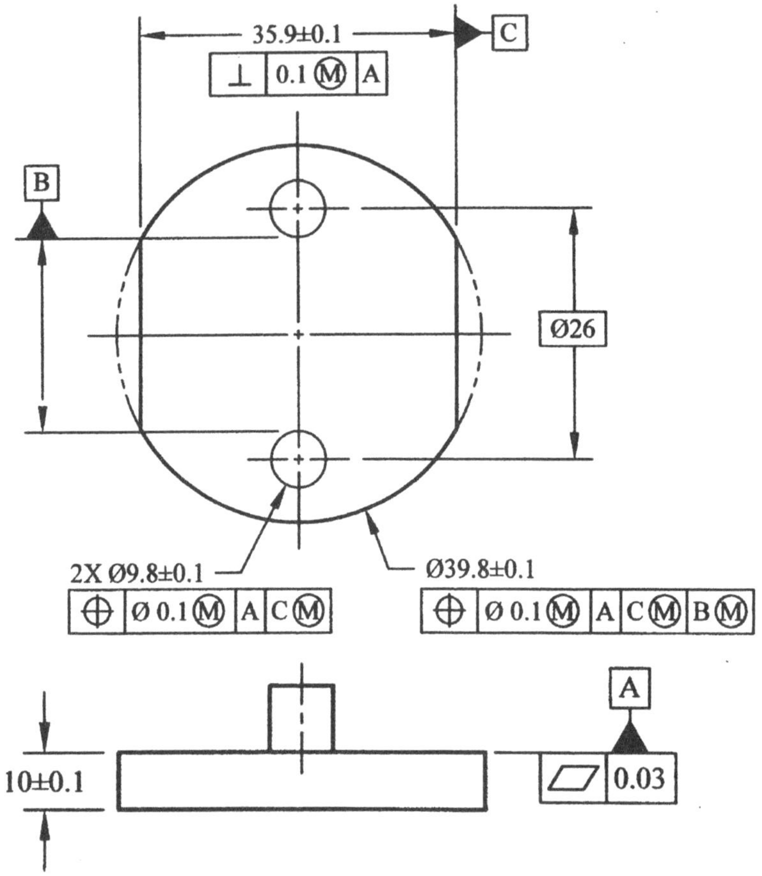

(b) [Datum feature simulator B]

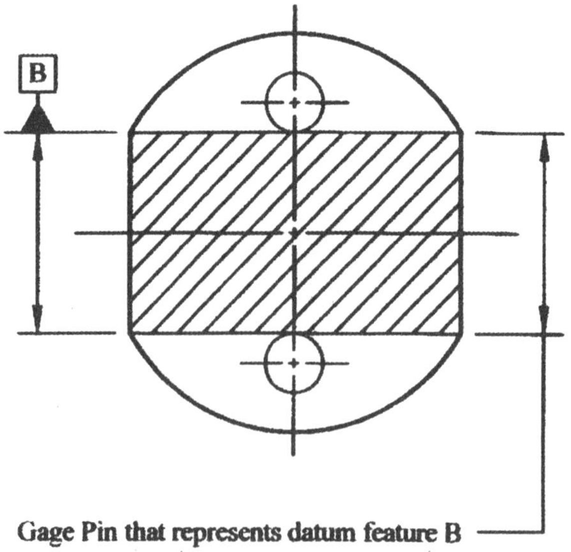

43. Using the correct datum feature designations, feature control frames and local notes:

a. Designate each of the 12 millimeter holes as a datum feature using one datum feature symbol (the letter D).

b. Locate the smaller 6 millimeter holes to the left and right of each large hole to that 12 millimeter hole only, using datum plane A for perpendicularity, datum axis D for location and datum plane B for angular orientation.

c. Use the local note 4X INDIVIUALLY twice.

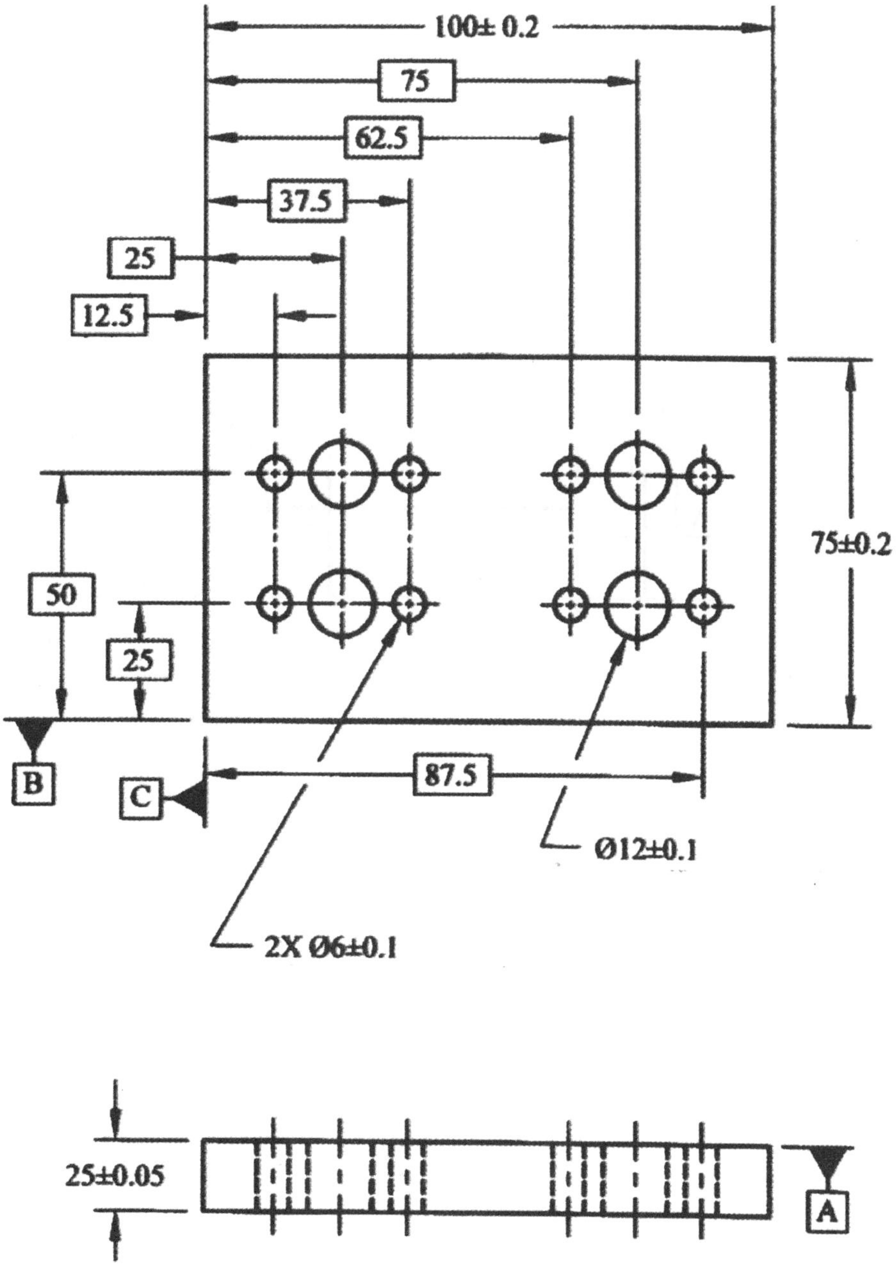

44. Moveable Datum Target Symbol. Please design a fixture to simulate the datum features shown in the drawing below.

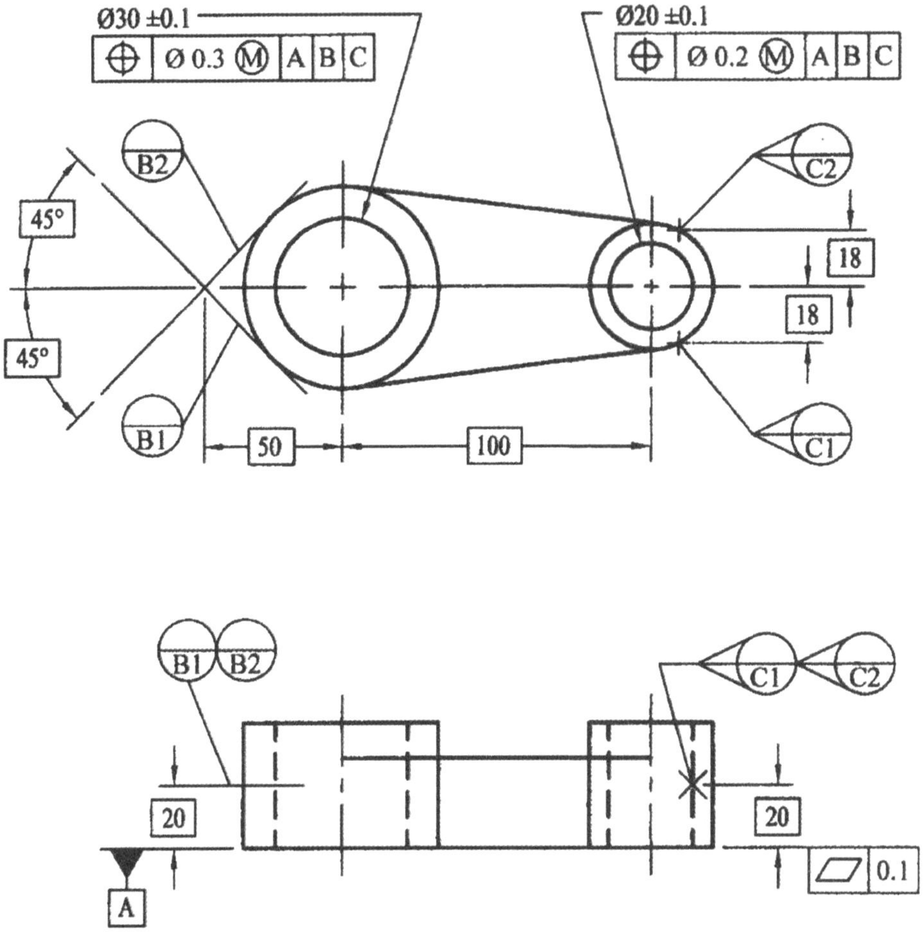

45. Spherical Moveable Datum Targets. Please assign a stationary spherical datum target of a diameter of 24 on the right end of this part and a moveable spherical datum target of a diameter of 20 on the left end of this part.

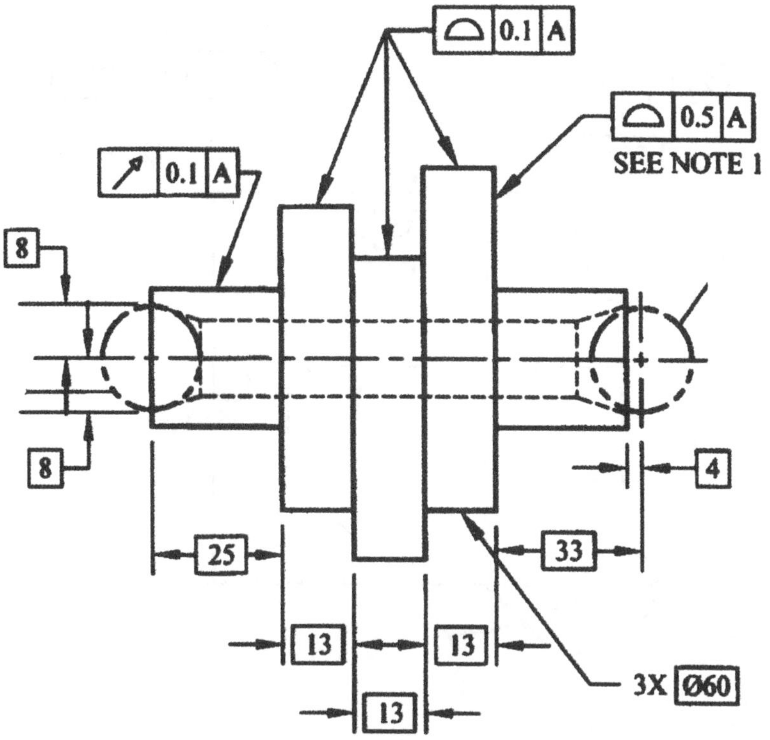

NOTE 1:

THIS CONTROL APPLIES TO ALL PLANER SURFACES LOCATED BY BASIC DIMENSIONS IN THIS VIEW.

46. Please dimension the size and location of the gage pins for the functional gage shown below the part drawing.

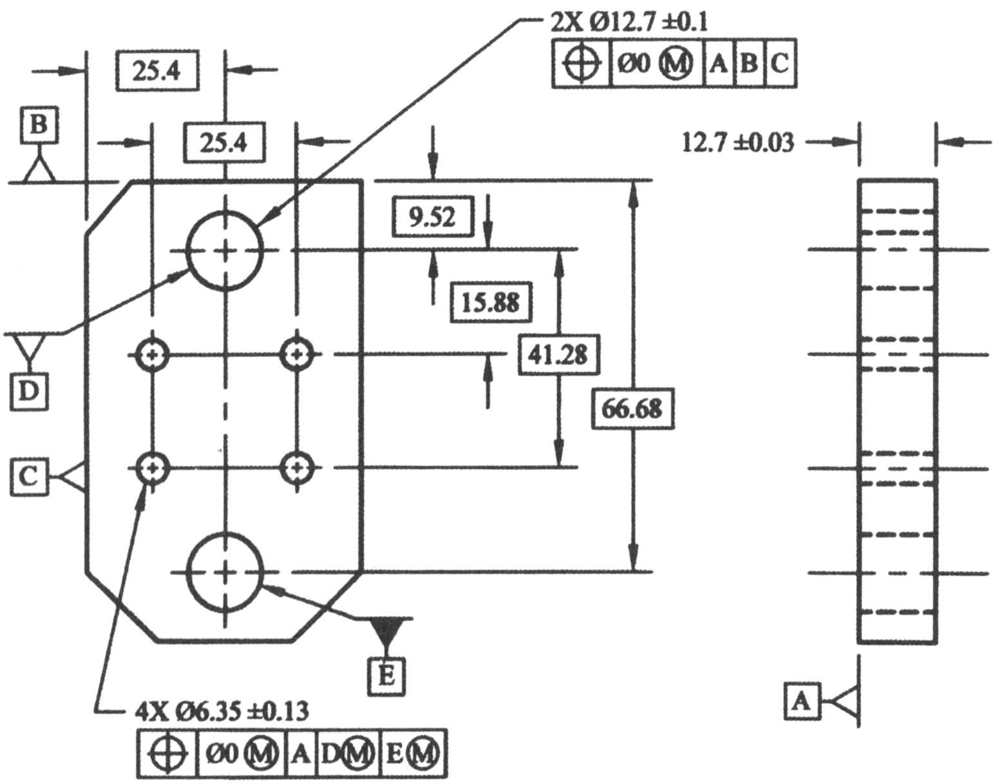

MEANS THIS:

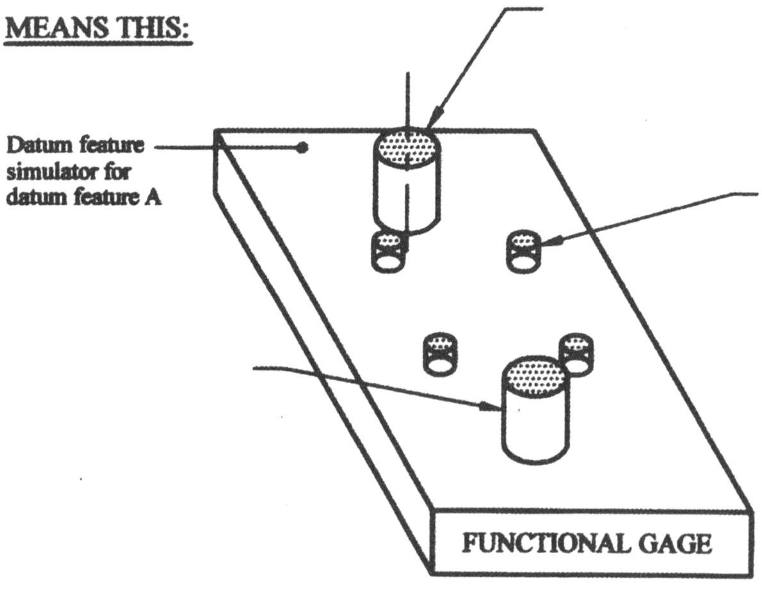

47. Please dimension and tolerance the functional gage shown beneath the part drawing. Use 10% of the part tolerance and the Practical Absolute method of dimensioning and tolerancing the gage.

Part Detail Drawing

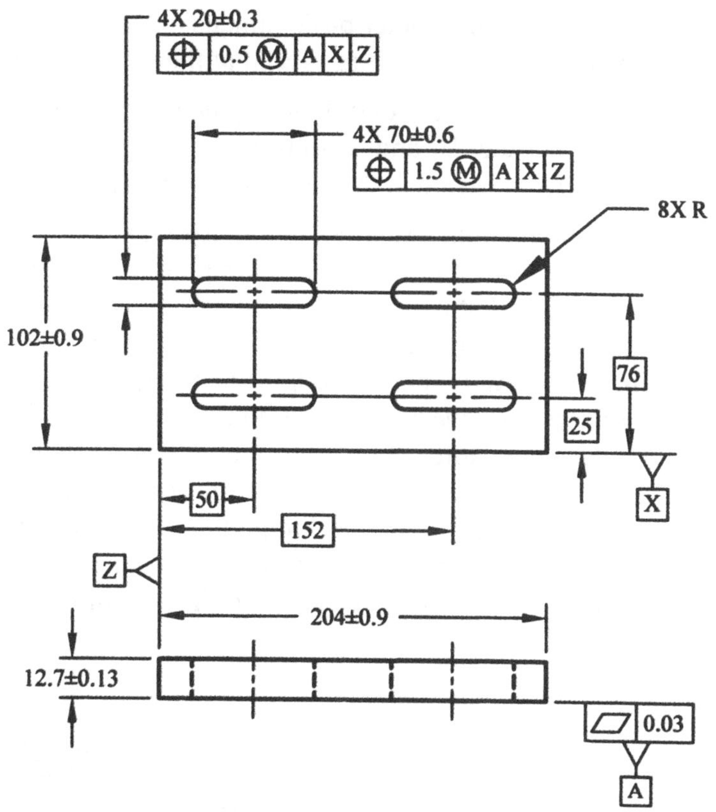

Functional Gage Drawing

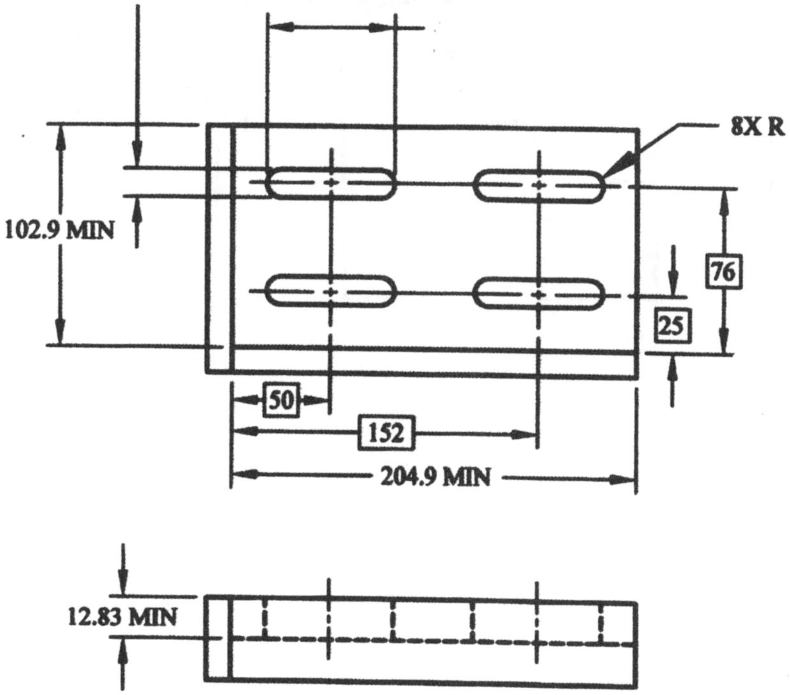

48. Please apply a three-level composite positional tolerancing control to the three-hole pattern to mean the following:

1. The top level control locates the three holes to datums B and C. Regardless of the material boundary of datum feature B, the three holes must be centered to datum centerplane B and hold the 19 millimeter basic dimension from datum plane C to within a diameter of 3 millimeters if the holes are produced at their maximum material condition.
2. The middle level control is ultimately only holding the orientation of the three holes (as a pattern) to datum centerplane B and datum plane C to within a diameter of 1.5 millimeters if the holes are produced at their maximum material condition.
3. The third level control is aligning the three holes' axes to one another for coaxiality to within a diameter of 0.3 millimeters if the holes are produced at their maximum material condition. It is worth noting that no datum feature references are used here, in that, none are necessary to position the three holes to each other. This tolerance is the one that allows the 8 millimeter screw to pass through the three holes. It generates a virtual condition (inner boundary) for the three holes that is calculated by subtracting the 0.3 millimeter diameter position tolerance from the 8.3 millimeter maximum material condition of the holes. 8.3 minus 0.3 equals an 8 millimeter diameter virtual condition. This is the worst mating condition and is compatible with the 8 millimeter screw that must pass through. Since 8 millimeter screws are actually a little smaller than 8 millimeters, there is even a little clearance built into this assembly condition.

Detail Drawing

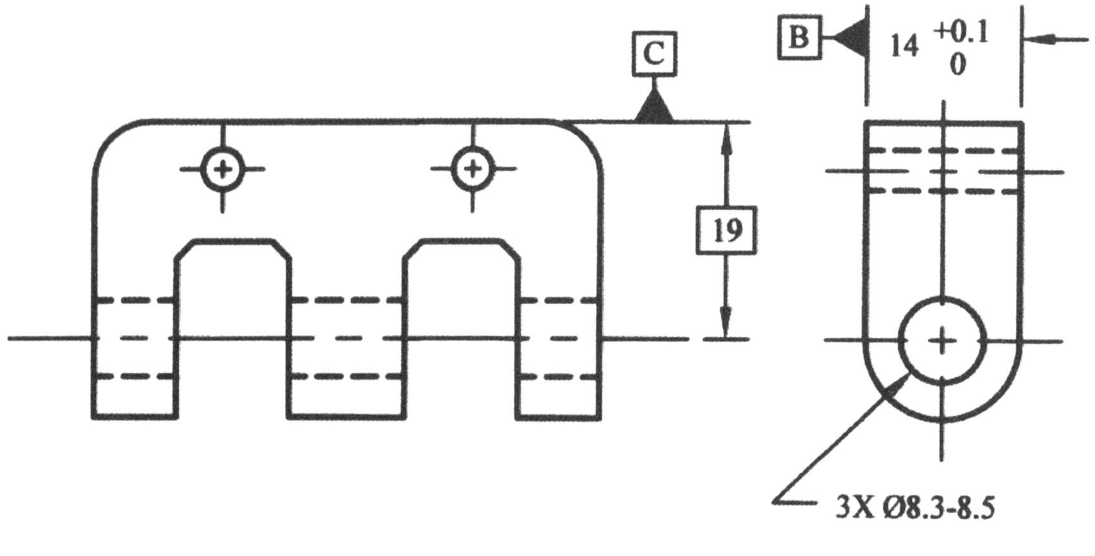

49. Please dimension the missing radii on the fixture at the bottom of the page to suit the part detail drawing shown on the top of the page.

Curved Surface as a Datum Feature

Detail Drawing

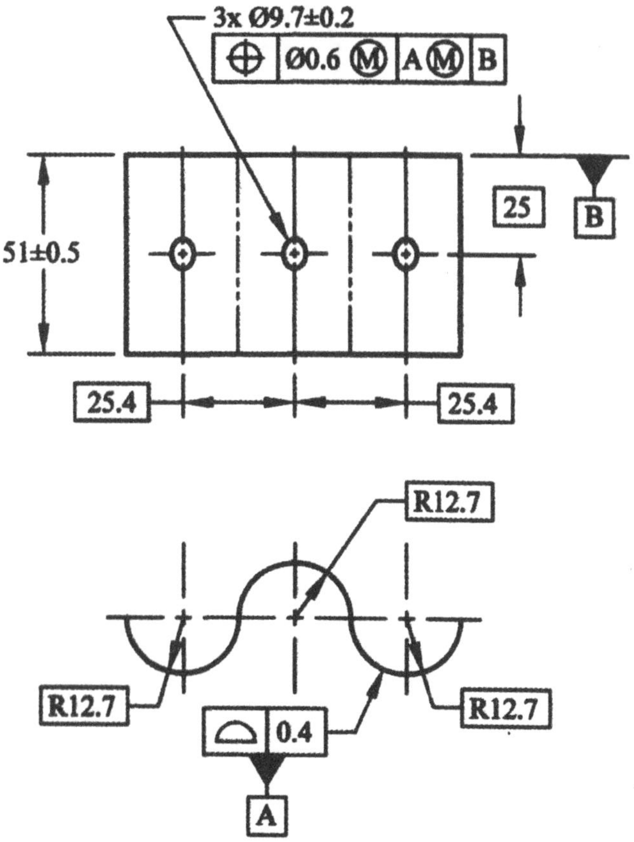

NOTES:
1. MATERIAL THICKNESS 2.5 MILLIMETERS
2. UNLESS OTHERWISE SPECIFIED ALL ANGLES ARE ±1 DEGREE
3. ALL RADII REFER TO DATUM FEATURE A

Fixture

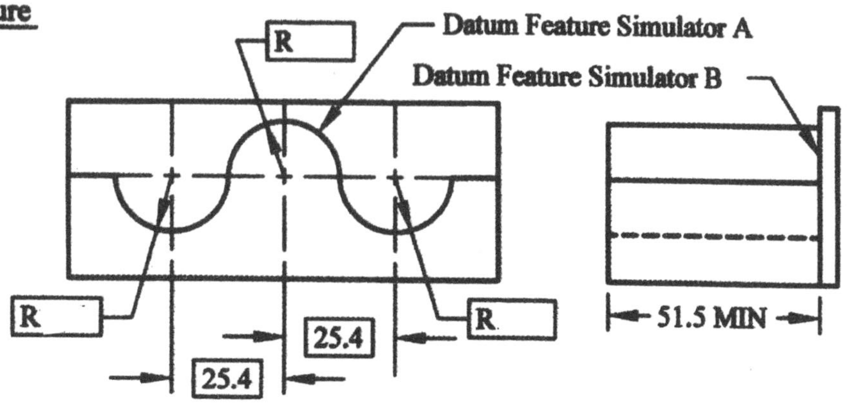

V. ANSWERBOOK SECTION
for
Geometric Dimensioning and Tolerancing
[per ASME Y14.5-2009]

1.

	Feature	Feature Control Frame	Outer Boundary	Inner Boundary
1)	ø20±2 Hole	⌖ | ⌀ 2 Ⓜ | A | B | C	22+6=28	18−2=16
2)	ø20±2 Hole	⌖ | ⌀ 2 Ⓛ | A | B | C	22+2=24	18−6=12
3)	ø20±2 Hole	⌖ | ⌀ 2 | A | B | C	22+2=24	18−2=16
4)	ø20^{+3}_{-2} Shaft	⌖ | ⌀ 0.5 Ⓜ | A | B | C	23.0+0.5=23.5	18.0−5.5=12.5
5)	ø15−16 Shaft	⌖ | ⌀ 0.4 Ⓛ | A | B | C	16.0+1.4=17.4	15.0−0.4=14.6
6)	ø$30^{+0.4}_{-0.2}$ Shaft	⌖ | ⌀ 0.3 | A | B | C	30.4+0.3=30.7	29.8−0.3=29.5
7)	$28.5^{+0.2}_{-0.1}$ Slot	⌖ | 0.8 Ⓜ | X | Y | Z	28.7+1.1=29.8	28.4−0.8=27.6
8)	23.08−24.04 Tab	⌖ | 0.5 Ⓜ | T | E | C	24.04+0.50=24.54	23.08−1.46=21.62
9)	75±2 Slot	⌖ | 1 Ⓛ | C | A | B	77+1=78	73−5=68
10)	70±0.5 Tab	⌖ | 0.5 Ⓛ | D | E | F	70.5+1.5=72	69.5−0.5=69
11)	12−13 Slot	⌖ | 0.3 | G | A | F	13+0.3=13.3	12−0.3=11.7
12)	9.5−10.6 Tab	⌖ | 1.2 | H | J | K	10.6+1.2=11.8	9.5−1.2=8.3

1. (continued)

①

$$\begin{array}{r} 28 = \text{O.B.} \\ +\ 16 = \text{I.B.} \\ \hline 44 = \text{Sum} \end{array} \qquad \begin{array}{r} 28 = \text{O.B.} \\ -\ 16 = \text{I.B.} \\ \hline 12 = \text{Diff.} \end{array}$$

$$\frac{44}{2} = 22 = \text{Mean} \quad \& \quad \frac{12}{2} = 6 = \pm\text{Tol.}$$

So; 22±6

②

$$\begin{array}{r} 24 = \text{O.B.} \\ +\ 12 = \text{I.B.} \\ \hline 36 = \text{Sum} \end{array} \qquad \begin{array}{r} 24 = \text{O.B.} \\ -\ 12 = \text{I.B.} \\ \hline 12 = \text{Diff.} \end{array}$$

$$\frac{36}{2} = 18 = \text{Mean} \quad \& \quad \frac{12}{2} = 6 = \pm\text{Tol.}$$

So; 18±6

③

$$\begin{array}{r} 24 = \text{O.B.} \\ +\ 16 = \text{I.B.} \\ \hline 40 = \text{Sum} \end{array} \qquad \begin{array}{r} 24 = \text{O.B.} \\ -\ 16 = \text{I.B.} \\ \hline 8 = \text{Diff.} \end{array}$$

$$\frac{40}{2} = 20 = \text{Mean} \quad \& \quad \frac{8}{2} = 4 = \pm\text{Tol.}$$

So; 20±4

④

$$\begin{array}{r} 23.5 = \text{O.B.} \\ +\ 12.5 = \text{I.B.} \\ \hline 36\ = \text{Sum} \end{array} \qquad \begin{array}{r} 23.5 = \text{O.B.} \\ -\ 12.5 = \text{I.B.} \\ \hline 11\ = \text{Diff.} \end{array}$$

$$\frac{36}{2} = 18 = \text{Mean} \quad \& \quad \frac{11}{2} = 5.5 = \pm\text{Tol.}$$

So; 18±5.5

⑤

$$\begin{array}{r} 17.4 = \text{O.B.} \\ +\ 14.6 = \text{I.B.} \\ \hline 32\ = \text{Sum} \end{array} \qquad \begin{array}{r} 17.4 = \text{O.B.} \\ -\ 14.6 = \text{I.B.} \\ \hline 2.8 = \text{Diff.} \end{array}$$

$$\frac{32}{2} = 16 = \text{Mean} \quad \& \quad \frac{2.8}{2} = 1.4 = \pm\text{Tol.}$$

So; 16±1.4

⑥

$$\begin{array}{r} 30.7 = \text{O.B.} \\ +\ 29.5 = \text{I.B.} \\ \hline 60.2 = \text{Sum} \end{array} \qquad \begin{array}{r} 30.7 = \text{O.B.} \\ -\ 29.5 = \text{I.B.} \\ \hline 1.2 = \text{Diff.} \end{array}$$

$$\frac{60.2}{2} = 30.1 = \text{Mean} \quad \& \quad \frac{1.2}{2} = 0.6 = \pm\text{Tol.}$$

So; 30.1±0.6

1. (continued)

⑦

$$\begin{array}{r} 29.8 = \text{O.B.} \\ +\ 27.6 = \text{I.B.} \\ \hline 57.4 = \text{Sum} \end{array} \qquad \begin{array}{r} 29.8 = \text{O.B.} \\ -\ 27.6 = \text{I.B.} \\ \hline 2.2 = \text{Diff.} \end{array}$$

$$\frac{57.4}{2} = 28.7 = \text{Mean} \quad \& \quad \frac{2.2}{2} = 1.1 = \pm \text{Tol.}$$

So; 28.7±1.1

⑧

$$\begin{array}{r} 24.54 = \text{O.B.} \\ +\ 21.62 = \text{I.B.} \\ \hline 46.16 = \text{Sum} \end{array} \qquad \begin{array}{r} 24.54 = \text{O.B.} \\ -\ 21.62 = \text{I.B.} \\ \hline 2.92 = \text{Diff.} \end{array}$$

$$\frac{46.16}{2} = 23.08 = \text{Mean} \quad \& \quad \frac{2.92}{2} = 1.46 = \pm \text{Tol.}$$

So; 23.08±1.46

⑨

$$\begin{array}{r} 78 = \text{O.B.} \\ +\ 68 = \text{I.B.} \\ \hline 146 = \text{Sum} \end{array} \qquad \begin{array}{r} 78 = \text{O.B.} \\ -\ 68 = \text{I.B.} \\ \hline 10 = \text{Diff.} \end{array}$$

$$\frac{146}{2} = 73 = \text{Mean} \quad \& \quad \frac{10}{2} = 5 = \pm \text{Tol.}$$

So; 73±5

⑩

$$\begin{array}{r} 72 = \text{O.B.} \\ +\ 69 = \text{I.B.} \\ \hline 141 = \text{Sum} \end{array} \qquad \begin{array}{r} 72 = \text{O.B.} \\ -\ 69 = \text{I.B.} \\ \hline 3 = \text{Diff.} \end{array}$$

$$\frac{141}{2} = 70.5 = \text{Mean} \quad \& \quad \frac{3}{2} = 1.5 = \pm \text{Tol.}$$

So; 70.5±1.5

⑪

$$\begin{array}{r} 13.3 = \text{O.B.} \\ +\ 11.7 = \text{I.B.} \\ \hline 25 = \text{Sum} \end{array} \qquad \begin{array}{r} 13.3 = \text{O.B.} \\ -\ 11.7 = \text{I.B.} \\ \hline 1.6 = \text{Diff.} \end{array}$$

$$\frac{25}{2} = 12.5 = \text{Mean} \quad \& \quad \frac{1.6}{2} = 0.8 = \pm \text{Tol.}$$

So; 12.5±0.8

⑫

$$\begin{array}{r} 11.8 = \text{O.B.} \\ +\ 8.3 = \text{I.B.} \\ \hline 20.1 = \text{Sum} \end{array} \qquad \begin{array}{r} 11.8 = \text{O.B.} \\ -\ 8.3 = \text{I.B.} \\ \hline 3.5 = \text{Diff.} \end{array}$$

$$\frac{20.1}{2} = 10.05 = \text{Mean} \quad \& \quad \frac{3.5}{2} = 1.75 = \pm \text{Tol.}$$

So; 10.05±1.75

2.

Part #1

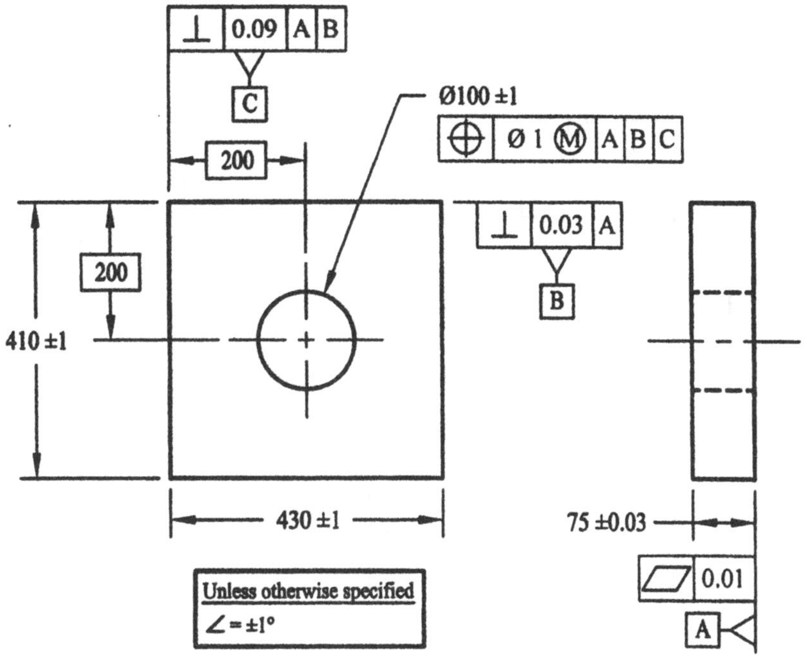

Part #2

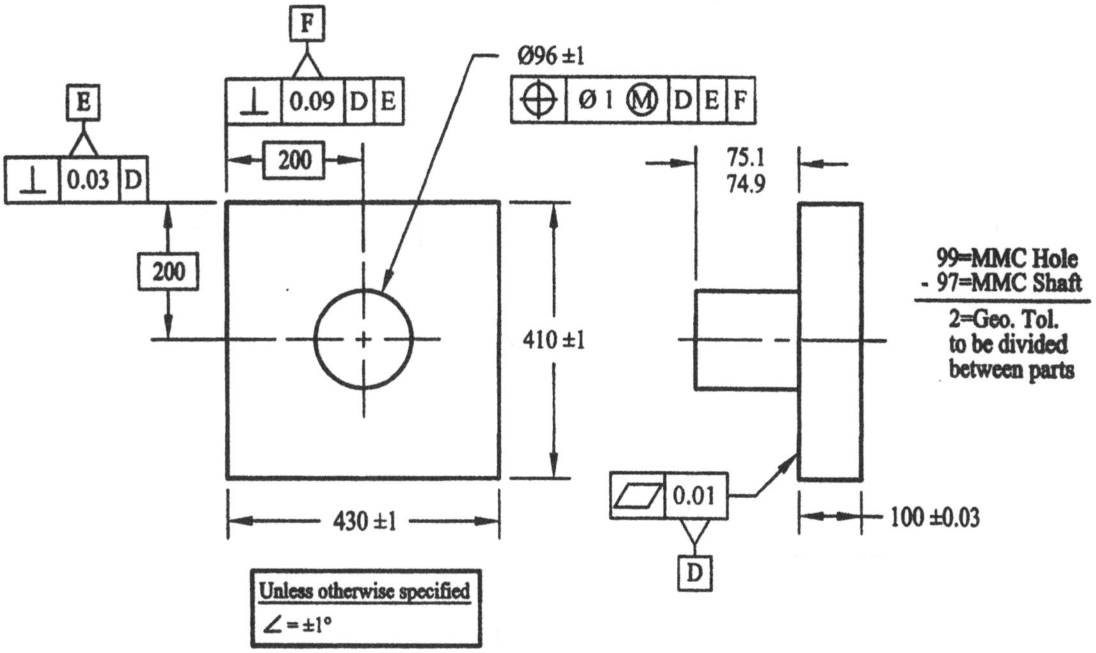

3. Practical Absolute Functional Gage for Part #1

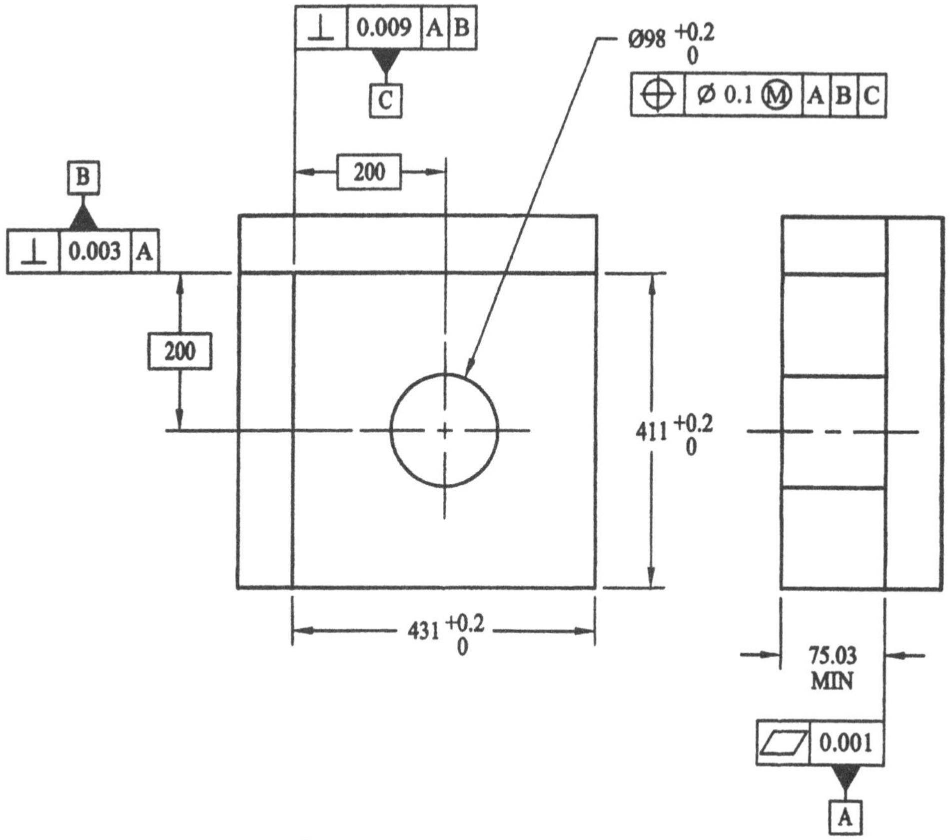

4.

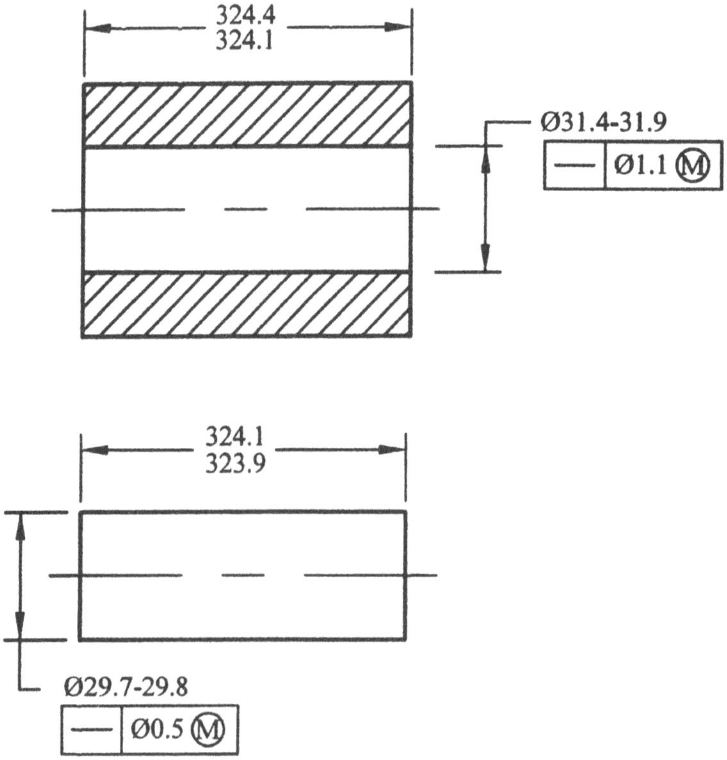

5.

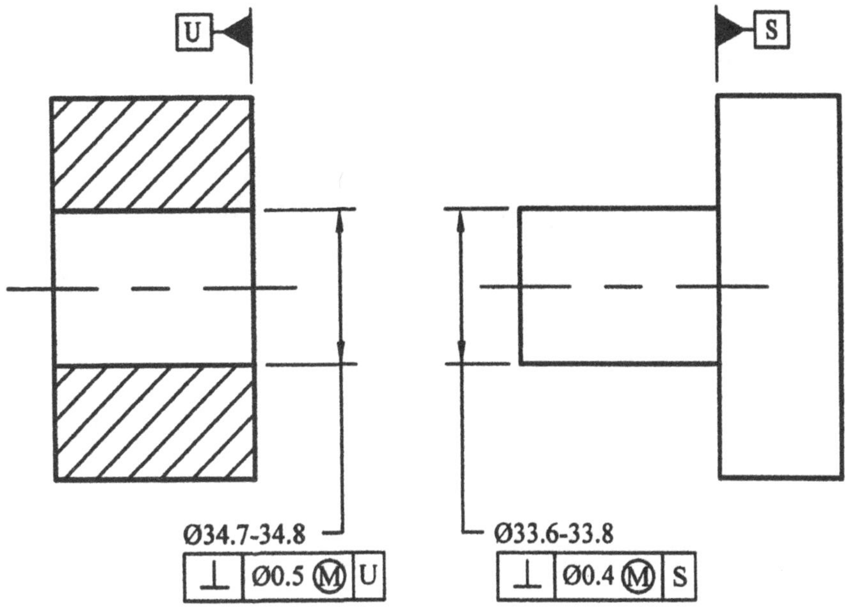

6.

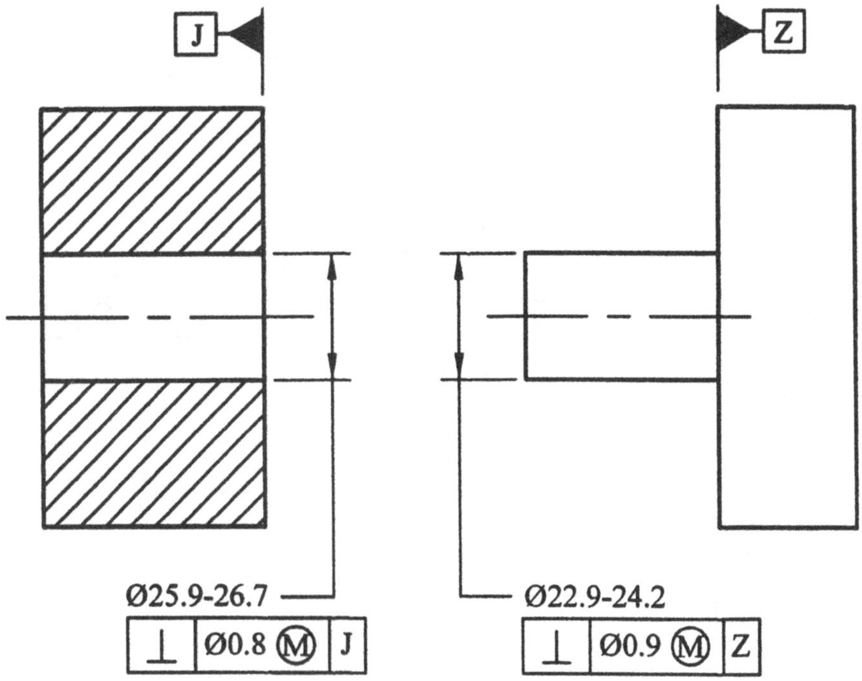

7.

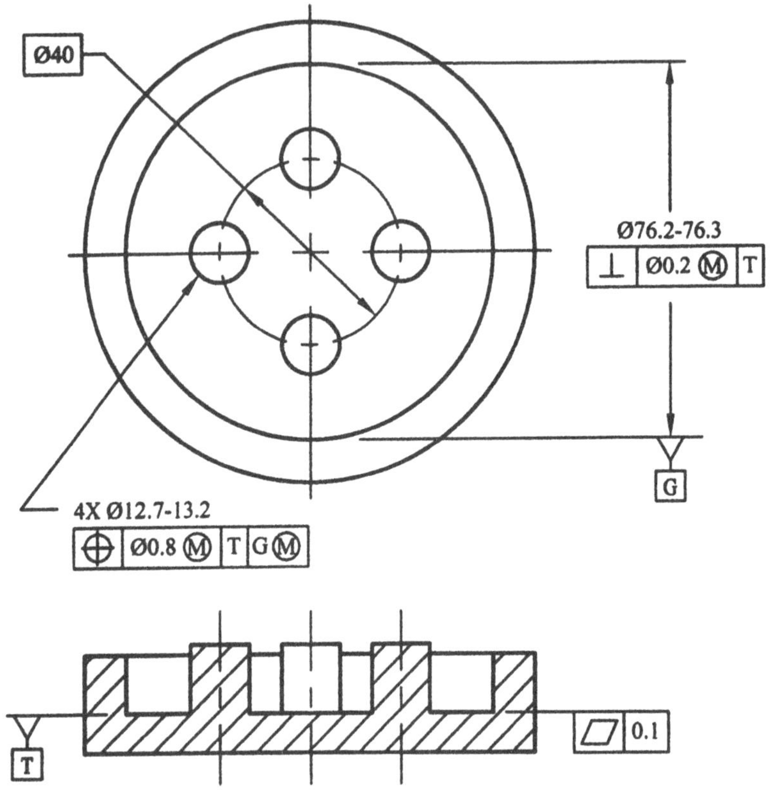
Ø40
Ø76.2-76.3
⊥ | Ø0.2 Ⓜ | T
G
4X Ø12.7-13.2
⌖ | Ø0.8 Ⓜ | T | GⓂ
T
▱ | 0.1

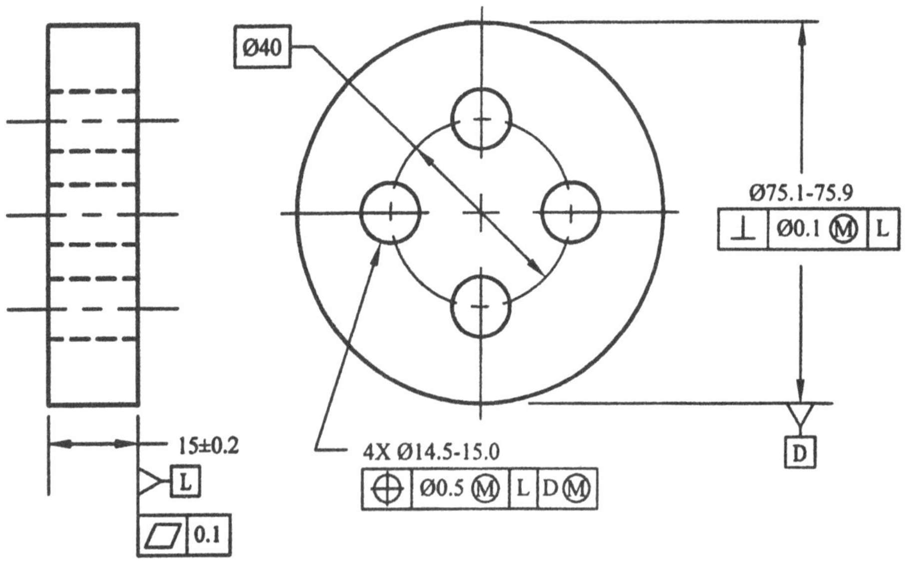
Ø40
Ø75.1-75.9
⊥ | Ø0.1 Ⓜ | L
D
15±0.2
L
▱ | 0.1
4X Ø14.5-15.0
⌖ | Ø0.5 Ⓜ | L | DⓂ

8.

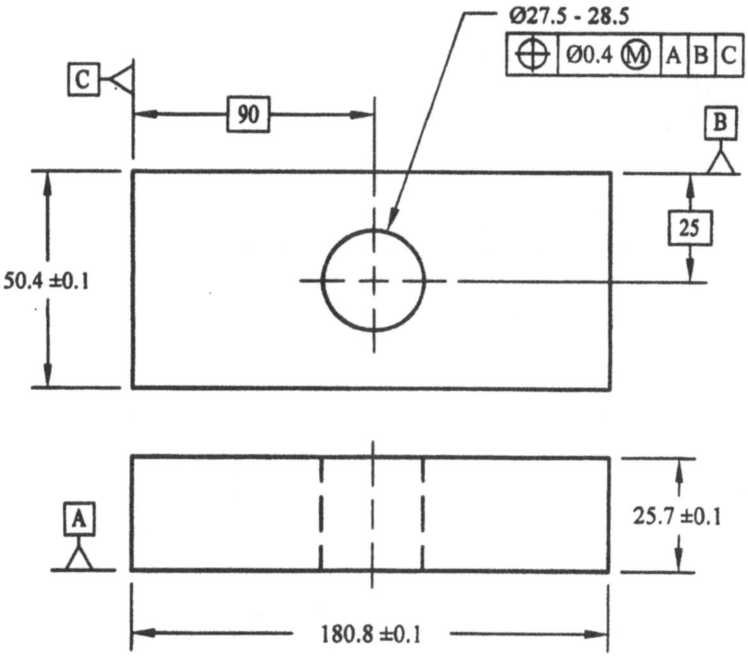

27.5 MMC
\- 0.4
① 27.1 Virt. Cond. to protect
② 27.75 protected
③ Good Part

AND

① 28.25 MIC
-27.50 MMC
0.75 Bonus Geo. Tol.
\+ 0.40 Orig. Geo. Tol.
Ø 1.15 Total Geo. Tol.

② 28.25 MIC
-27.75 MIC at ⌖
Ø 0.50 out of ⌖

③ Good Part

Part 1 as produced being inspected:

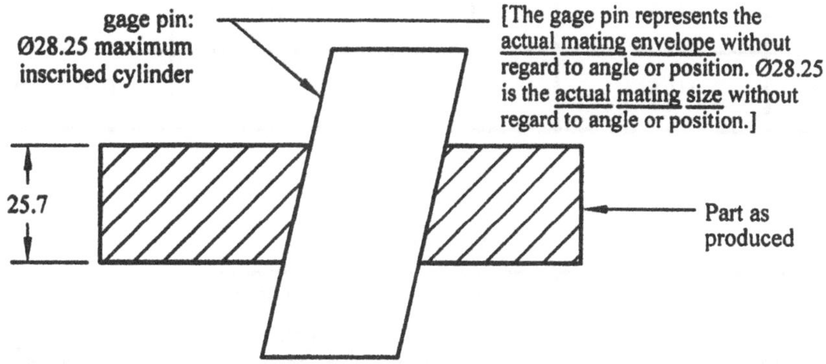

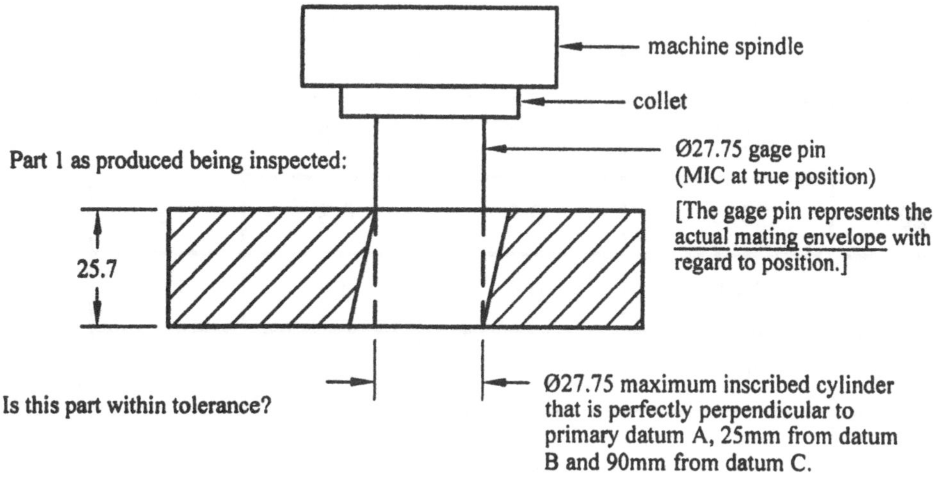

1. Is this part within tolerance?

9.

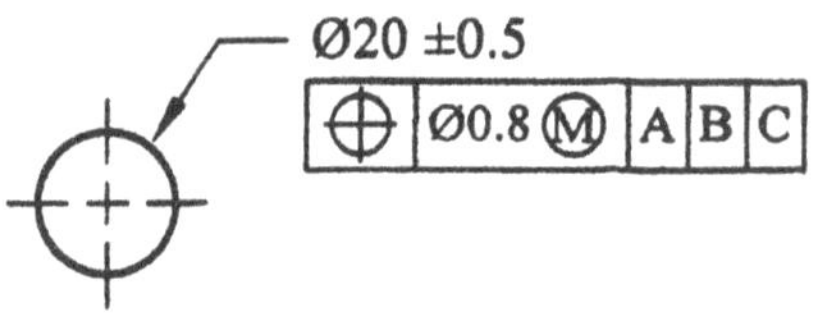

IF A HOLE

	Actual	Less MMC	Bonus Equals	Original Pos. Tol.	Total
A.	20.5	19.5	1	0.8	1.8
B.	20.3	19.5	0.8	0.8	1.6
C.	20.1	19.5	0.6	0.8	1.4
D.	20	19.5	0.5	0.8	1.3
E	19.8	19.5	0.3	0.8	1.1
F.	19.6	19.5	0.1	0.8	0.9
G.	19.5	19.5	0	0.8	0.8

IF A SHAFT

	MMC	Less Actual	Bonus Equals	Original Pos. Tol.	Total
H.	20.5	19.5	1	0.8	1.8
I.	20.5	19.6	0.9	0.8	1.7
J.	20.5	19.8	0.7	0.8	1.5
K.	20.5	20	0.5	0.8	1.3
L.	20.5	20.1	0.4	0.8	1.2
M.	20.5	20.3	0.2	0.8	1
N.	20.5	20.5	0	0.8	0.8

10.

Actual Feature Size	26	25.8	25.6	25.4	25.1
⌖ | Ø0.5 Ⓜ | A | B | C	A 0.5	B 0.7	C 0.9	D 1.1	E 1.4
// | Ø0.4 Ⓜ | A	F 0.4	G 0.6	H 0.8	I 1	J 1.3
⌖ | Ø0.5 | A | B | C	K 0.5	L 0.5	M 0.5	N 0.5	O 0.5
⊥ | Ø0 Ⓜ | A	P 0	Q 0.2	R 0.4	S 0.6	T 0.9
⌖ | Ø0.2 Ⓜ | L | D | C	U 0.2	V 0.4	W 0.6	X 0.8	Y 1.1
— | Ø0.3 Ⓜ	Z 0.3	AA 0.5	BB 0.7	CC 0.9	DD 1.2
⌭ | 0.05	EE 0.05	FF 0.05	GG 0.05	HH 0.05	II 0.05
⌖ | Ø0.08 Ⓜ | J | C | E	JJ 0.08	KK 0.28	LL 0.48	MM 0.68	NN 0.98

11.

Dimension from Datum C	Dimension from Datum B	Actual Hole Size	Answers		
			A	B	C
(1) 1.901	.924	.500	.015	.0028	YES
(2) 1.905	.929	.505	.020	.0128	YES
(3) 1.902	.926	.510	.025	.0045	YES
(4) 1.925	.932	.502	.017	.0519	NO
(5) 1.895	.922	.500	.015	.0117	YES
(6) 1.890	.925	.507	.022	.020	YES
(7) 1.900	.910	.509	.024	.030	NO
(8) 1.901	.926	.511	violates LMC	.0028	NO
(9) 1.870	.909	.509	.024	.068	NO

12.

as produced:

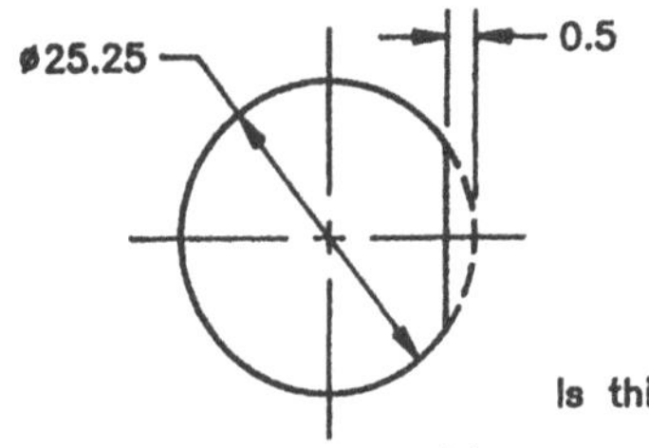

Is this part within tolerance ? **Yes**

as produced:

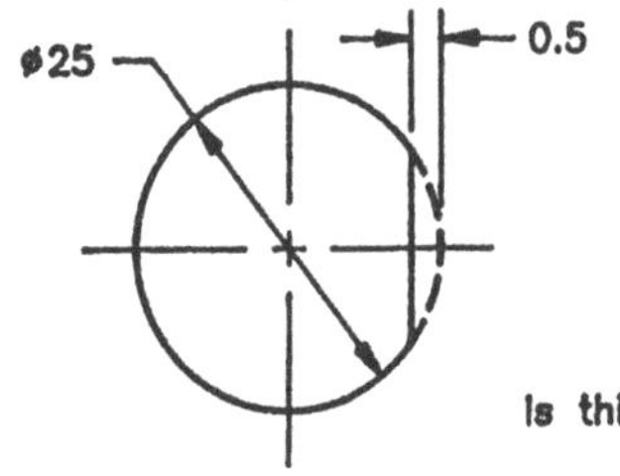

Is this part within tolerance ?

No – violates LMC

as produced:

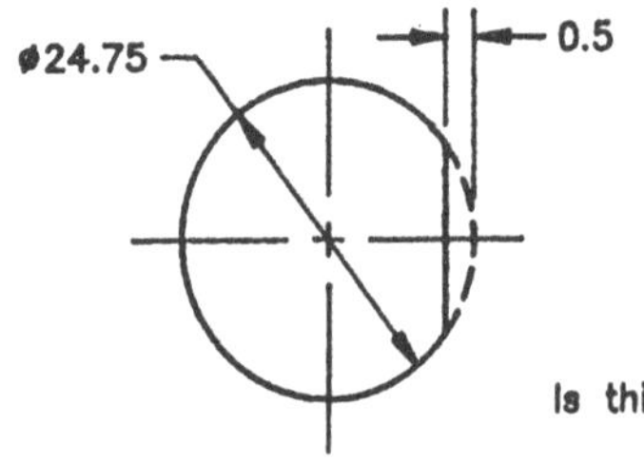

Is this part within tolerance ?

No – violates LMC

13.

as produced:

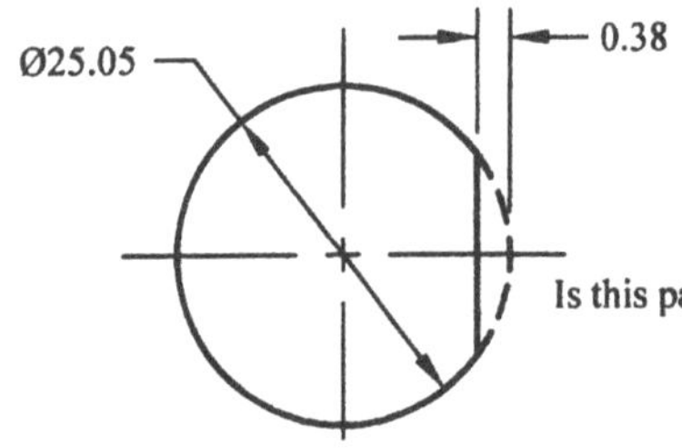

Is this part within tolerance ? No

Flat is deeper than
0.35 & violates LMC

as produced:

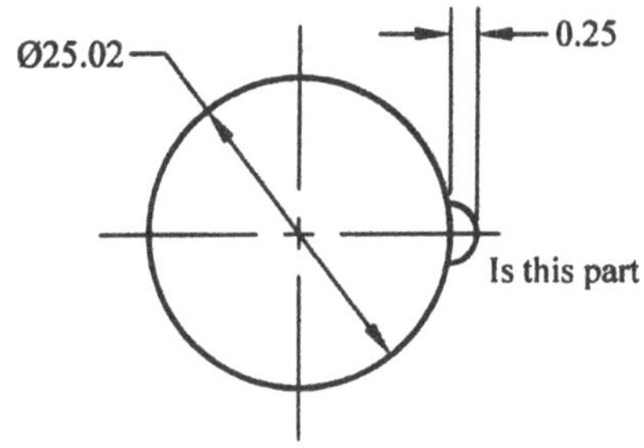

Is this part within tolerance ? No

Part is round enough,
but part violates MMC

25.02
\+ 0.25
Ø25.27 ← Actual

as produced:

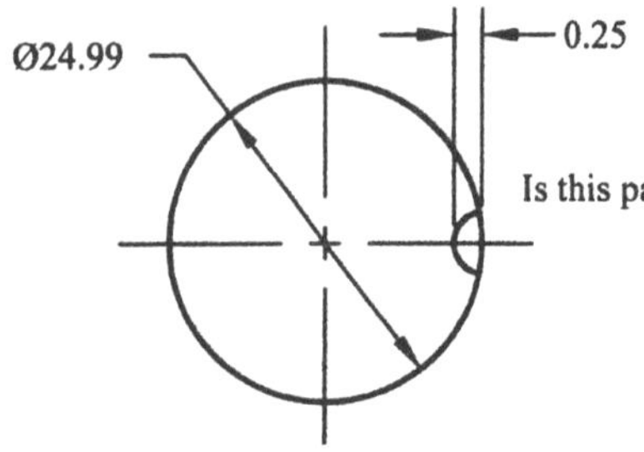

Is this part within tolerance ? No

It violates LMC

24.99
\- 0.25
Ø24.74 ← Actual

14.

No. Part is OK. Allowed to taper 0.03 per side, or a diameter of 0.06. It only tapers ∅0.04.

15.

as produced:
1st part

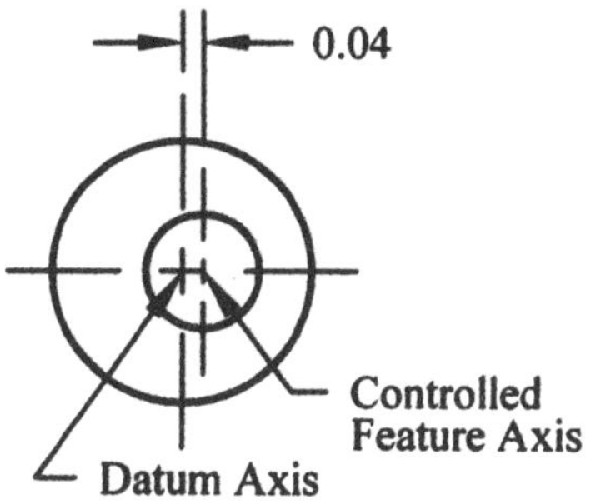

Is the part good?
Why, or why not?

No. Part is only allowed to be off center 0.035.

2nd part

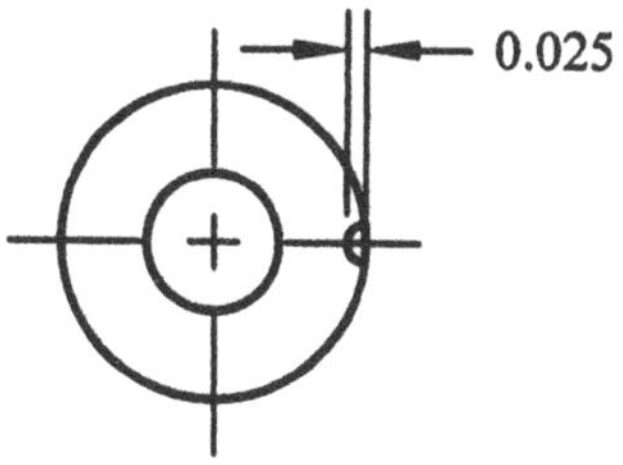

Is the part good?
Why, or why not?

No. It violates the size tolerance.

3rd part

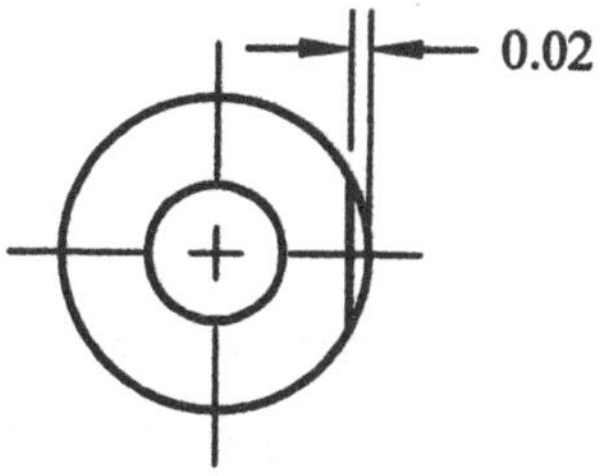

Is the part good?
Why, or why not?

No. It violates circularity control.

16.

1) Size is okay.
2) ○ is okay. It doesn't control taper.
3) — is okay. Don't see any pits, bumps or curves.
4) ⌰ is bad. Small diameter is only allowed to taper ∅0.16, but actually tapers ∅0.2.

17.

No. The cylindricity control allows the part to taper 0.01 on each side for a total of 0.02 on the diameter. Circular runout does not control taper, and the size tolerance on the small diameter has not been violated. So, the part does not violate any of the tolerances depicted.

18.

1) Size is okay.
2) ⌭ is okay. Allowed to taper ∅0.04 and only tapers ∅0.03.
3) ↗ is okay. It doesn't control taper.

19.

1) Size is okay.
2) ⌭ is bad. Allowed to taper ∅0.08, but does taper ∅0.11.
3) ⌰ is bad. Allowed to taper ∅0.14, but does taper ∅0.2.

20.

1st Part: Yes, FIM is 0.07.
2nd Part: No. Violates size and cylindricity tolerance.
3rd Part: Yes. FIM is 0.01.

21. Part 12135 – Bushing

- Outside diameter measures:
  .525 at its largest **OK**
  .519 at its smallest **OK**
  .515
  .519
  ∅.006 actual taper |⌭|.005| **allows taper of .005 per side (a ∅.010)**

- Part outside diameter is round to within .0002. **OK**

- FIM of I.D. while centered on and rotating about datum axis D is .002 at each cross section. **OK**

- Inside diameter is:
  .2550 at its largest **OK**
  .2549 at its smallest **OK** **.002 must be off center (eccentric) because inside diameter indicates shape is good.**

- Part thickness varies between .196 and .195. It is rigid. **OK**

- Part thickness has a slight uniform bow. It has a centerplane straightness of .003 **No. Exceeds MMC envelope.**

22.

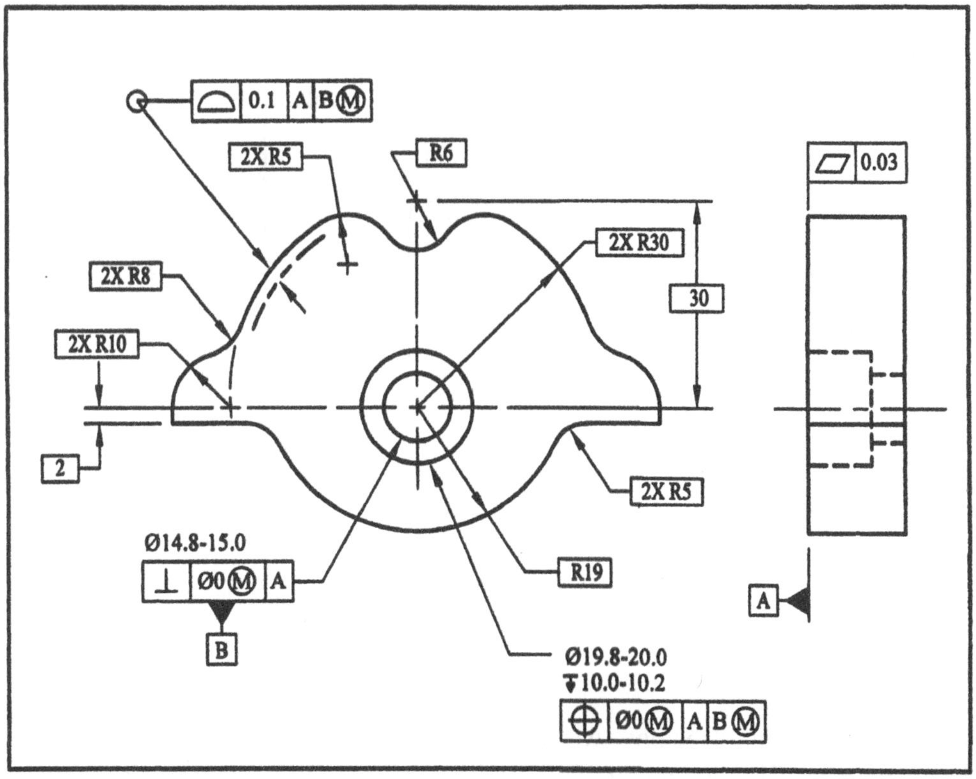
0.1 A B Ⓜ
2X R5
R6
0.03
2X R30
2X R8
30
2X R10
2
2X R5
Ø14.8-15.0
Ø0 Ⓜ A
B
R19
A
Ø19.8-20.0
10.0-10.2
Ø0 Ⓜ A B Ⓜ

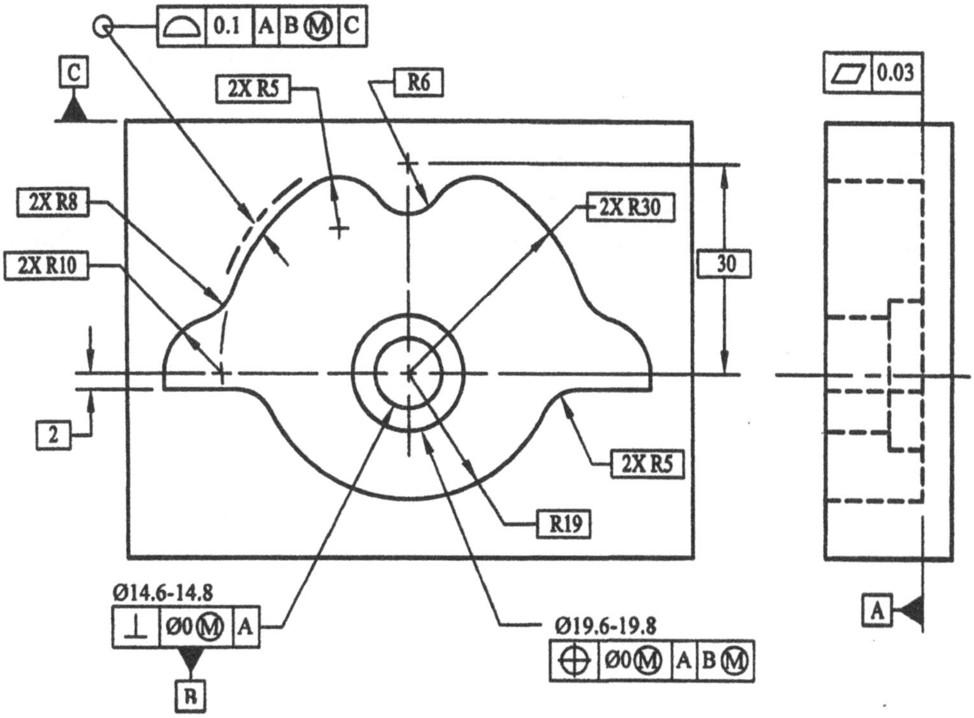
0.1 A B Ⓜ C
C
2X R5
R6
0.03
2X R8
2X R30
2X R10
30
2
2X R5
R19
Ø14.6-14.8
Ø0 Ⓜ A
B
Ø19.6-19.8
Ø0 Ⓜ A B Ⓜ
A

23.

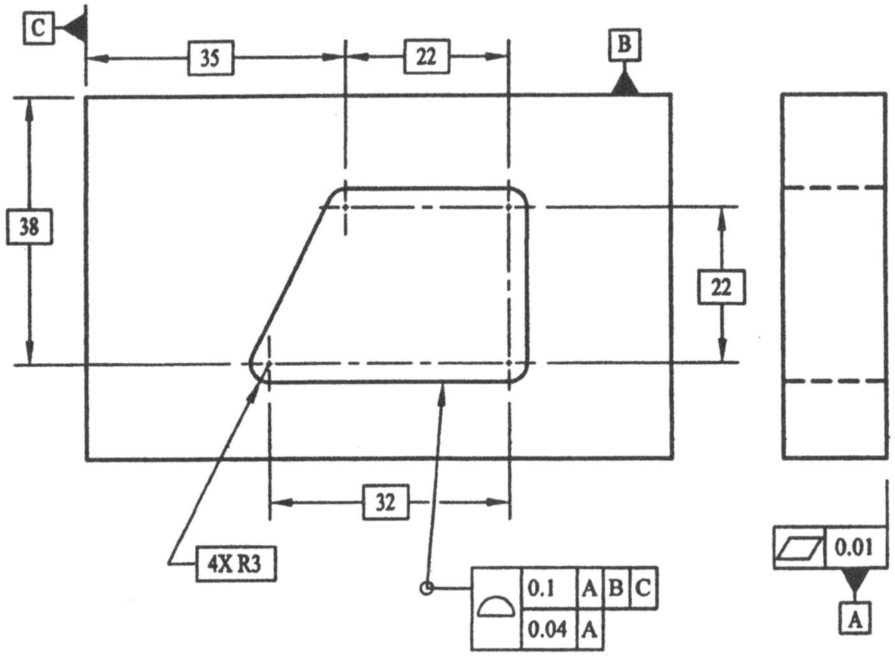

This composite profile control has two separately verifiable levels of geometric tolerance. The upper level of profile tolerance of 0.1mm locates the oddly shaped hole, while the lower level of control tolerances to within 0.04 per surface the size and shape (3-D form) of the oddly configured hole. Since the primary datum A is repeated in the lower level control, the perpendicularity of the surface of the hole is held within the 0.04 tolerance zone which is oriented to A.

The 0.04 level of tolerance controls the 3-dimensional form of the hole automatically since that is its main function. The fact that this profile of a surface tolerance extends all-around in this view means that it will also control the size of the oddly shaped hole. The size tolerance is 0.04 per surface. So, if we wanted to know how that affects the basic dimensions on the part, we could just add them up and apply the profile tolerance of 0.04 per surface.

24.

1) Yes
2) No 1.963
   <u>+ .015</u>
   1.978 MAX allowed
3) Yes
4) No. 1.123
   <u>- .020</u>
   1.103 MIN allowed
5) No. Allowed ∅.015 MAX; therefore, part is okay.
6) Not necessarily. With pattern shift the part may be good.
7) Yes.
8) No. The minimum allowed over the pin measurement is calculated as follows:

MIN	**MAX**
1.963 basic dimension	1.963 basic dimension
+ .500 (1/2 of .495 + .505 = 1.000)	+ .500 (1/2 of .495 + .505 = 1.000)
2.463	2.463
- .010 (1/2 of ∅.020)	+ .010 (1/2 of ∅.020)
2.453	2.473
- .005 (1/2 of ∅.010)	+ .005 (1/2 of ∅.010)
2.448 MIN allowed over the pins measurement	2.478 MAX allowed over the pins measurement

-Allowed range 2.448 to 2.478

-Actual measurement of 2.458 is within the allowed range; therefore, it is not indicated that the tolerance has been exceeded.

25.

1) How flat must datum feature A be when produced? **.002**

2) For the four holes that become datum pattern D, what are they being positioned to besides datum feature A? **each other**

3) How many holes does the profile control tie the profile to? **all seven**

4) What acts as the origin of measurement for location for the three hole pattern? **the central axis of datum pattern D as generated by its four virtual condition boundaries**

5) What characteristic of geometry does datum A tolerance when used in each feature control frame? **perpendicularity**

6) Is the profile control a total wide zone of .01 or is it ±.01? **a total wide zone of .010**

7) What is the virtual condition of the holes in the three hole pattern? **∅.510**

8) How many gages would be necessary to gage the MMC of all seven holes, the two positional controls and the profile control (if only the outer tolerance zone boundary of the profile control was to be gaged)? **one**

9) If the two outermost .51-.52 holes are produced at a diameter of .514 and their axes a distance apart of 1.126, are they within tolerance? **1.120 + .002 (1/2 of .004) + .002 (1/2 of .004) = 1.124 allowed; therefore, 1.126 is bad**

10) If the surface profile has a bump on the upper left radius that exceeds the basic radius shown by .005, is the surface in violation of the tolerance zone under all circumstances? **with pattern shift, no**

11) If the largest gage pin that can be inserted into the middle .510-.520 hole measures a diameter of .518 and the hole just to its left can receive a gage pin no larger than .512 and a vernier caliper measurement from the outermost points on the gage pin diameters is a distance of 1.080, would that indicate that they exceed their positional tolerance? **.560 + .259 (1/2 of .518) + .256 (1/2 of .512) + .004 (1/2 of .008) + .001 (1/2 of .002) equals 1.080 MAX over the pins meas. allowed**

12) How does the gaging you visualized when answering Question 8 differ from the gaging that would be used for the part on Page 50? **It would be the same.**

26.

1) Centerplane of part at .999

2)
```
   .624
 + .621
  1.245
```

3)
```
  1.998
- 1.245
   .753 Size of Slot
```

4) .753 ÷ 2 = .3765 Half of Slot

5)
```
   .6240
 + .3765
  1.0005 Slot Centerplane from Top of Part
```

6)
```
   .6210
 + .3765
   .9975 Slot Centerplane from Bottom of Part
```

7)
```
   .999       .999       .999± .001
 + .001     - .001
  1.000       .998       .998 – 1.000 Extent of Centered Tolerance Zone
```

8) No

9) Slot must be reworked. Slot is .0005 too close to bottom of part. Remove minimum of .001 from top of slot:

```
   .623
 + .621
  1.244

  1.998
- 1.244
   .754 Size of Slot
```

.754 ÷ 2 = .377 Half of Slot

```
   .623
 + .377
  1.000 Slot Centerplane from Top of Part

   .621
 + .377
   .998 Slot Centerplane from Bottom of Part
```

.999± .001 or .998 – 1.000

Centerplane of Slot <u>In</u> Tolerance Zone

27.

1) If the holes on the 1.600 diameter bolt circle have been produced at a distance from datum axis B of .820, under what conditions might the holes meet their positional tolerance? **If the holes are produced at ∅.590.**

2) If the 2 holes opposite each other on the bolt circle are produced at a distance apart of 1.630 and the holes measure a diameter of .578, do they meet their positional tolerance? **∅.578 holes get ∅.028 positional tolerance each. So, 1.600 + (1/2 of .028) + (1/2 of .028) = 1.628 allowed; therefore, 1.630 is a NO.**

3) If the center hole is produced at a diameter of .586 and is off center of datum axis B a radius of .017, it is within its positional tolerance? **At ∅.586, the positional tolerance is ∅.036 (R.018), so .017 is okay. Yes.**

4) If the part thickness measures .626 with a vernier caliper and curves like a banana .006 in the right side view, does it meet its size requirements? **.626 + .006 = .632 which violates MMC envelope, so the answer is no.**

5) If datum feature B is produced at a diameter of 2.976 and is perfectly perpendicular to datum A, how much pattern shift would be allowed for the 5-hole pattern? **2.994 – 2.976 = ∅.018 or R.009**

6) If the outside diameter is produced at a size that will fit into a hole that is 2.980 and seat on datum feature A with three high points of contact, how much perpendicularity tolerance is the O.D. allowed? **2.994 – 2.980 = ∅.014**

7) If none of the allowed perpendicularity tolerance in question #6 was used by the O.D., what could the tolerance be used for instead? **Pattern shift of the 5 holes.**

8) If gage pins were inserted into two opposite holes on the 1.600 bolt circle and a measurement taken over the pins was 2.230, could the holes be within their positional tolerance? **Yes.** Under what circumstances? **If holes are made at ∅.590, then 1.600 + (1/2 of .590) + (1/2 of .590) + (1/2 of .040) + (1/2 of .040) = 2.230 allowed.**

9) If two holes on the bolt circle were produced at a diameter of .580, what would their maximum allowed axial separation be? **Positional tolerance at ∅.580 = ∅.030, so 1.600 + (1/2 of .030) + (1/2 of .030) = 1.630**

10) What is the maximum amount of pattern shift for the five hole pattern that may ever occur on this part? **2.994 – 2.960 = ∅.034 or R.017**

28.

FIXED FASTENER SITUATION

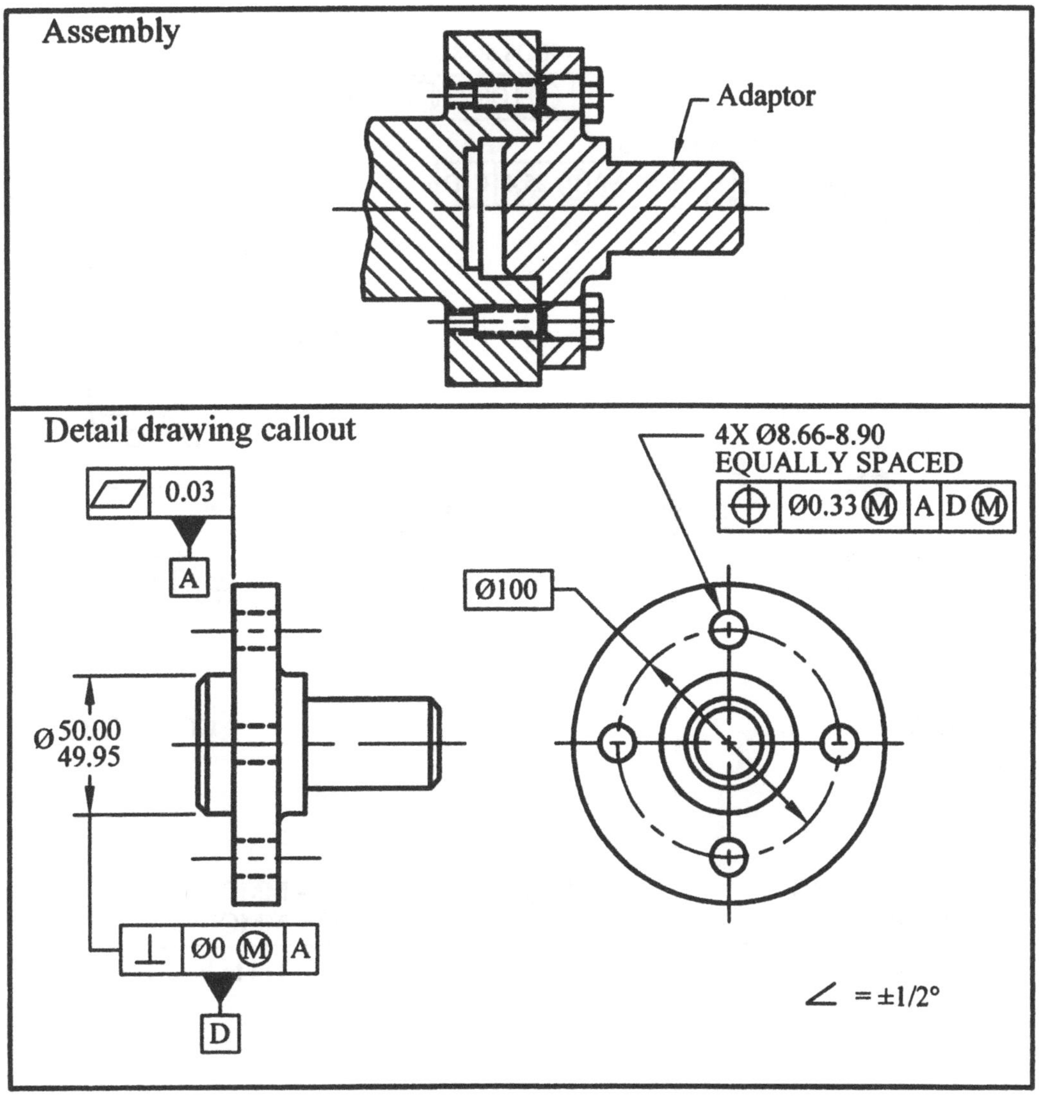

1. Locate the four-hole pattern. To calculate the positional tolerance, use an 8mm screw.

$$\begin{array}{rl} 8.66 & = \text{MMC Hole} \\ -\ 8.00 & = \text{MMC Fastener} \\ \hline 0.66 & \end{array}$$

$$2\overline{)0.66} = 0.33$$

Extra information: The bore that mates with the 49.95-50.00 diameter has a virtual condition of 50.00.

Sequence reminder: 1) flatness 2) perpendicularity 3) position

$$\begin{array}{r} 50 \\ -\ 50 \\ \hline 0 \end{array}$$

29.

Detail drawing callouts

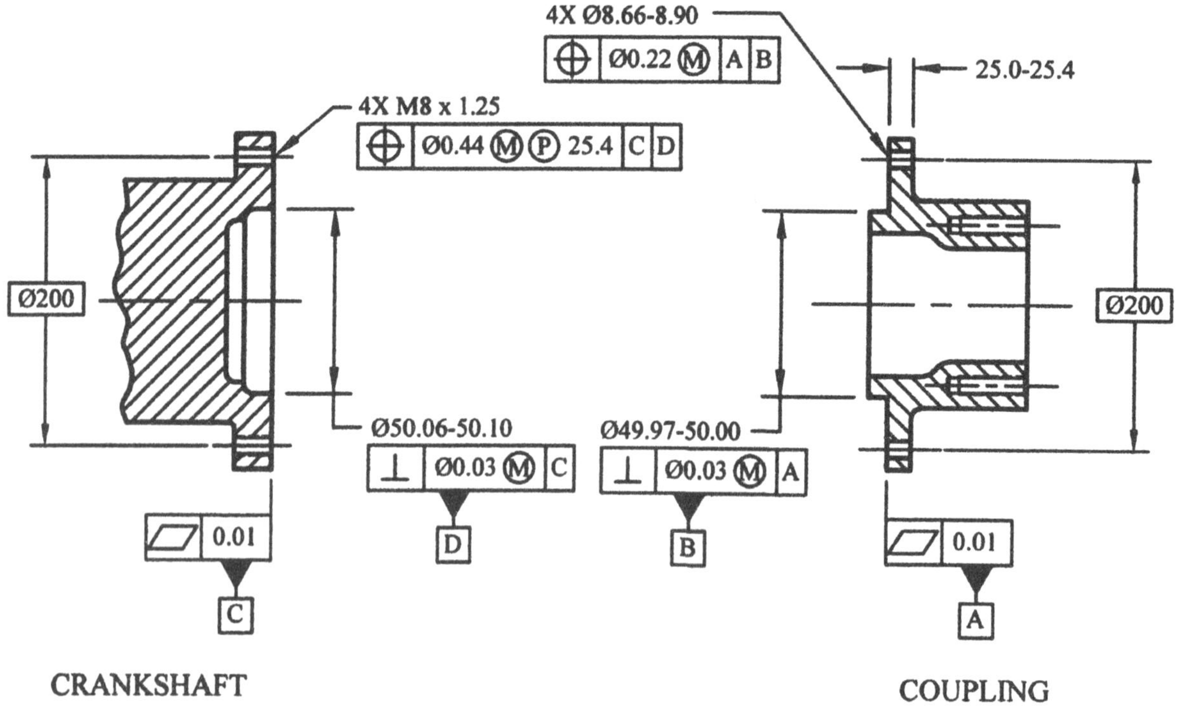

50.06 = MMC	0.03 for ⊥	8.66 = MMC
50.00 = MMC	2) 0.06	8.00 = MMC
0.06		0.66 = Tol to be divided

30.

Assembly

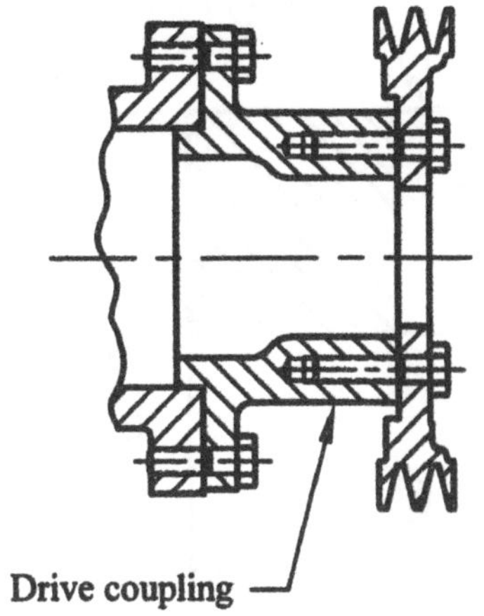

Detail part and gage for checking position of M8 x 1.25 tapped holes

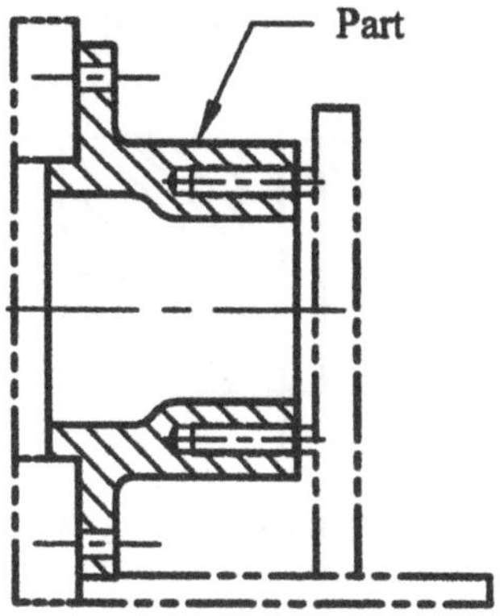

Detail drawing callouts

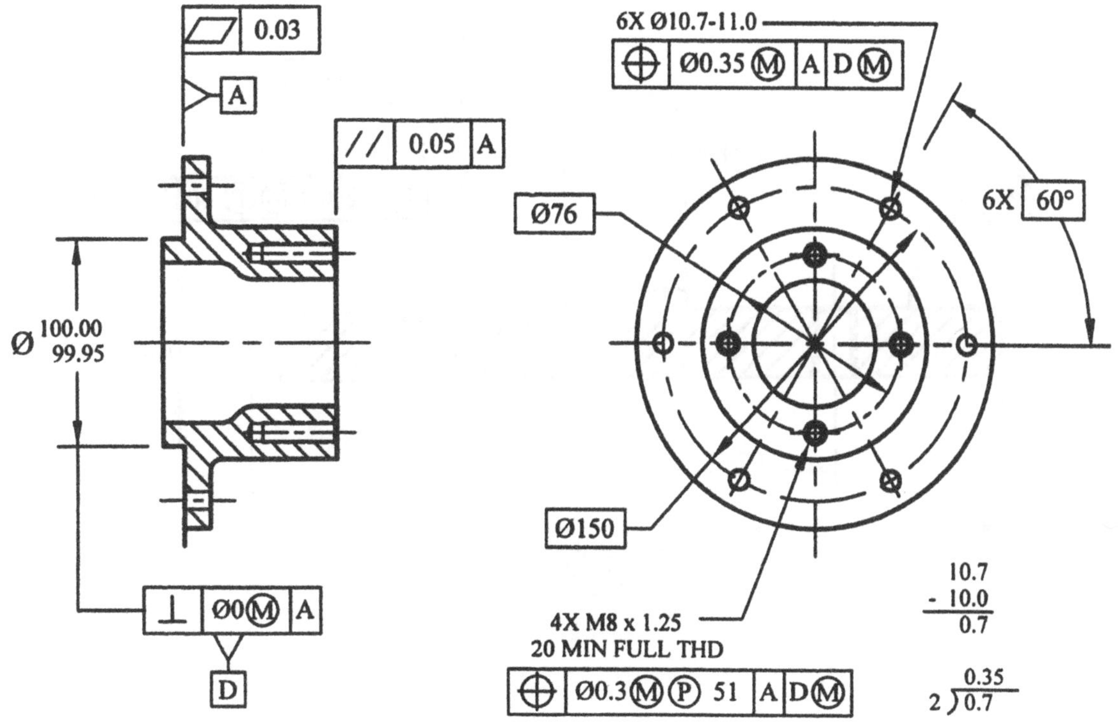

31.

Detail drawing callouts

PART 1

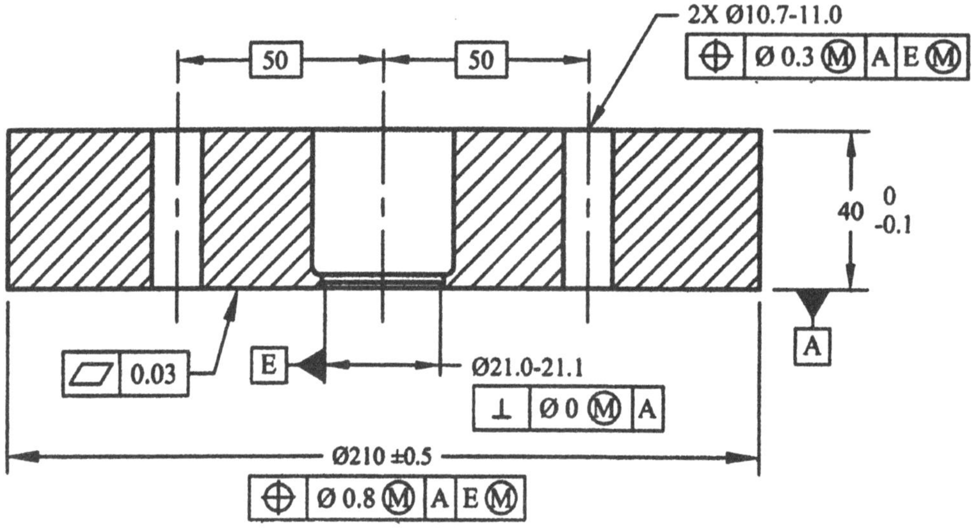

PART 2

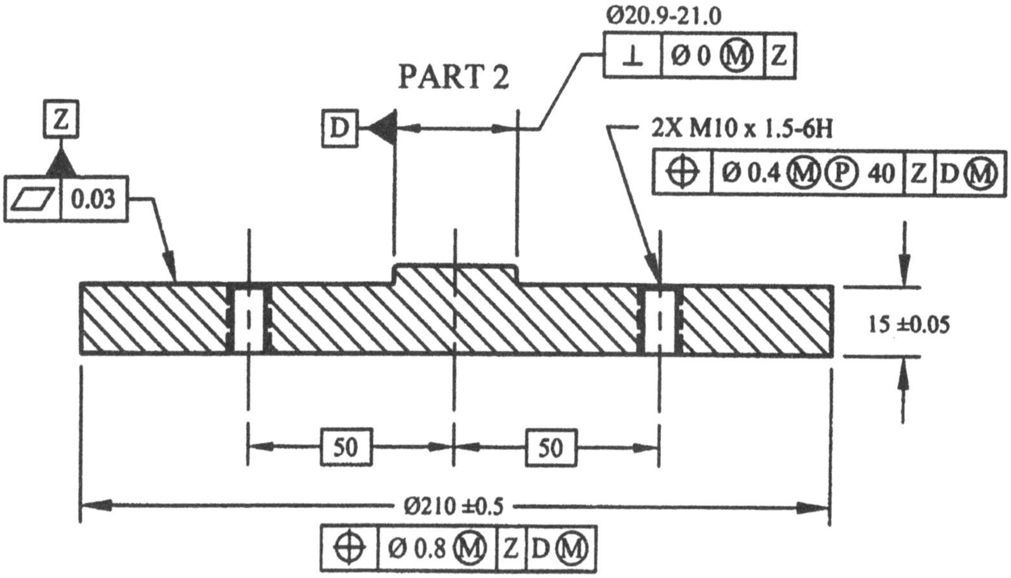

32.

Part 1

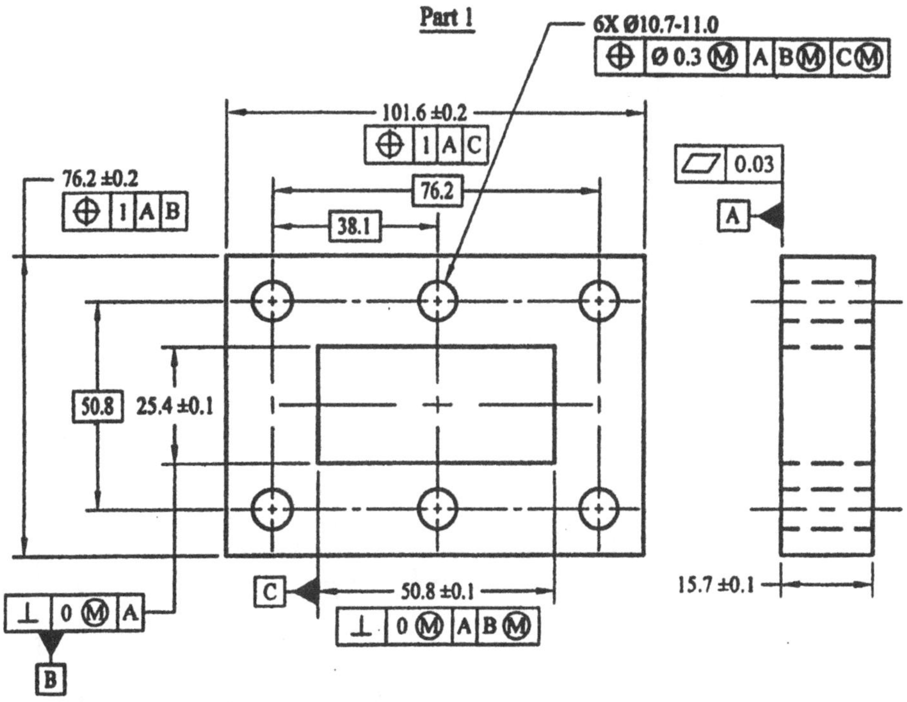

UNLESS OTHERWISE SPECIFIED ALL ANGLES ARE ±1°

Part 2

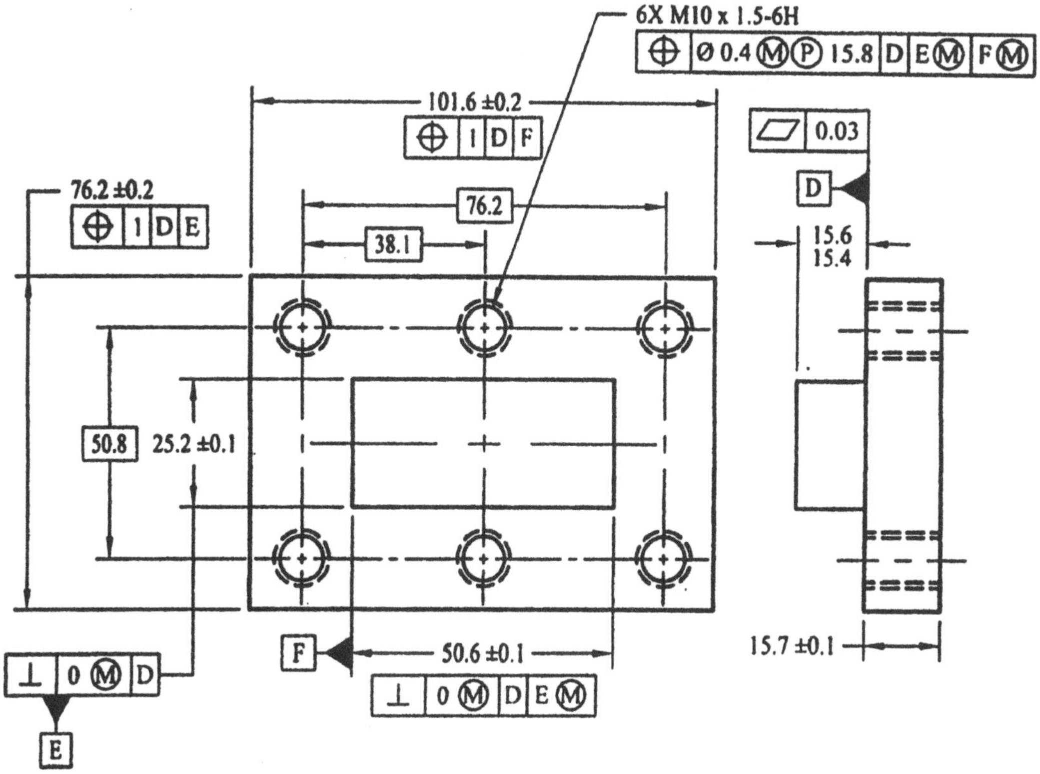

UNLESS OTHERWISE SPECIFIED ALL ANGLES ARE ±1°

33.

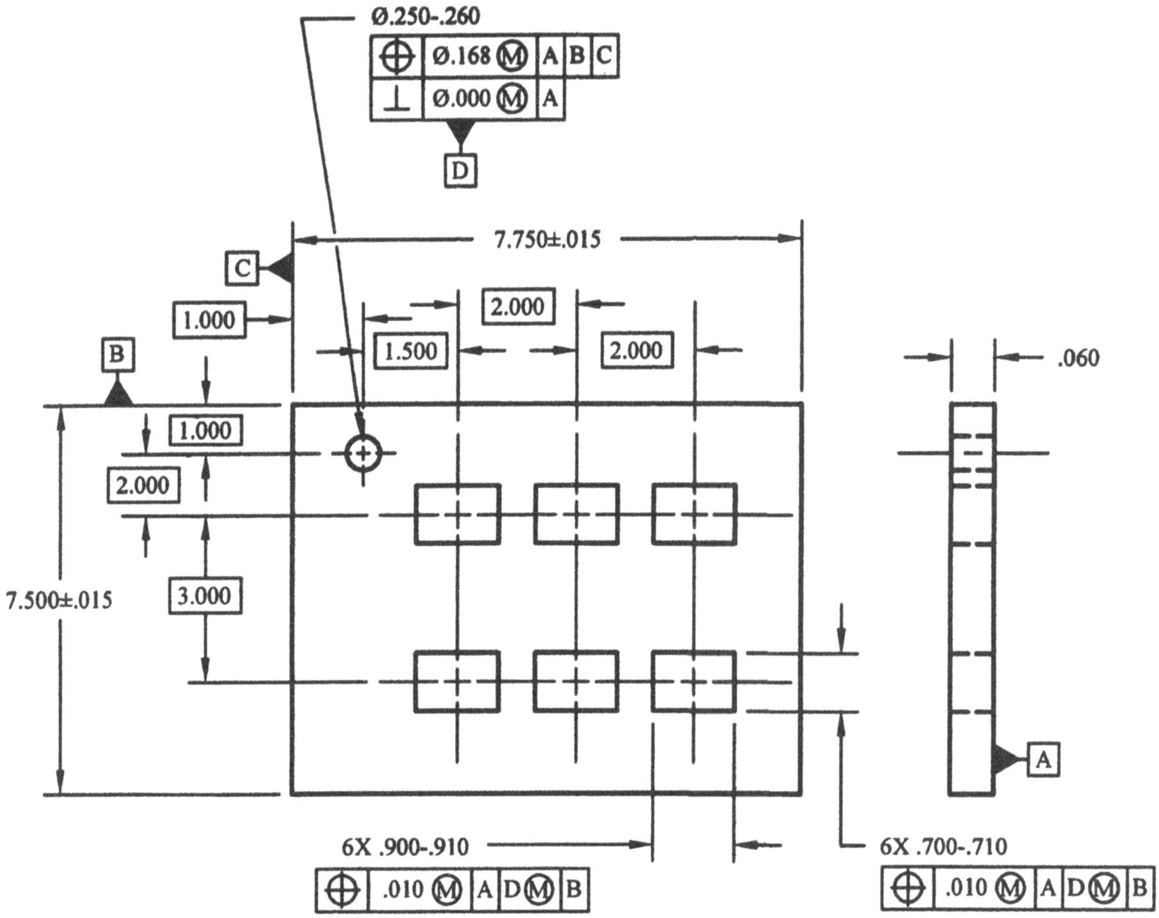

.X = ±.06
.XXX = ±.005
∠ = ±1°

34.

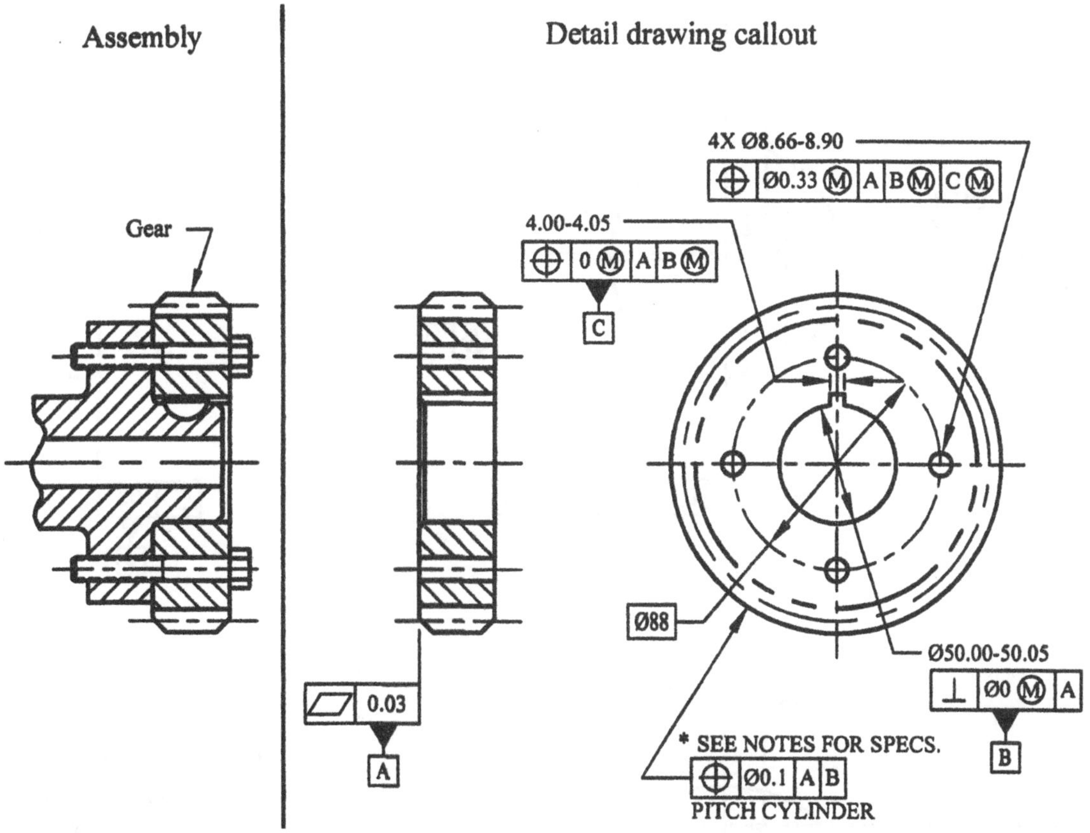

8.66 = MMC
\- 8.00 = MMC
0.66 = Tol. to be divided

$$2\,\overline{)0.66}\quad = 0.33$$

50 = MMC Hole
\- 50 = MMC Shaft
0 = Tol. (Allowance)

35.

1) None, except ⊕ and ⊥ might be easier to interpret. There is only one hole. No need for composite position.
2) Total wide tolerance zone.
3) Angular orientation – it stops pattern rotation.
4) Pattern could rotate with the only control being the general tolerance of ± 5°.

36.

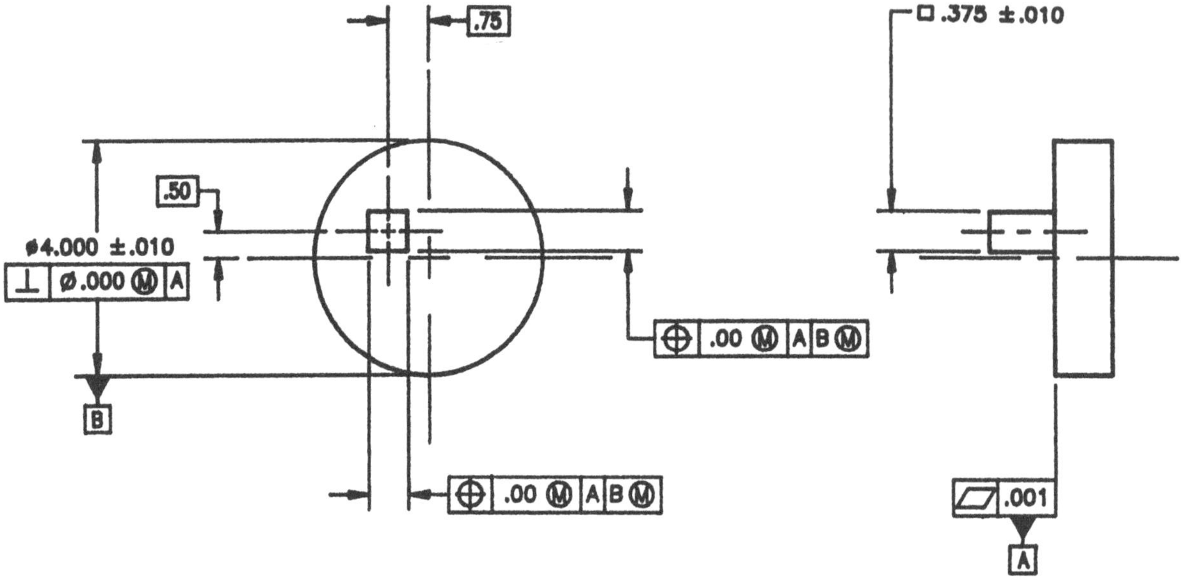

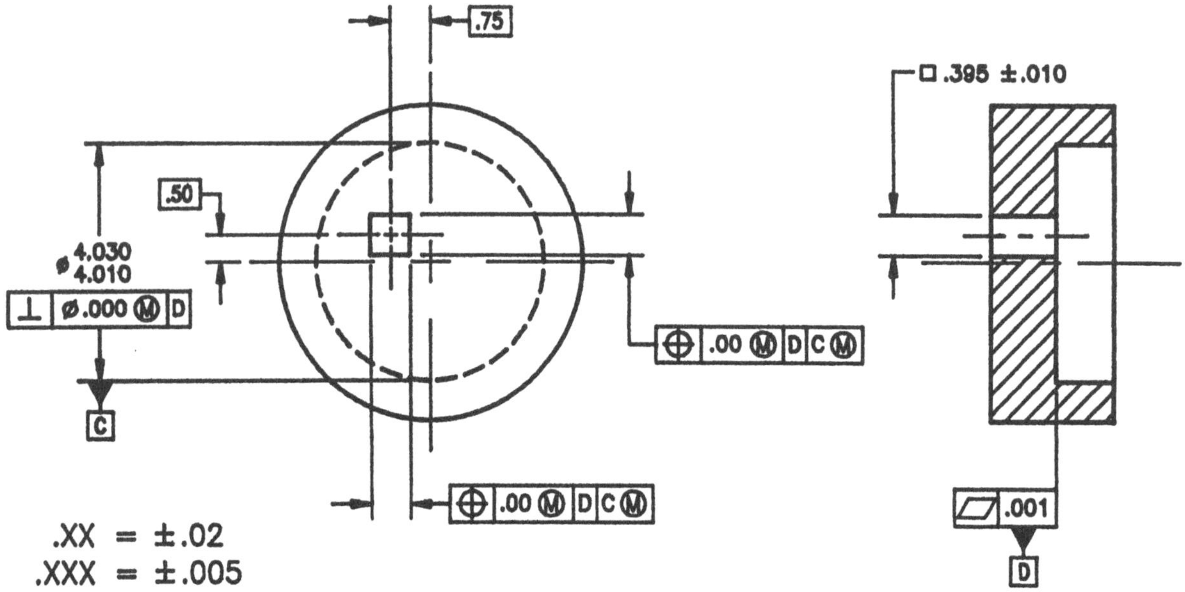

.XX = ±.02
.XXX = ±.005
∠ = ±1°

37. Problem #1

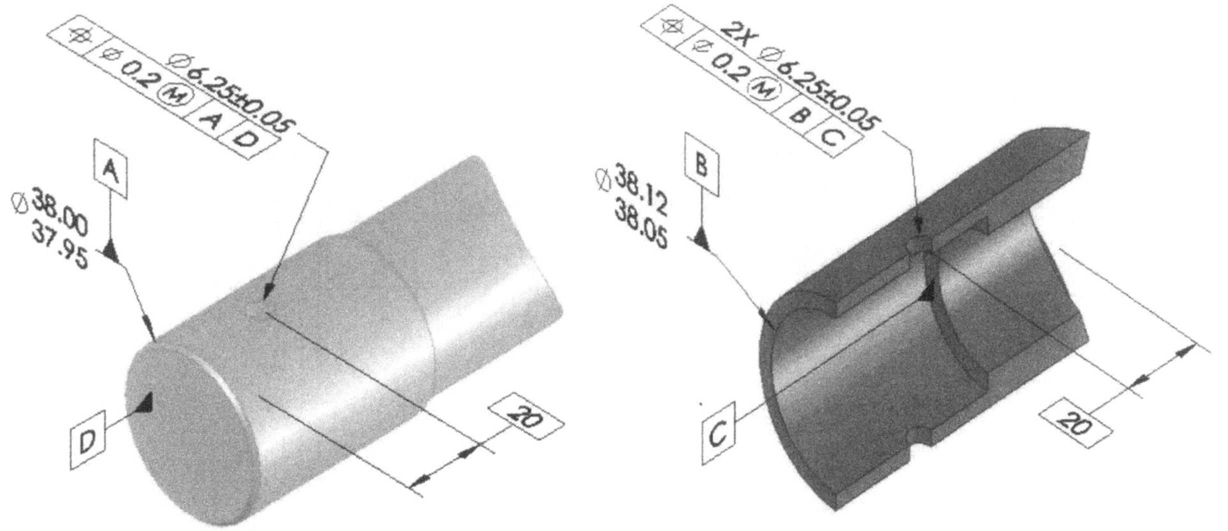

a) Inner shaft

b) Outer shaft

SolidWorks Corporation

37. Problem #2

Assembly

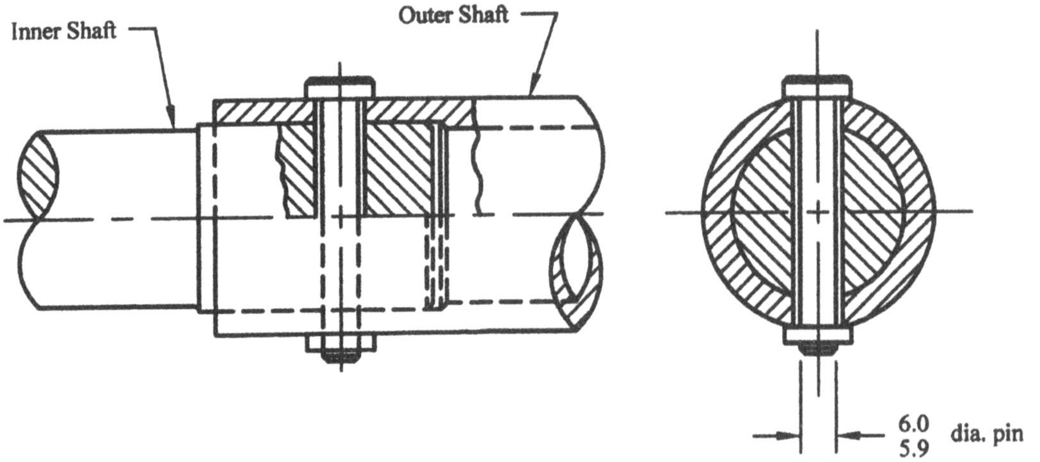

Detail drawing callouts

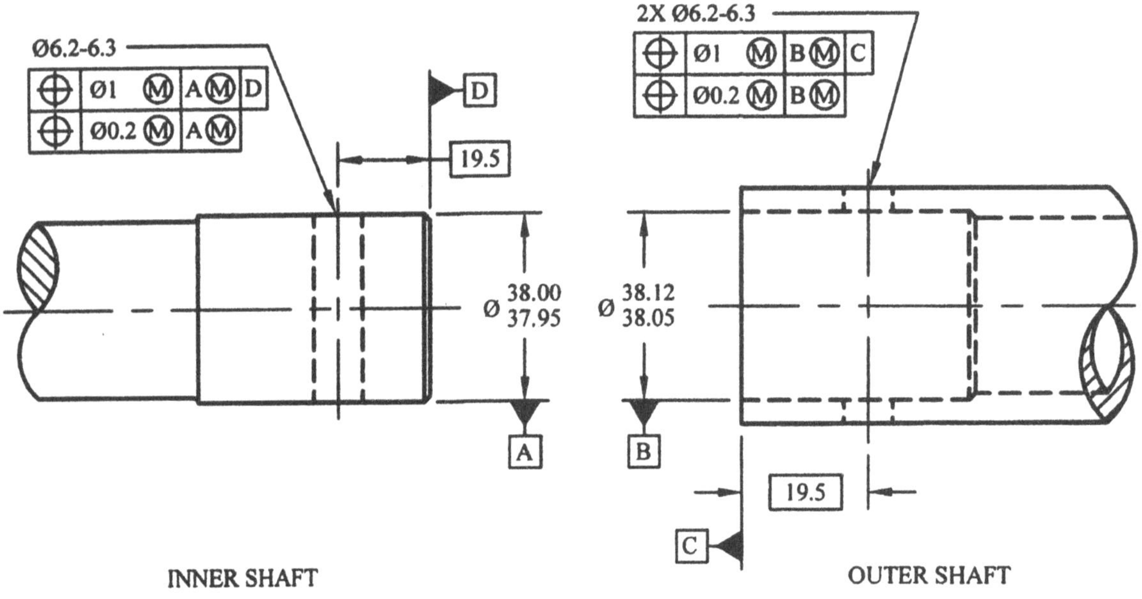

∠ = ±1°

38.

Assembly

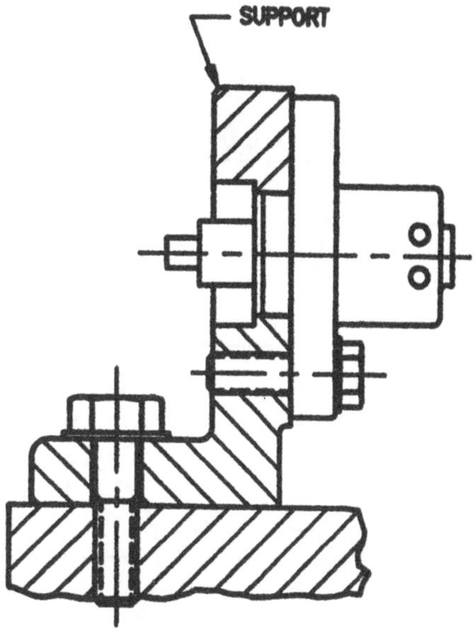

Detail drawing callout

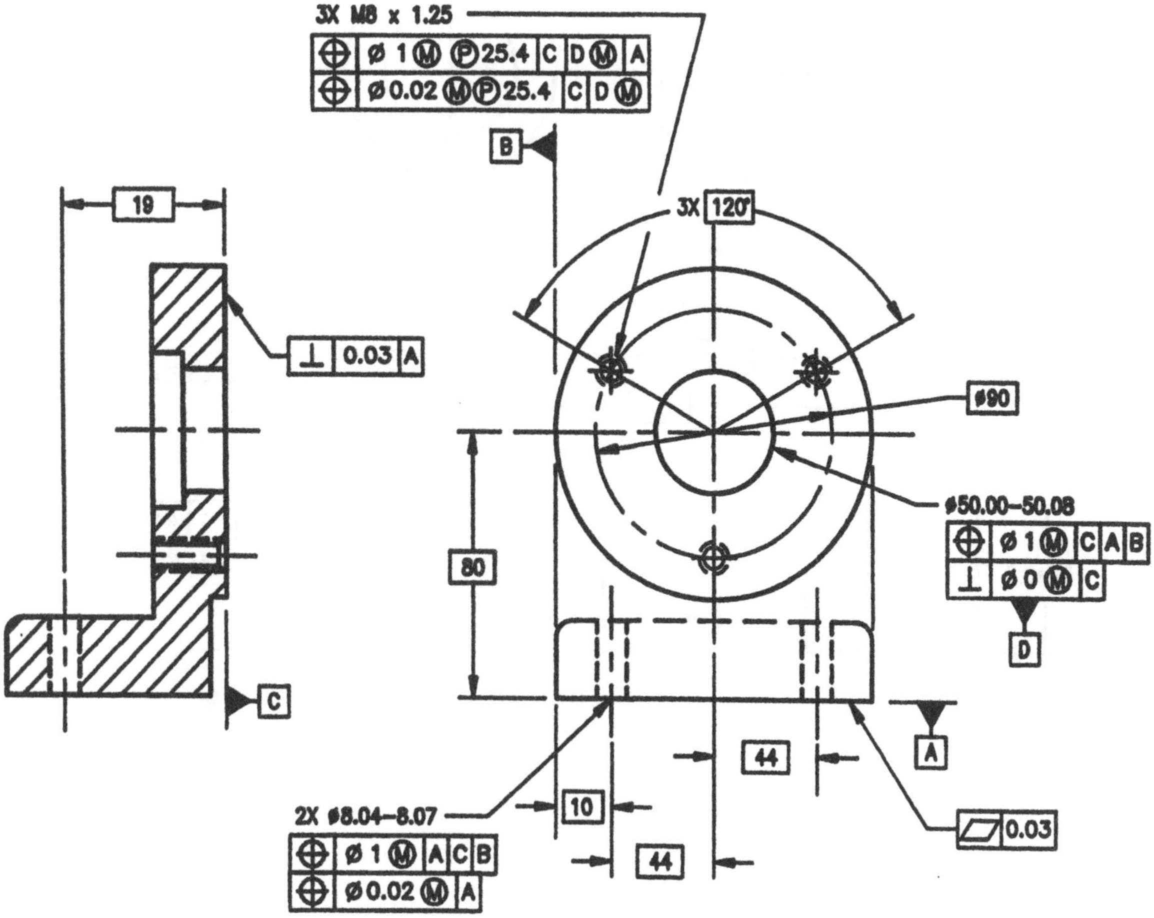

39.

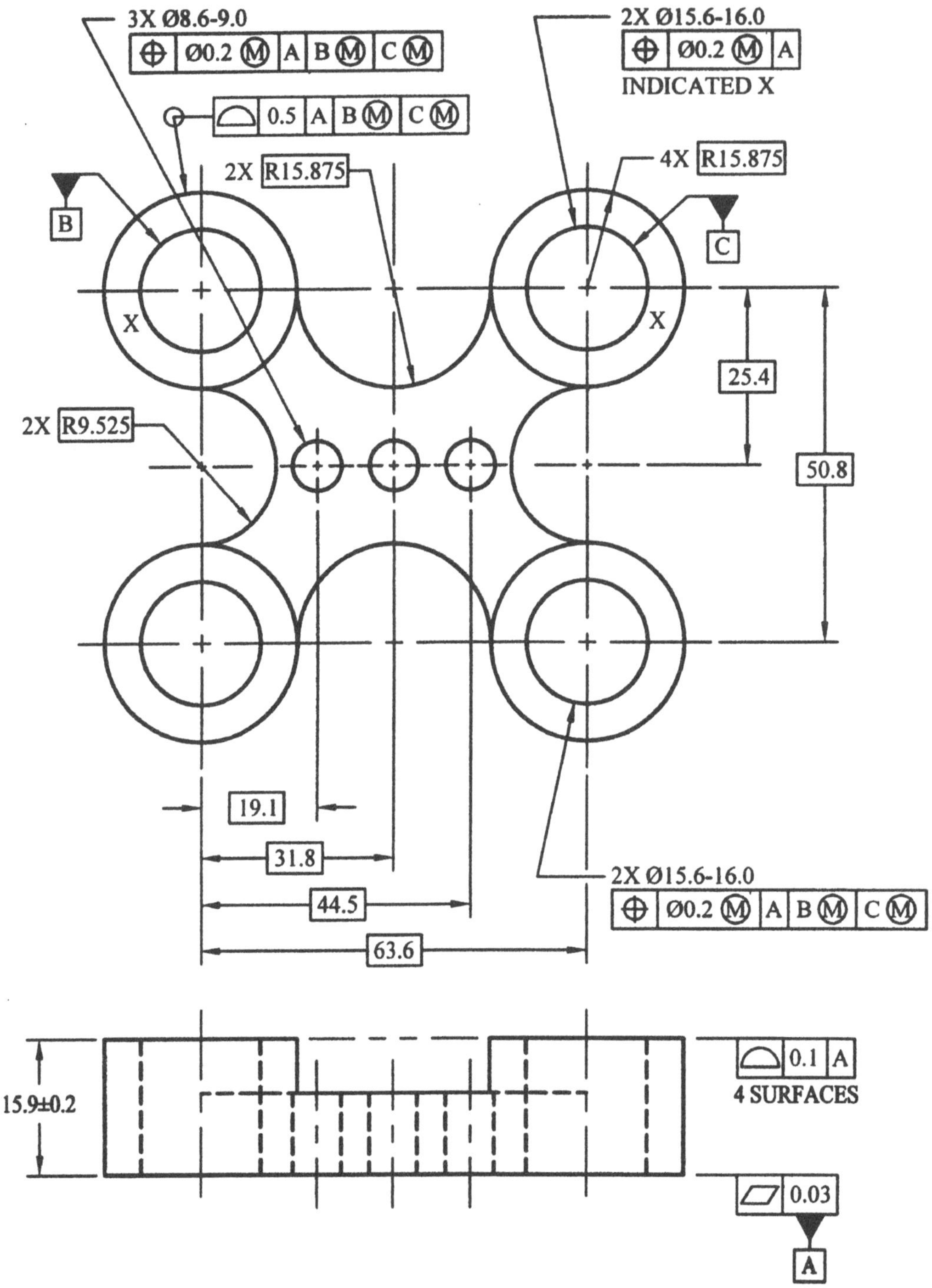

3X Ø8.6-9.0
⊕ Ø0.2 Ⓜ A B Ⓜ C Ⓜ
2X Ø15.6-16.0
⊕ Ø0.2 Ⓜ A
INDICATED X
0.5 A B Ⓜ C Ⓜ
2X R15.875
4X R15.875
B
C
X
X
25.4
50.8
2X R9.525
19.1
31.8
44.5
63.6
2X Ø15.6-16.0
⊕ Ø0.2 Ⓜ A B Ⓜ C Ⓜ
15.9±0.2
0.1 A
4 SURFACES
0.03
A

40.

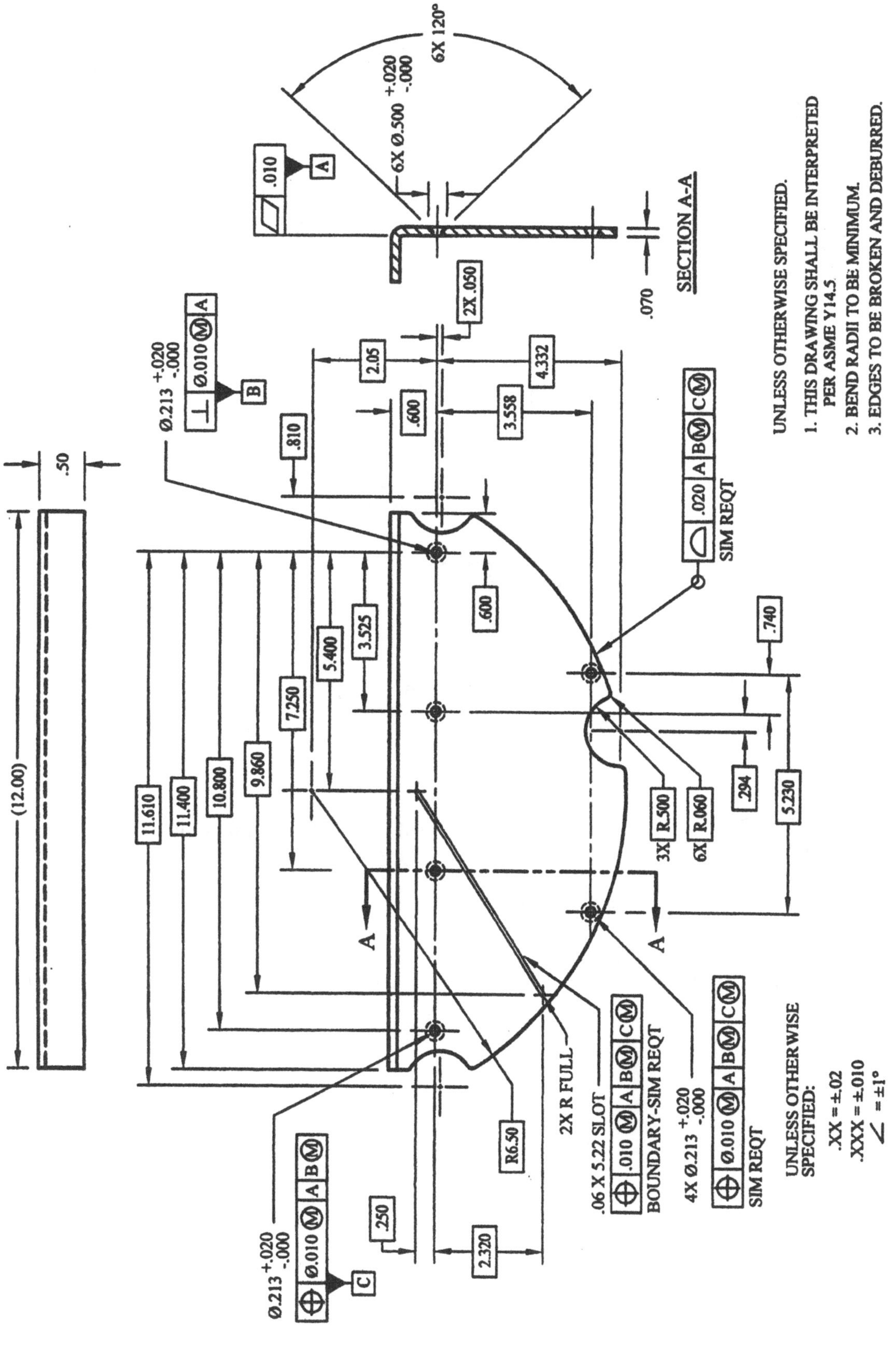

.50
(12.00)
Ø.213 +.020 -.000
⊥ Ø.010 Ⓜ A
B
.810
2.05
.600
2X.050
3.558
4.332
.010
A
6X Ø.500 +.020 -.000
6X 120°
.070
SECTION A-A
.020 A B Ⓜ C Ⓜ
SIM REQT
UNLESS OTHERWISE SPECIFIED.
1. THIS DRAWING SHALL BE INTERPRETED PER ASME Y14.5
2. BEND RADII TO BE MINIMUM.
3. EDGES TO BE BROKEN AND DEBURRED.
11.610
11.400
10.800
9.860
7.250
5.400
3.525
.600
.740
.294
5.230
3X R.500
6X R.060
A
A
R6.50
2X R FULL
.06 X 5.22 SLOT
.010 Ⓜ A B Ⓜ C Ⓜ
BOUNDARY-SIM REQT
4X Ø.213 +.020 -.000
Ø.010 Ⓜ A B Ⓜ C Ⓜ
SIM REQT
UNLESS OTHERWISE SPECIFIED:
.XX = ±.02
.XXX = ±.010
∠ = ±1°
Ø.213 +.020 -.000
Ø.010 Ⓜ A B Ⓜ
C
.250
2.320

41.

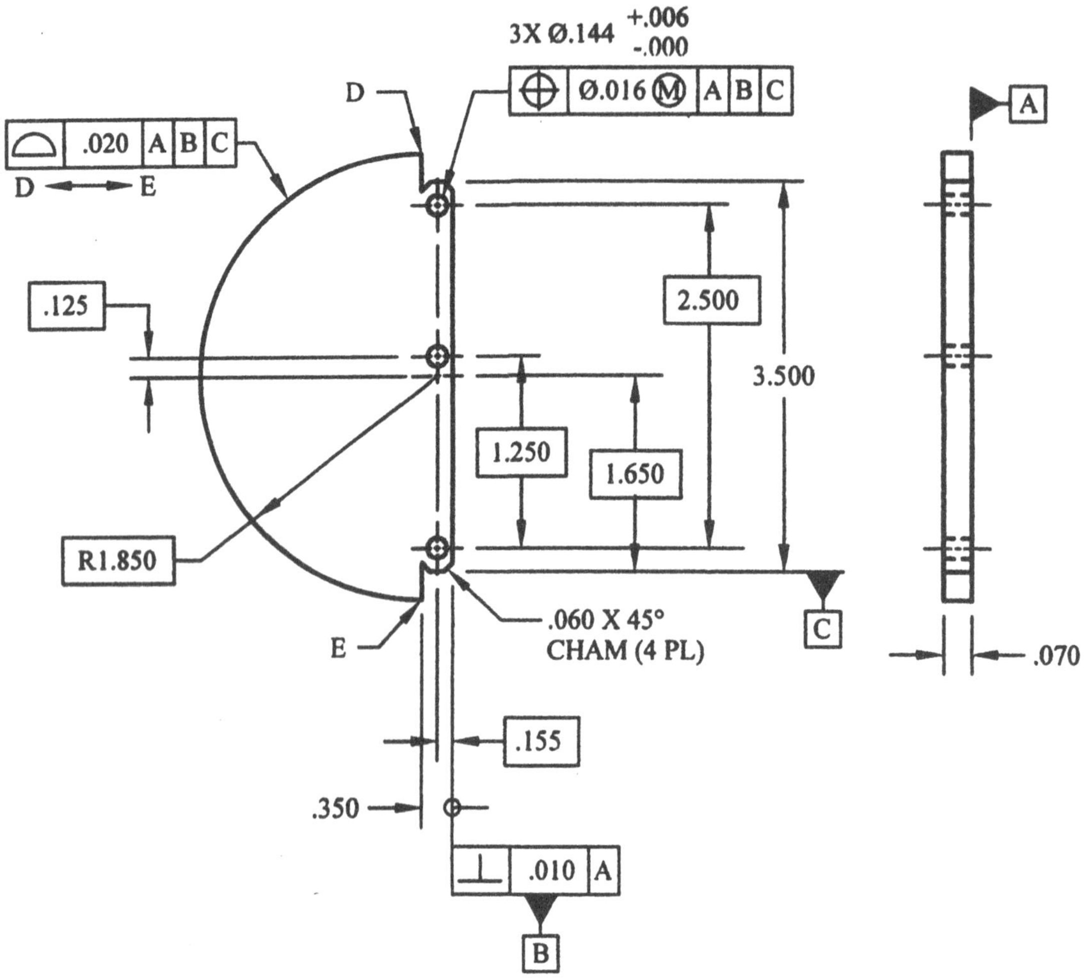

NOTES:

1. TO BE INTERPRETED PER ASME Y14.5
2. ALL GEOMETRIC TOLERANCES TO BE INSPECTED WHILE PART IS RESTRAINED ON PRIMARY DATUM.

UNLESS OTHERWISE SPECIFIED:

.XXX = ±.010

∠ = ±1°

42.

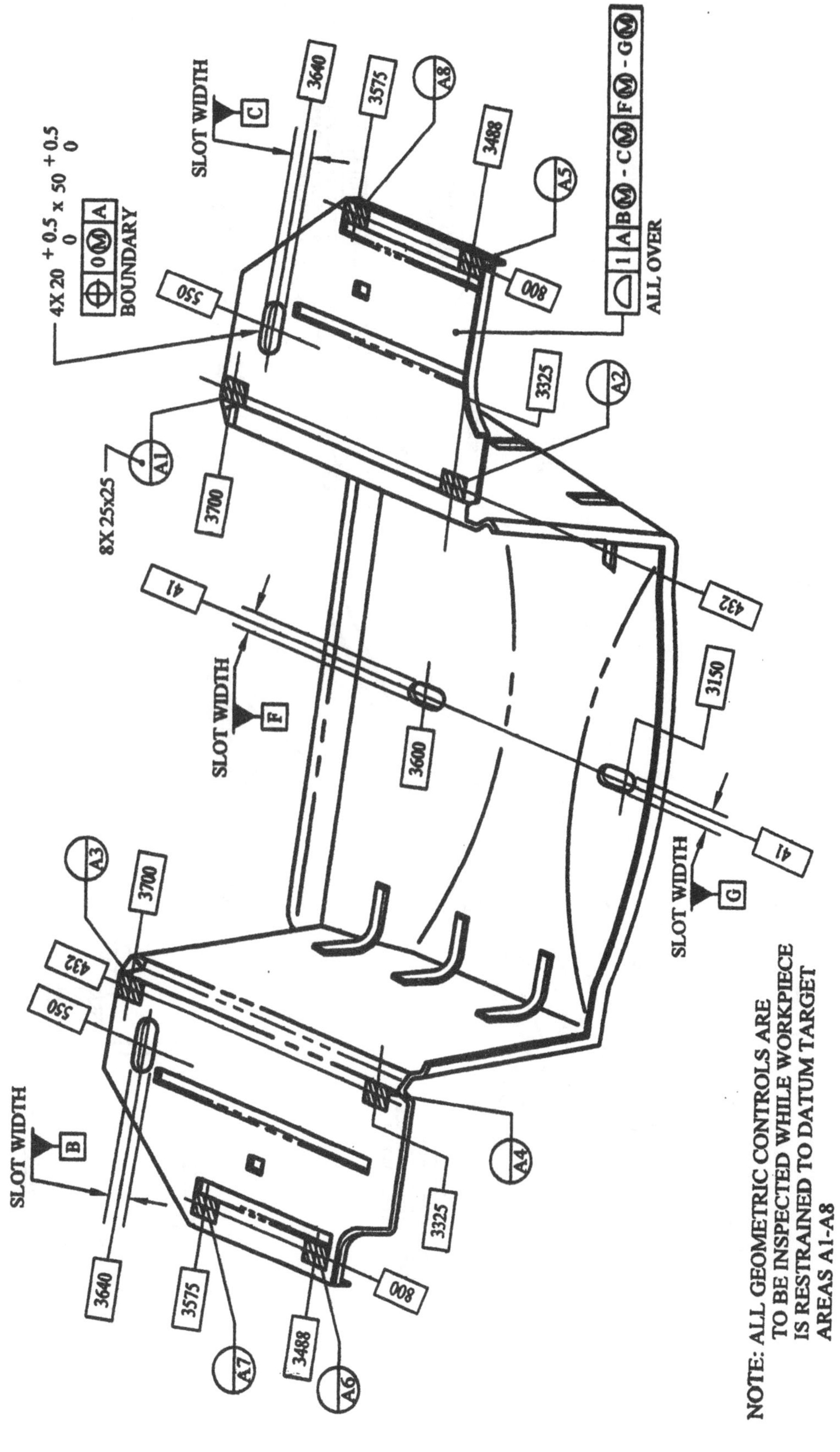
4X 20 +0.5 0 x 50 +0.5 0
BOUNDARY
SLOT WIDTH
C
B
F
G
8X 25x25
ALL OVER
A1
A2
A3
A4
A5
A6
A7
A8
3640
3575
3488
3325
3700
3600
3150
550
008
432
41
NOTE: ALL GEOMETRIC CONTROLS ARE
TO BE INSPECTED WHILE WORKPIECE
IS RESTRAINED TO DATUM TARGET
AREAS A1-A8

43.

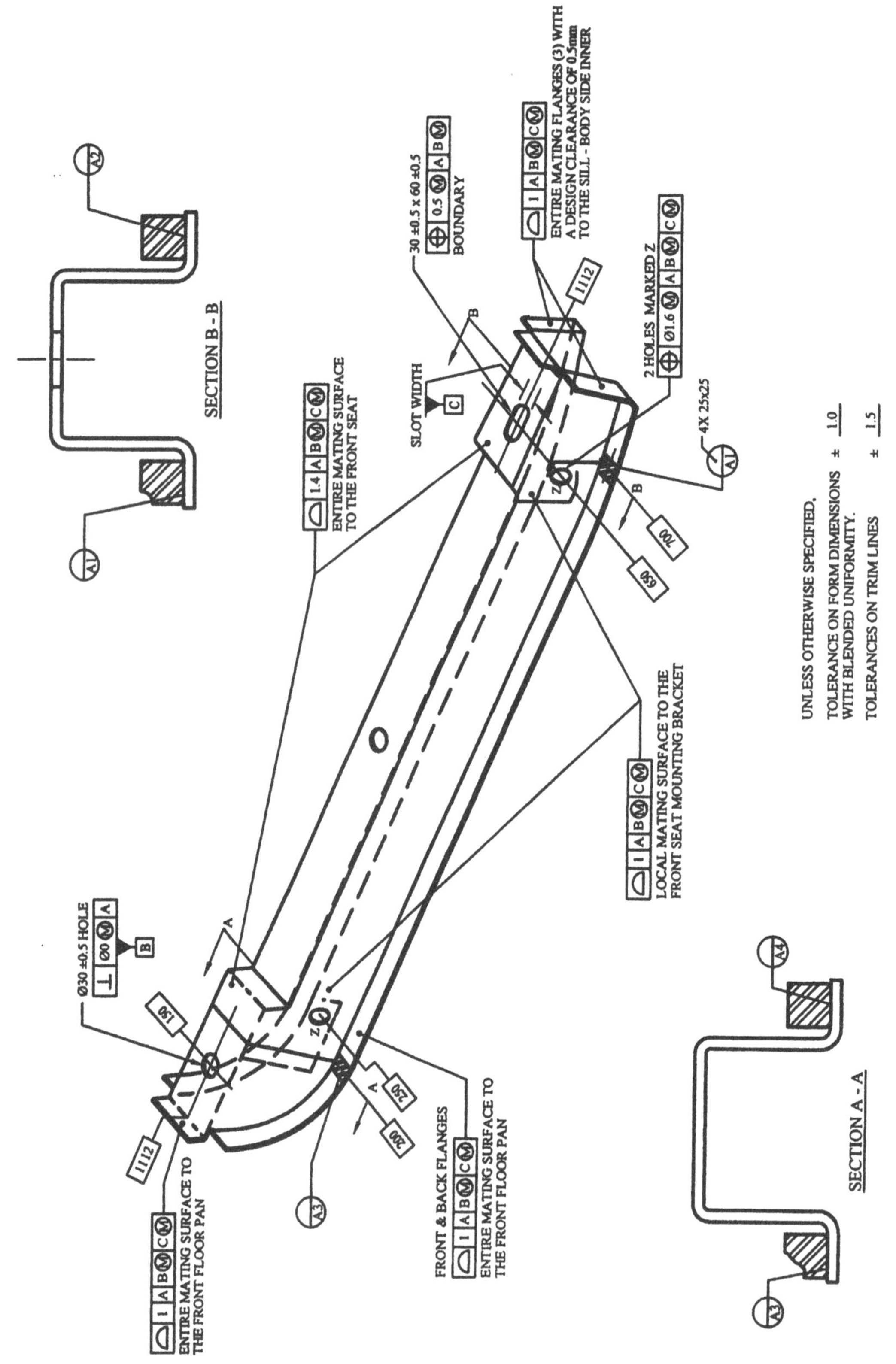
SECTION B - B
SECTION A - A
30 ±0.5 x 60 ±0.5
BOUNDARY
ENTIRE MATING FLANGES (3) WITH A DESIGN CLEARANCE OF 0.5mm TO THE SILL - BODY SIDE INNER
SLOT WIDTH
2 HOLES MARKED Z
4X 25x25
ENTIRE MATING SURFACE TO THE FRONT SEAT
LOCAL MATING SURFACE TO THE FRONT SEAT MOUNTING BRACKET
Ø30 ±0.5 HOLE
ENTIRE MATING SURFACE TO THE FRONT FLOOR PAN
FRONT & BACK FLANGES
ENTIRE MATING SURFACE TO THE FRONT FLOOR PAN
UNLESS OTHERWISE SPECIFIED.
TOLERANCE ON FORM DIMENSIONS ± 1.0
WITH BLENDED UNIFORMITY.
TOLERANCES ON TRIM LINES ± 1.5

44.

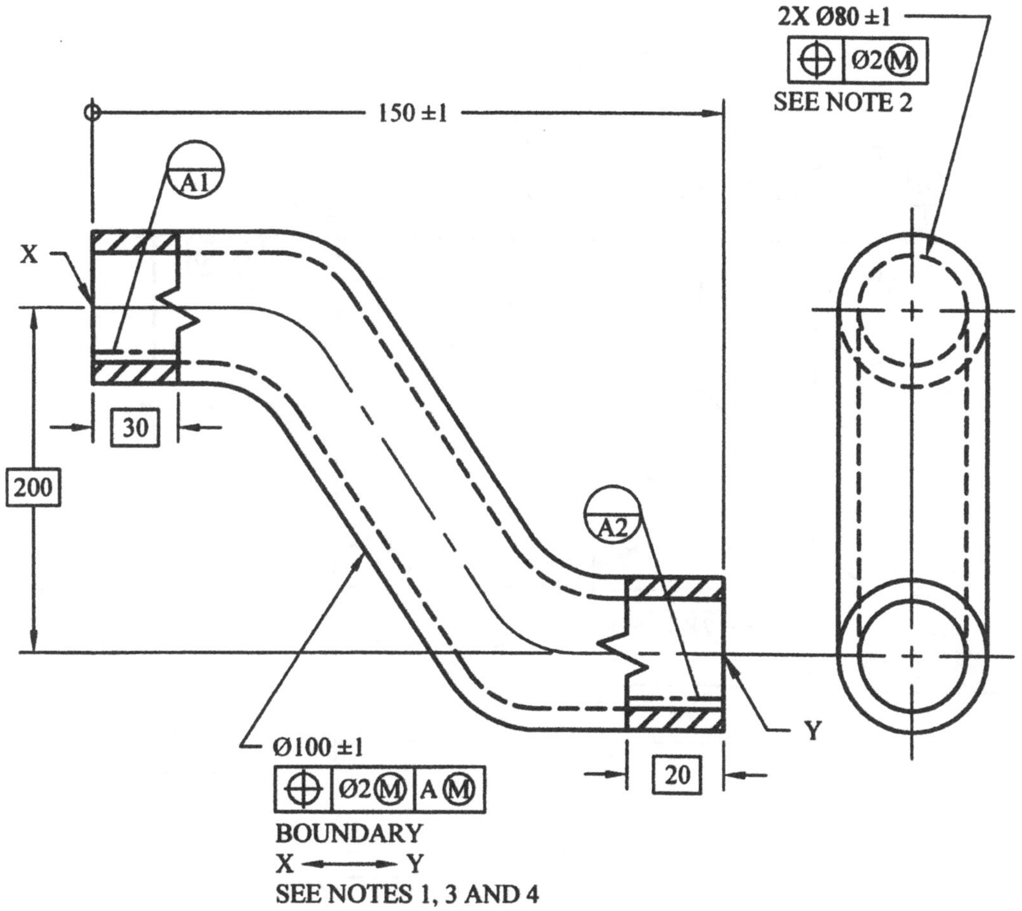

NOTE 1: FOR COMPLETE DEFINITION OF THIS PART REFER TO THE CAD MODEL.

NOTE 2: THIS POSITION CONTROL APPLIES TO DATUM TARGET DIAMETERS A1 AND A2 ONLY. IT POSITIONS THEM TO EACH OTHER.

NOTE 3: UNLESS OTHERWISE SPECIFIED, CONSIDER ALL CAD DIMENSIONS AS BASIC.

NOTE 4: DATUM AXIS A IS FORMED HALFWAY BETWEEN THE VIRTUAL CONDITION BOUNDARIES OF DATUM TARGET DIAMETERS A1 AND A2.

45.

Workpiece

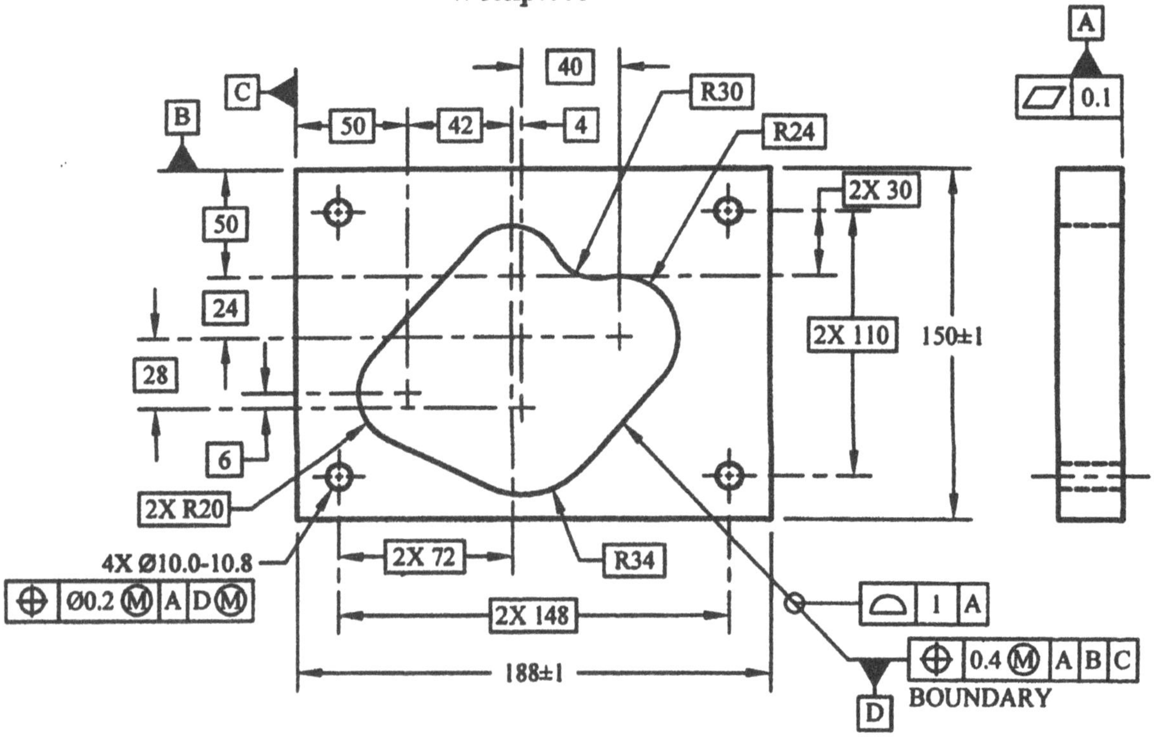

Gage

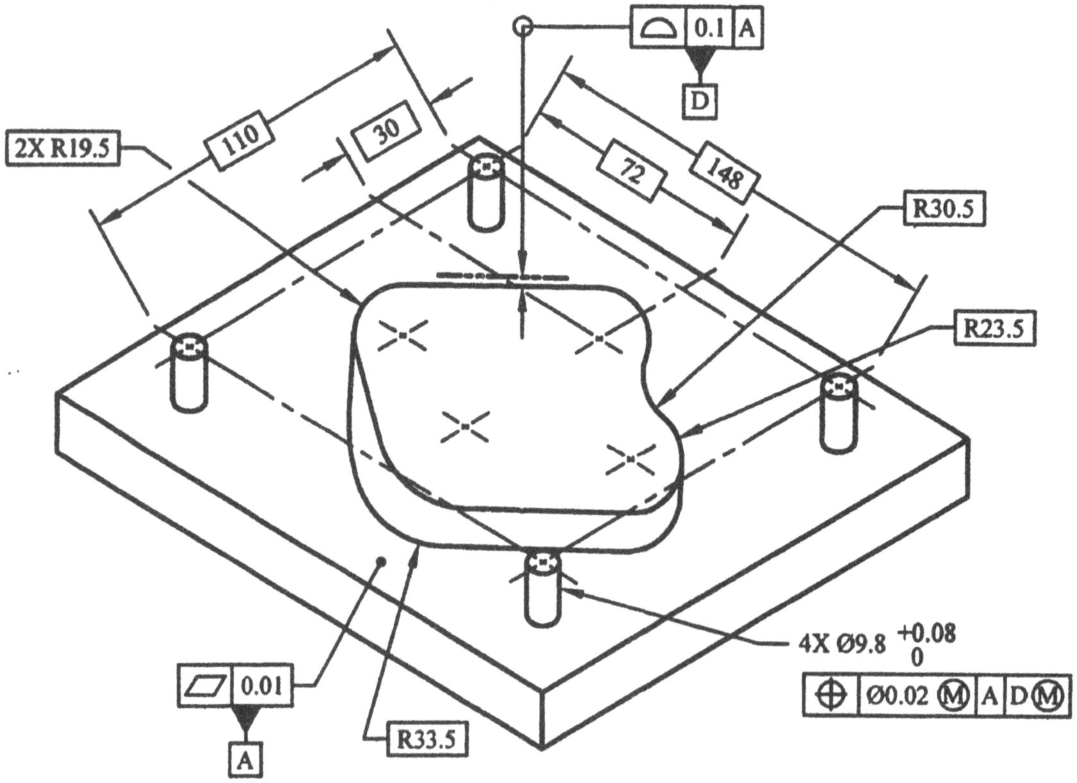

46.

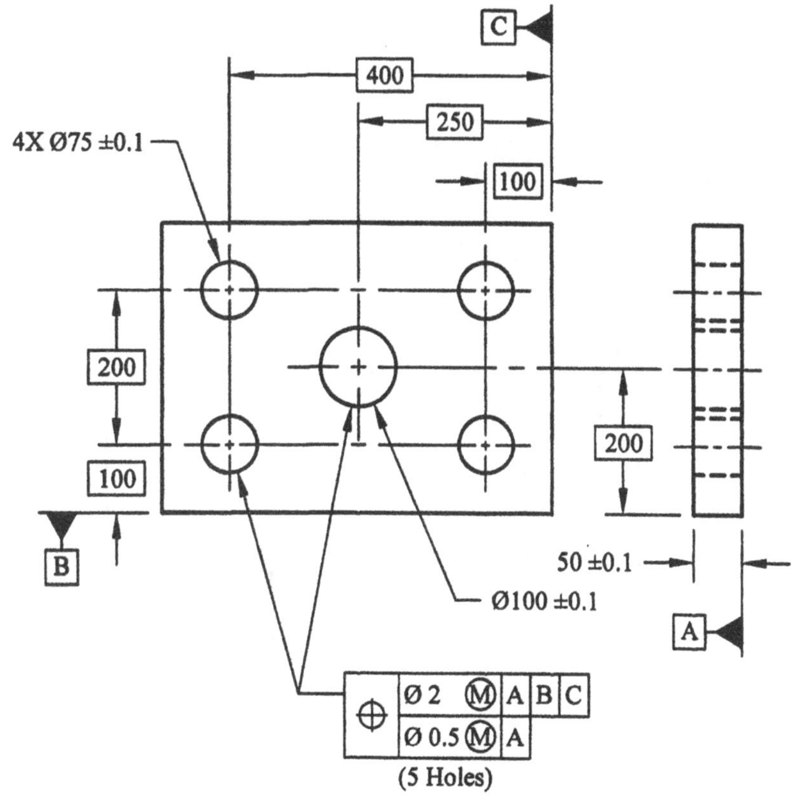
C
400
250
4X Ø75 ±0.1
100
200
100
B
200
50 ±0.1
Ø100 ±0.1
A
Ø2 Ⓜ A B C
Ø 0.5 Ⓜ A
(5 Holes)

47.

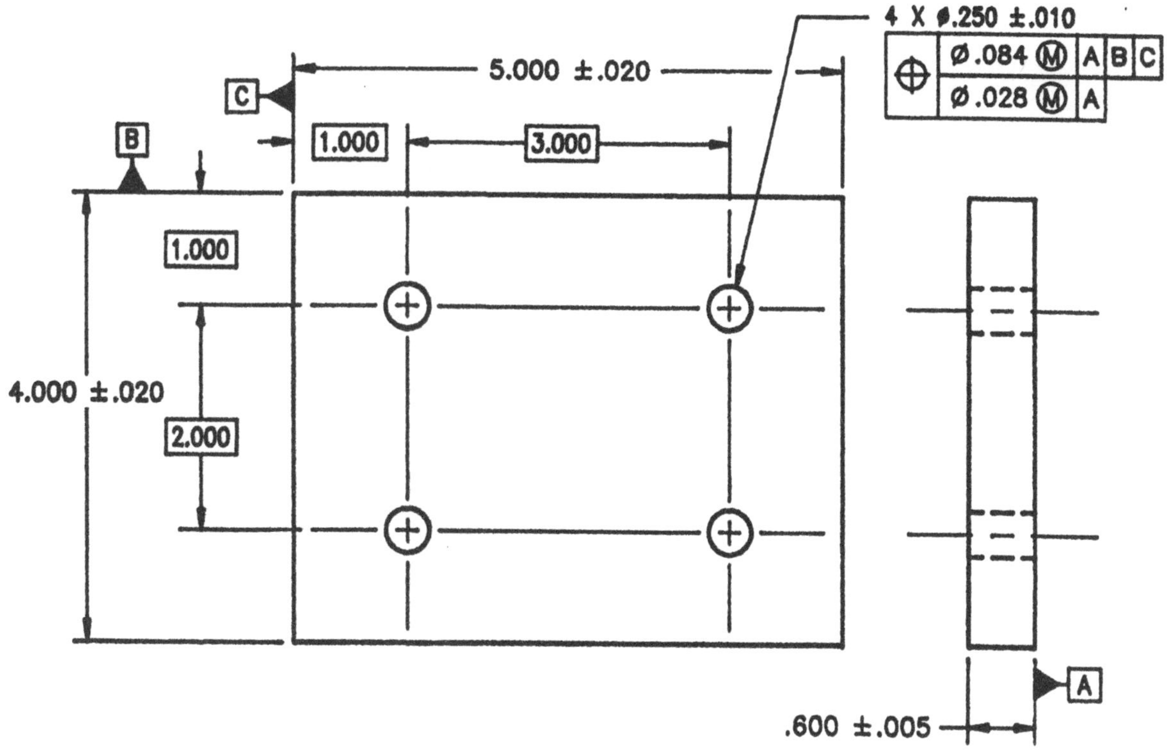
4 X Ø.250 ±.010
Ø.084 Ⓜ A B C
Ø.028 Ⓜ A
5.000 ±.020
C
B
1.000
3.000
1.000
4.000 ±.020
2.000
A
.600 ±.005

48.

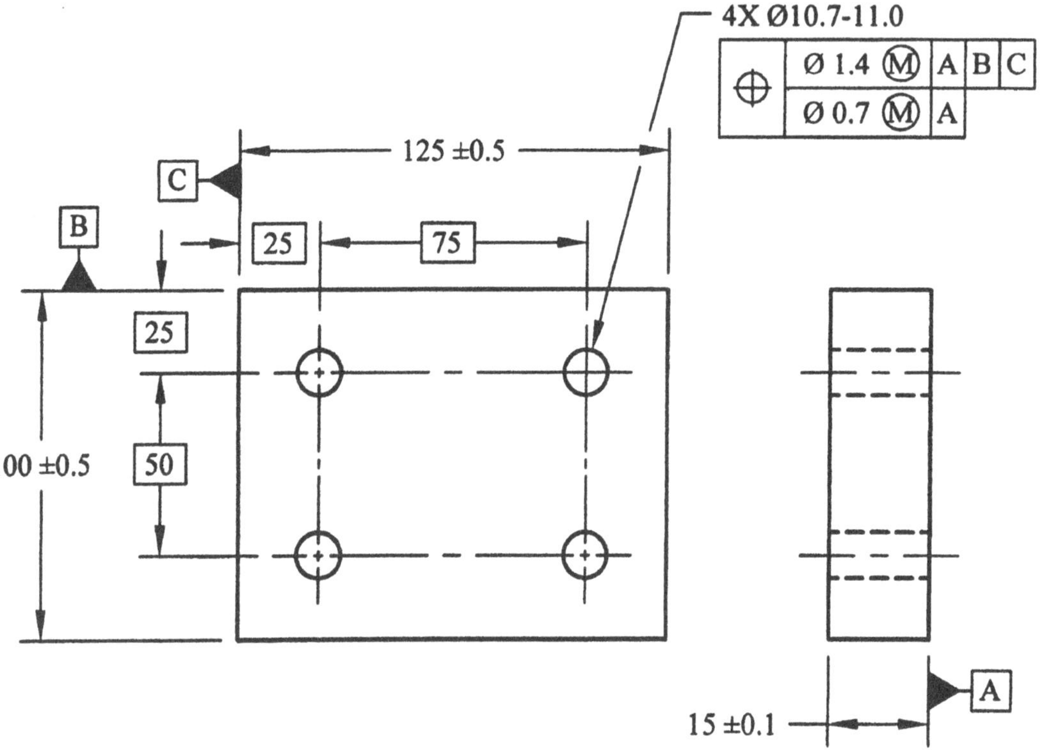

UNLESS OTHERWISE SPECIFIED
ALL ANGLES ARE ±1°

49.

Assembly

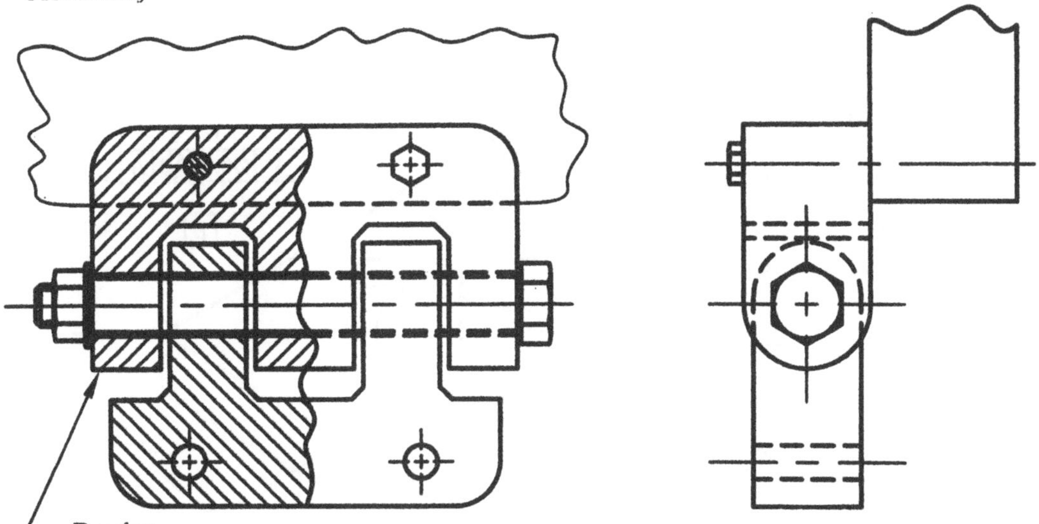

Detail Drawing Cutout

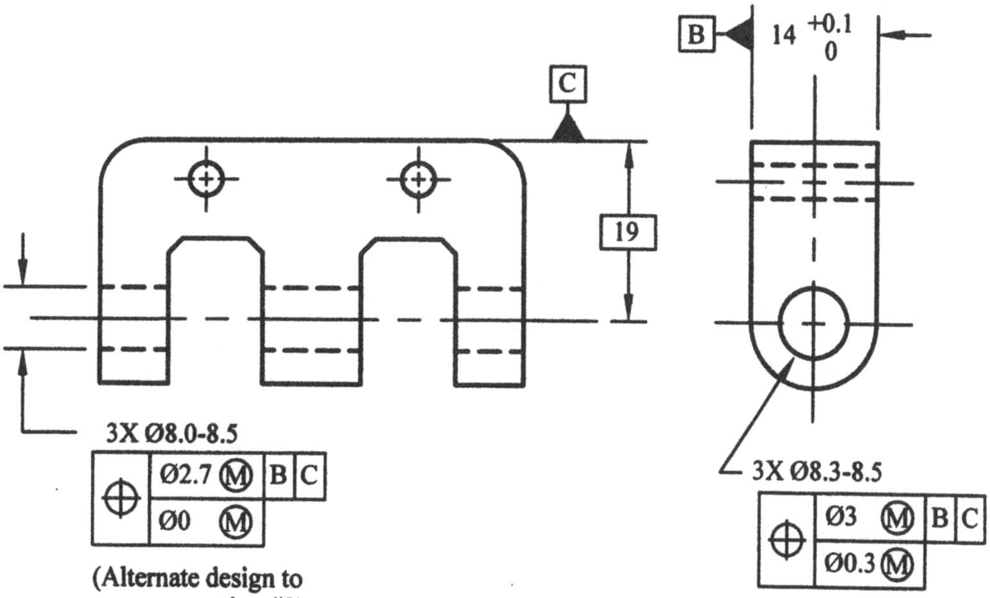

(Alternate design to answer question #2)

Answer to question #1

50.

GAGE

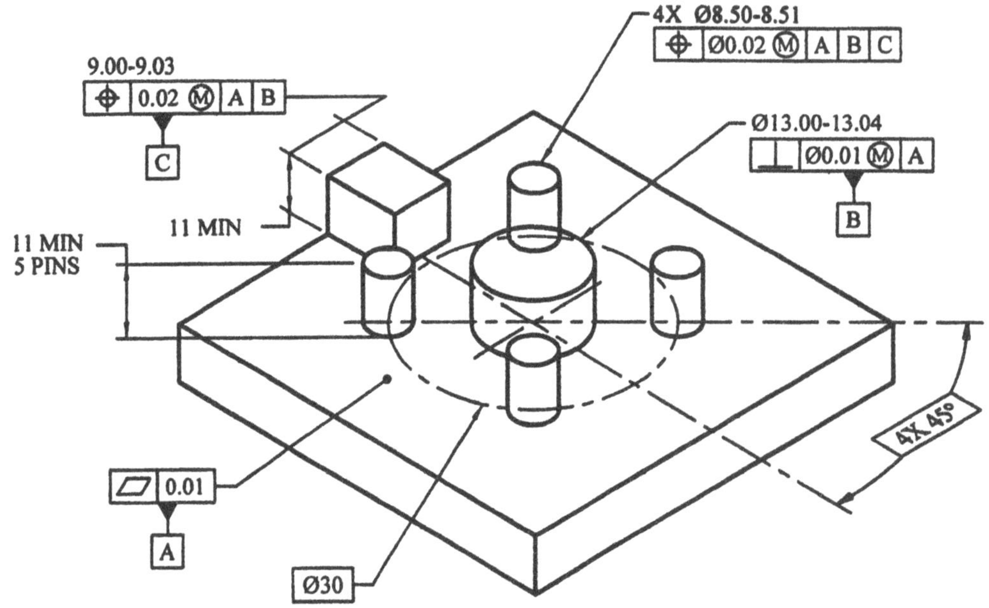

WORKPIECE APPLIED TO GAGE

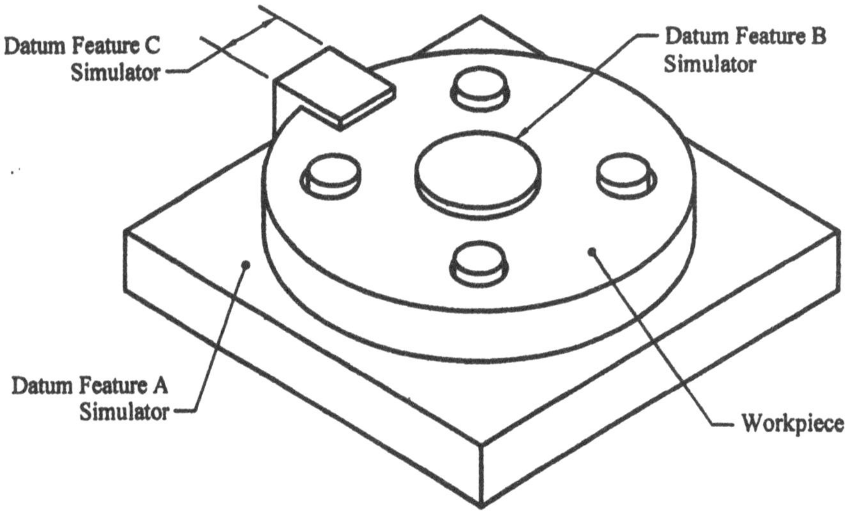

All gages in this section have used either the Absolute Tolerancing Method (shown on the GO and NOGO gages) or the Practical Absolute Tolerancing Methods (shown on both Functional Gages). These gaging practices use as the premise that all gage pins have plus tolerances and all gage holes have minus tolerances for all GO gages and Functional Gages. For NOGO gages, all gage pins have minus tolerances and all gage holes have plus tolerances. This is to achieve a gage that does not accept parts that are out of their tolerance ranges.

51.

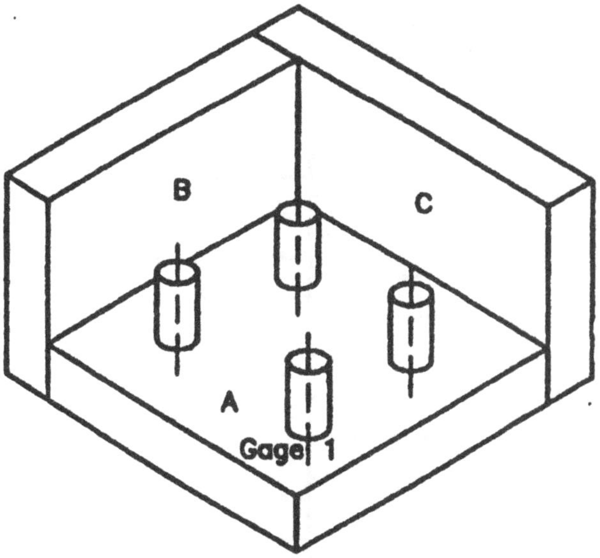

GAGE 1 – Datum to pattern
All four pins are located by the same basic dimensions (from the datum feature simulators) as used on the part drawing.

Pin sizes are ∅.355 $^{+.XXX}_{-.000}$

with | ⌖ | ∅.000 Ⓜ | A | B | C |

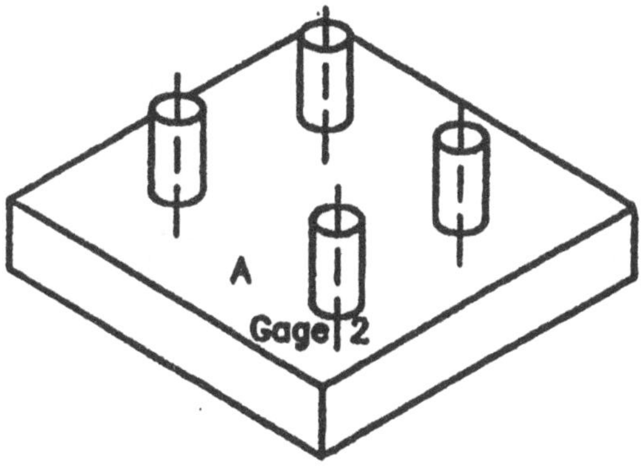

GAGE 2 – Feature to feature
pin axis to pin axis basic dimensions are the same as gage 1 (and part drawing).

Pin sizes are ∅.360 $^{+.XXX}_{-.000}$

with

⌖	∅ .XXX Ⓜ	A	
	∅.000 Ⓜ	A	

52.

Part #1

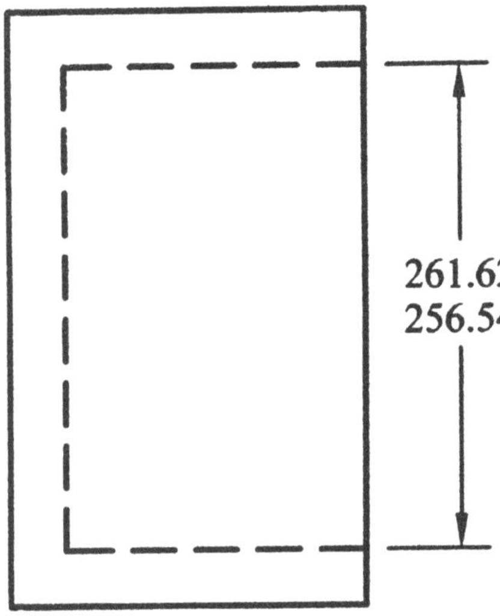

Part #2-#11

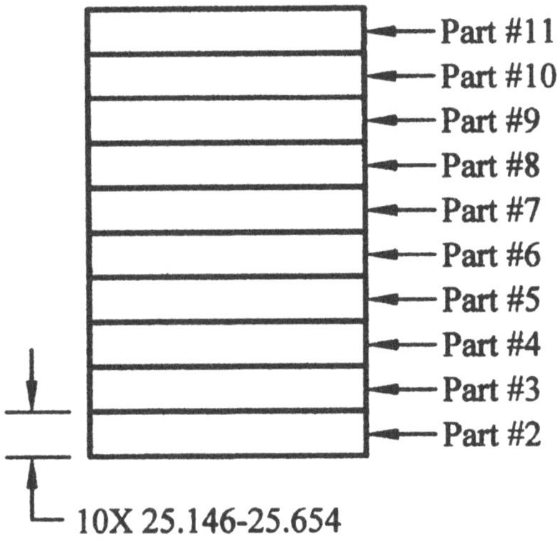

Part #1:

$$\begin{array}{r} 261.62 \\ +\ 256.54 \\ \hline 518.16 \end{array} \qquad \begin{array}{r} 261.62 \\ -\ 256.54 \\ \hline 5.08 \end{array}$$

1/2 of 518.16 = 259.08

1/2 of 5.08 = 2.54

So, expressed as an equal bilaterally toleranced dimension: 259.08±2.54

Part #2-#11:

$$\begin{array}{r} 25.654 \\ \text{X}\ 10 \\ \hline 256.54 \end{array} \qquad \begin{array}{r} 25.146 \\ \text{X}\ 10 \\ \hline 251.46 \end{array} \qquad \begin{array}{r} 256.54 \\ +\ 251.46 \\ \hline 508 \end{array} \qquad \begin{array}{r} 256.54 \\ -\ 251.46 \\ \hline 5.08 \end{array}$$

1/2 of 508 = 254

1/2 of 5.08 = 2.54

So, expressed as an equal bilaterally toleranced dimension: 254±2.54

or:

$$\begin{array}{r} 25.654 \\ +\ 25.146 \\ \hline 50.8 \end{array} \qquad \begin{array}{r} 25.654 \\ -\ 25.146 \\ \hline 0.508 \end{array}$$

50.8/2 = 25.4
0.508/2 = 0.254

So: 25.4±0.254---10 times

52. (continued)

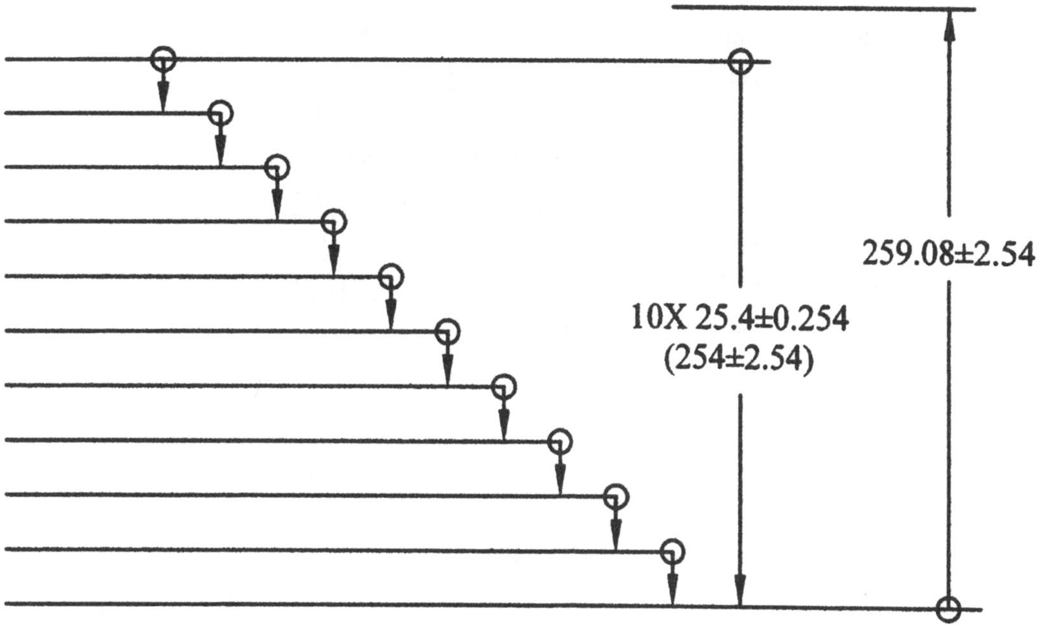

DIMENSIONS		±TOL
-	+	
254	259.08	2.54 2.54
254	259.08	5.08 TOTALS
259.08 - 254.00 5.08	5.08 + 5.08 10.16=MAX GAP	5.08 - 5.08 0=MIN GAP (zero interference)

Possible solutions to interference problems, should ever they occur:

1. Increase MMC size of cavity.
2. Decrease MMC size of plates.
3. Decrease tolerance on cavity (effectively increasing the MMC).
4. Decrease the tolerances on the plates (effectively decreasing their collective MMC).

53.

Resultant Condition of Hole #1 = <u>Ø6.0 + 1.4 = Ø7.4</u>
Virtual Condition of Hole #1 = <u>Ø5.0 - 0.4 = Ø4.6</u>
Resultant Condition of Hole #2 = <u>Ø6.0 + 1.4 = Ø7.4</u>
Virtual Condition of Hole #2 = <u>Ø5.0 - 0.4 = Ø4.6</u>
Resultant Condition of Hole #3 = <u>Ø6.0 + 1.0 = Ø7.0</u>
Virtual Condition of Hole #3 = <u>Ø5.0 - 0 = Ø5.0</u>
Resultant Condition of Hole #4 = <u>Ø6.0 + 1.0 = Ø7.0</u>
Virtual Condition of Hole #4 = <u>Ø5.0 - 0 = Ø5.0</u>

Resultant Condition of Hole #1 = Ø7.4 - Virtual Condition of Hole #1 = Ø4.6 Difference = Ø2.8 Resultant Condition of Hole #3 = Ø7 - Virtual Condition of Hole #3 = Ø5 Difference = Ø2 Hole #1: 1/2 of Difference = <u>Ø1.4</u> Hole #3: 1/2 of Difference = <u>Ø1</u>	Resultant Condition of Hole #2 = Ø7.4 - Virtual Condition of Hole #2 = Ø4.6 Difference = Ø2.8 Resultant Condition of Hole #4 = Ø7 - Virtual Condition of Hole #4 = Ø5 Difference = Ø2 Hole #2: 1/2 of Difference = <u>Ø1.4</u> Hole #4: 1/2 of Difference = <u>Ø1</u>
Resultant Condition of Hole #1 = Ø7.4 + Virtual Condition of Hole #1 = Ø4.6 Sum = Ø12 1/2 Sum of R.C. and V.C. of Hole #1 = <u>Ø6</u> 1/2 Sum ± 1/2 Difference of Hole #1 = Ø6±1.4 1/2 of 1/2 Sum ± 1/2 of 1/2 Difference of Hole #1 = R3±0.7	Resultant Condition of Hole #2 = Ø7.4 + Virtual Condition of Hole #2 = Ø4.6 Sum = Ø12 1/2 Sum of R.C. and V.C. of Hole #2 = <u>Ø6</u> 1/2 Sum ± 1/2 Difference of Hole #2 = Ø6±1.4 1/2 of 1/2 Sum ± 1/2 of 1/2 Difference of Hole #2 = R3±0.7
Resultant Condition of Hole #3 = Ø7 + Virtual Condition of Hole #3 = Ø5 Sum = Ø12 1/2 Sum of R.C. and V.C. of Hole #3 = <u>Ø6</u> 1/2 Sum ± 1/2 Difference of Hole #3 = Ø6±1 1/2 of 1/2 Sum ± 1/2 of 1/2 Difference of Hole #3 = R3±0.5	Resultant Condition of Hole #4 = Ø7 + Virtual Condition of Hole #4 = Ø5 Sum = Ø12 1/2 Sum of R.C. and V.C. of Hole #4 = <u>Ø6</u> 1/2 Sum ± 1/2 Difference of Hole #4 = Ø6±1 1/2 of 1/2 Sum ± 1/2 of 1/2 Difference of Hole #4 = R3±0.5

53. (continued) MIN GAP

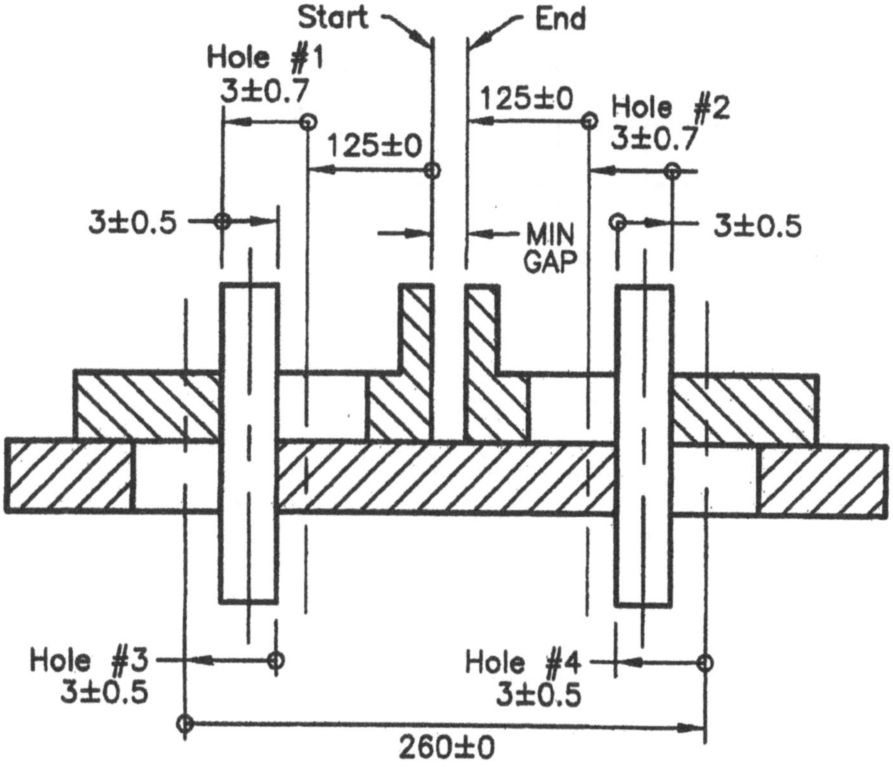

Loop Diagram

Right to Left	Left to Right		
−	+	±Tol	
125		0.0	Basic Dim
3		0.7	Hole #1
	3	0.5	Pin
3		0.5	Hole #3
	260	0.0	Basic Dim
3		0.5	Hole #4
	3	0.5	Pin
3		0.7	Hole #2
125		0.0	Basic Dim
262	266	3.4	Totals
266 −262 4		4.0 −3.4 0.6 MIN GAP	

53. (continued) MAX GAP

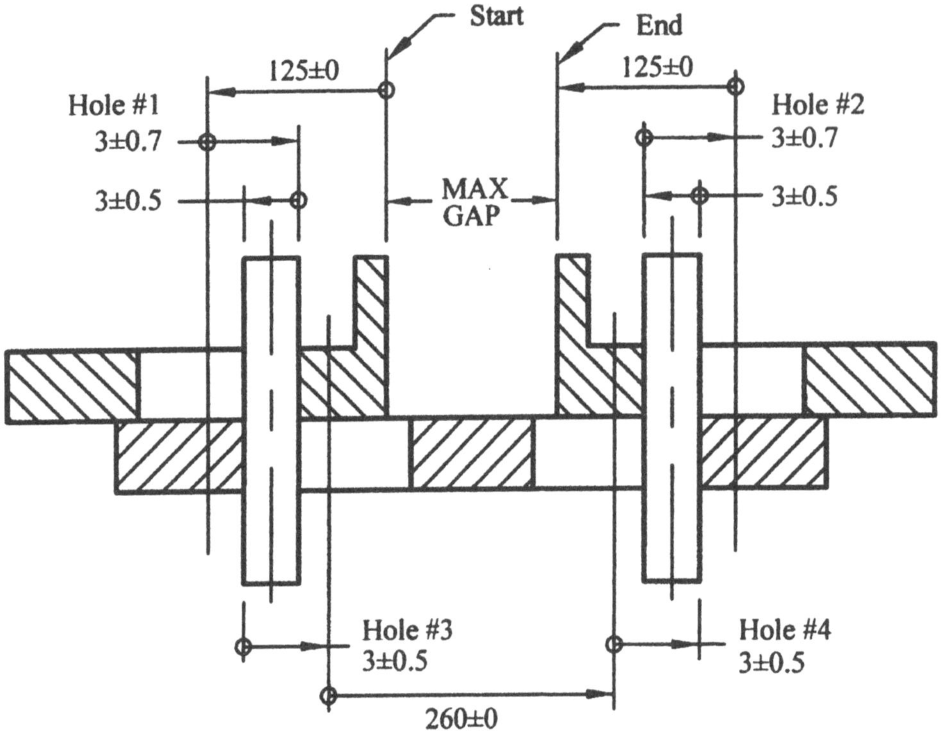

Loop Diagram

Right to Left	Left to Right		
-	+	±Tol	
125		0.0	Basic Dim
	3	0.7	Hole #1
3		0.5	Pin
	3	0.5	Hole #3
	260	0.0	Basic Dim
	3	0.5	Hole #4
3		0.5	Pin
	3	0.7	Hole #2
125		0.0	Basic Dim
256	272	3.4	Totals
272 - 256 16		16.0 + 3.4 19.4	MAX GAP

54.

–	+	±	± squared	Type
3		0.7	0.49	Hole #1
	3	0.5	0.25	Pin
3		0.5	0.25	Hole #3
3		0.5	0.25	Hole #4
	3	0.5	0.25	Pin
3		0.7	0.49	Hole #2
12	6	3.4	1.98	Totals

$$1.5\sqrt{1.98} = 2.110687$$

$$\frac{2.110687 \text{ (tol. likely to be consumed)}}{3.4 \text{ (arithmetically calculated tol.)}} = 0.6207902$$

So, 2.110687 is about 62% of 3.4.

$$\frac{1}{0.6207902} = 1.6108501$$ and $1.6108501 \times 2.110687 = 3.4$ (the originally calculated arithmetic tolerance)

So, we will increase the tolerances for the pertinent features to 161%.

54. (continued)

So, the ±0.7 tolerance for Hole #1 and Hole #2 becomes 1.61 x 0.7 = 1.127 or a tolerance for each that is ±1.127. The tolerance for Hole #3, Hole #4 and both pins becomes 1.61 x 0.5 = ±0.805. <u>This is the answer to the problem.</u>

To Reintegrate these tolerances into a combination of size and geometric tolerance, the following steps are performed:

±1.127 for Holes #1 and #2:

1) 3 x 2 = Ø6

2)
```
  1.127
    x 2
  -----
  2.254 = Tol
```

3)
```
  6.000                6.000
- 2.254              + 2.254
  -----                -----
  3.746 = I.B.         8.254 = O.B.
```

4) 0.4 = Original Geo. Tol. at MMC so;
```
  1.61   = Increase
x 0.4    = Geo. Tol. at MMC
  -----
  0.644  = new ⊕ Tol. at MMC ⟨ST⟩
```

⊕	Ø 0.644 Ⓜ ⟨ST⟩	A	B	C

5) 1.4 = Geo. Tol. at LMC so;
```
  1.61   = Increase
x 1.4    = Geo. Tol. at LMC
  -----
  2.254  = new Geo. Tol. ⊕ Tol. at LMC ⟨ST⟩
```

6) Add 0.644 to I.B. 3.746
```
  3.746
+ 0.644
  -----
  4.390 = MMC
```

7) Subtract 2.254 from O.B. 8.254
```
  8.254
- 2.254
  -----
  6.000 = LMC
```

8) Ø4.39 - 6.00 ⟨ST⟩

⊕	Ø 0.644 Ⓜ ⟨ST⟩	A	B	C

54. (continued)

±0.805 for Holes #3 and #4;

1) 3 x 2 = Ø6

2)

$$\begin{array}{r} 0.805 \\ \times\ 2 \\ \hline 1.610 \end{array} = \text{Tol}$$

3)

$$\begin{array}{r} 6.000 \\ -\ 1.610 \\ \hline 4.390 \end{array} = \text{I.B.} \qquad \begin{array}{r} 6.000 \\ +\ 1.610 \\ \hline 7.610 \end{array} = \text{O.B.}$$

4) 0 = Original Geo. Tol. so; 0 x 1.61 = 0

⊕	Ø 0 Ⓜ ⟨ST⟩	A	B	C

5) 1mm = Geo. Tol. at LMC so;

$$\begin{array}{r} 1.61 \\ \times\ 1 \\ \hline 1.61 \end{array}$$

= new Geo. Tol. ⊕ Tol. at LMC ⟨ST⟩

6) Add 0 to I.B. 4.390 so; new MMC = 4.390

7) Subtract 1.61 from O.B. 7.610

$$\begin{array}{r} 7.610 \\ -\ 1.610 \\ \hline 6.000 \end{array} = \text{LMC}$$

8) Ø4.39 - 6.00 ⟨ST⟩

⊕	Ø 0 Ⓜ ⟨ST⟩	A	B	C

Had we used the RSS formula <u>without the correction factor</u> of 1.5, we would have found the following;

$$\sqrt{1.98} = 1.4071247$$

$$\frac{1.4071247 \text{ (tol. likely to be consumed)}}{3.4 \text{ (arithmetically calculated tol.)}} = 0.4138602$$

so, 1.4071247 is about 41% of 3.4.

$$\frac{1}{0.4138602} = 2.4162748 \quad \text{and} \quad \begin{array}{r} 2.4162748 \\ \times\ 1.4071247 \\ \hline 3.4 \end{array} \text{ (the original calculated arithmetic tolerance)}$$

So, had we used the RSS formula without the 1.5 correction factor, the tolerance for the pertinent features would have been increased by 242% instead of 161%.

54. (continued)

Using the Simpler Methodology to Keep the Same Size Mean Dimension instead of the Same Functional Boundary Mean

Holes 1 & 2-With this method, the original size mean dimension on the holes of Ø5.5 is kept. The original size tolerance of ±0.5 is multiplied by 1.61 to get 0.805. So, the new size of the slot is **Ø5.5±0.805 Statistical Tolerance** using the Root Sum Square method with a safety factor of 1.5. The original position tolerance of Ø0.4 at MMC is multiplied by 1.61 to become **Ø0.644 at MMC Statistical Tolerance** using the Root Sum Square method with a safety factor of 1.5.

Holes 3 & 4-With this method, the original size mean dimension on the holes of Ø5.5 is kept. The original size tolerance of ±0.5 is multiplied by 1.61 to get 0.805. So, the new size of the slot is **Ø5.5±0.805 Statistical Tolerance** using the Root Sum Square method with a safety factor of 1.5. Since the original position tolerance for holes 3 & 4 was zero at MMC, mulitplying it by anything leaves it unchanged. So, the position tolerance is still **Ø0 at MMC.**

55.

Part #4

Step 1: Outside Diameter

263.0 = MMC
+ 0.2 = Geo. Tol.
Ø263.2 = Outer Boundary

$\frac{263.2}{2}$ = R131.6 Shaft

Step 2: Clearance Hole

Ø9.0 = LMC

$\frac{9.0}{2}$ = R4.5 Hole

Part #3

Step 1: Mounted Screw (in threaded hole)

7.76 = LMC Screw
- 0.40 = Geo. Tol.

7.36 = Inner Boundary Screw
- 0.06 = Pattern Shift due to D Ⓜ Reference

Ø7.30 = Inner Boundary with Pattern Shift

Step 2:

Ø79.9 = LMC of D

$\frac{Ø79.9}{2}$ = R39.95 Shaft

Part #2

Step 1:

Ø80.08 = LMC of D
+ 0.20 = Position Tol. of D
Ø80.28 = Outer Boundary of D

$\frac{80.28}{2}$ = R40.14 Hole

Step 2:

Ø109.00 = LMC of E

$\frac{109}{2}$ = R54.5 Shaft

Part #1

Step 1: Central Hole

111.00 = LMC Hole
+ 1.20 = Geo. Tol. at LMC
112.20 = Outer Boundary Hole

$\frac{112.2}{2}$ = R56.1 Hole

MMC of C = 264.6 = R132.3

55. (continued)

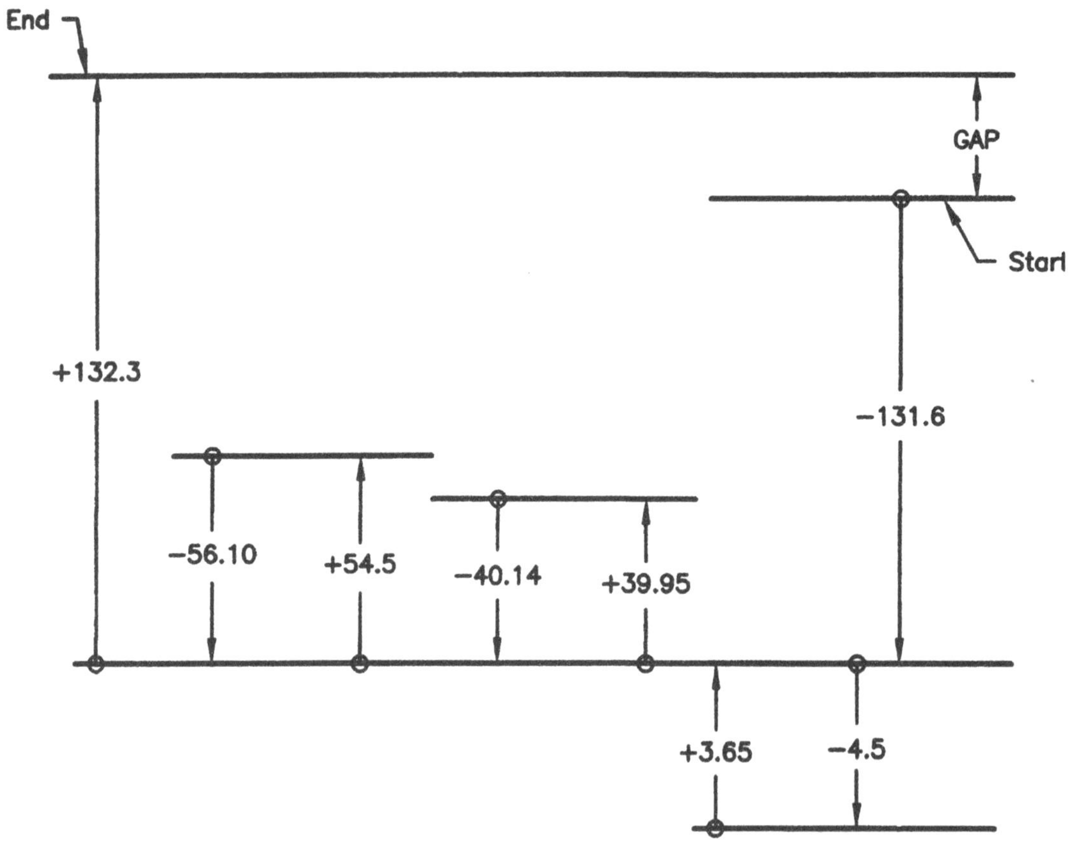

−	+
131.60	
4.50	
	3.65
	39.95
40.14	
	54.50
56.10	
	132.30
−232.34	+230.40

Totals

$$\begin{array}{r} +230.40 \\ + \; -232.34 \\ \hline -1.94 \end{array}$$

1.94 = MAX. Interference

56. Problem #1

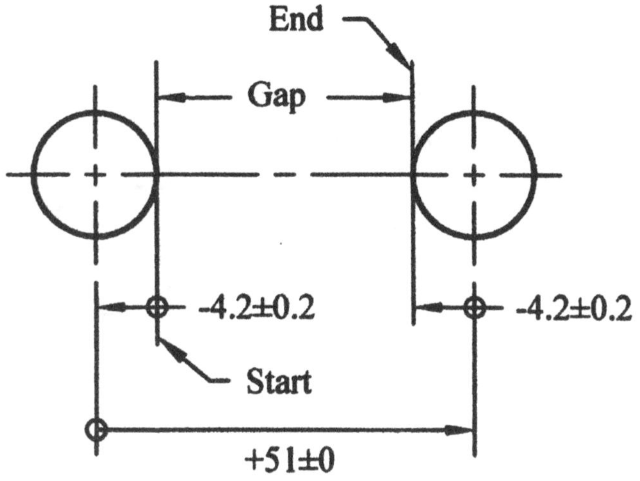

Inner Boundary of Hole = 8.0 (MMC) - 0 (Geo. Tol. at MMC) = Ø8.0
Outer Boundary of Hole = 8.4 (LMC) +0.4 (Geo. Tol. at LMC) = Ø8.8

8.8 = O.B. of Hole	8.8 = O.B. of Hole
- 8.0 = I.B. of Hole	+ 8.0 = I.B. of Hole
0.8 = Difference	16.8 = Sum

$\frac{0.8}{2}$ = 0.4 = ±Tolerance $\frac{16.8}{2}$ = 8.4 = Mean Dimension

Divide both by 2 to convert to radii = 4.2±0.2

-	+	±Tol	
4.2		0.2	
	51	0.0	
4.2		0.2	
8.4	51	0.4	Totals

+51.0	42.6	42.6
+ -8.4	- 0.4	+ 0.4
+42.6	42.2 = MIN GAP	43.0 = MAX GAP

56. Problem #2

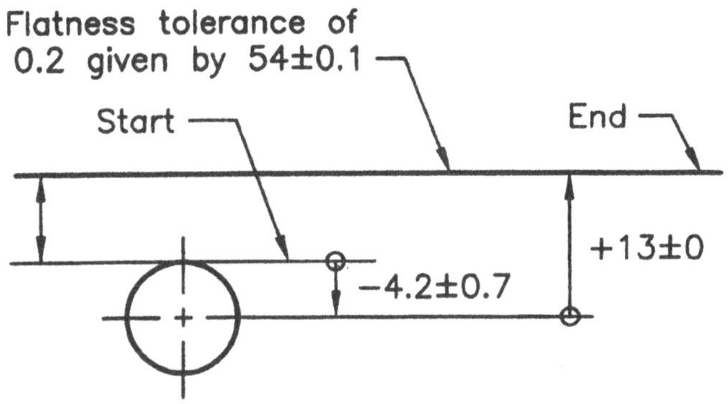

Ø8.4 (LMC of Hole) + 1.4 (Geo. Tol. at LMC to D) = Ø9.8 = O.B. Hole
Ø8.0 (MMC of Hole) - 1.0 (Geo. Tol. at MMC to D) = Ø7.0 = I.B. Hole

9.8 = O.B. of Hole	9.8 = O.B. of Hole
- 7.0 = I.B. of Hole	+ 7.0 = I.B. of Hole
2.8 = Difference	16.8 = Sum

$\frac{2.8}{2}$ = 1.4 = ±Tolerance $\frac{16.8}{2}$ = 8.4 = Mean Dimension

Divide both by 2 to convert to radii = 4.2±0.7

13.0	8.8	8.8
- 4.2	+ 0.7	- 0.7
8.8	9.5 = MAX WALL	8.1 = MIN WALL to Datum Plane D
		- 0.2 = Flatness Tol. of Surface D (Rule #1)
		7.9 = MIN WALL to Surface D (Low Points)

56. Problem #3

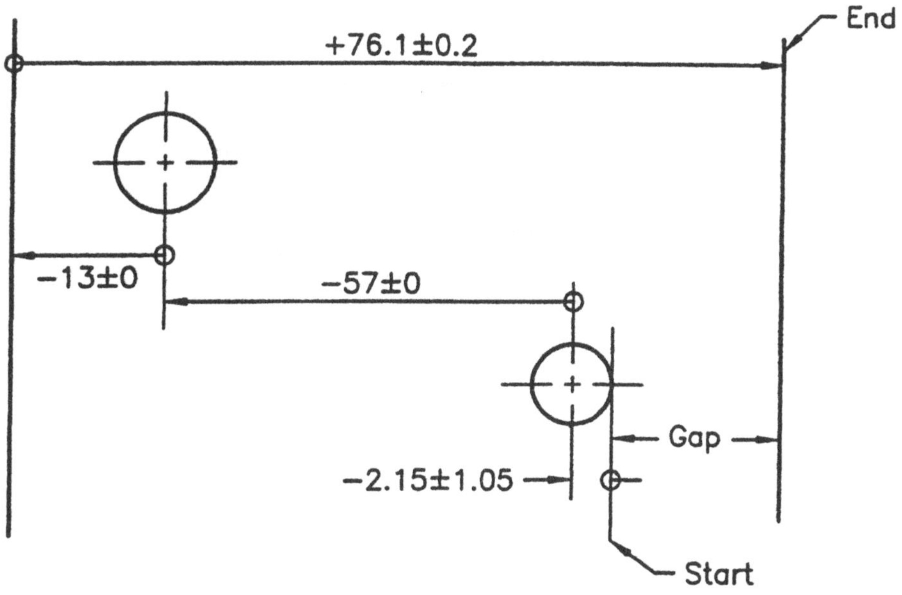

Ø4.3 (LMC) + 0.3 (Geo. Tol. at LMC) = Ø4.6 = O.B. Hole
+1.4 = Pos. Tol. of B at LMC
+0.4 = Pattern Shift (B referenced at MMC)
Ø6.4 = Hole's outer boundary plus all tolerances that apply

Ø4.0 (MMC) - 0 (Geo. Tol. at MMC) = Ø4.0 = I.B. Hole
-1.4 = Pos. Tol. of B at LMC
- 0.4 = Pattern Shift (B referenced at MMC)
Ø2.2 = Hole's outer boundary minus all tolerances that apply

Ø6.4 - Ø2.2 = Ø4.2 = Diff. Ø6.4 + Ø2.2 = Ø8.6 = Sum Divide both by 2. $\frac{Ø4.2}{2}$ = Ø2.1 & $\frac{Ø8.6}{2}$ = Ø4.3

So, Ø4.3±2.1 converted to a radius is R2.15±1.05

Overall dimension of part is 76.3 = O.B. and 75.9 = I.B. [76.1 - 0.2 (G.T. at LMC)]

76.3 - 75.9 = 0.4 difference and 76.3 + 75.9 = 152.2 sum

Divide both by 2. $\frac{0.4}{2} = 0.2$ and $\frac{152.2}{2} = 76.1$. So; 76.1±0.2

56. Problem #3 (continued)

-	+	±Tol	
2.15		1.05	
57.00		0.00	
13.00		0.00	
	76.10	0.20	
72.15	76.10	1.25	Totals

$$\begin{array}{r} 76.10 \\ -\ 72.15 \\ \hline 3.95 \end{array} \qquad \begin{array}{r} 3.95 \\ -\ 1.25 \\ \hline 2.70 \end{array} = \text{MIN WALL} \qquad \begin{array}{r} 3.95 \\ +\ 1.25 \\ \hline 5.20 \end{array} = \text{MAX WALL}$$

The bottom edge can be calculated using the following set of numbers:

First we must calculate the outer and inner boundary of the overall dimension in the top to bottom direction. 54.1 + 0 = 54.1 outer boundary and 53.9 - 0.2 = 53.7 inner boundary. 54.1 - 53.7 = 0.4 difference. 54.1 + 53.7 = 107.8 sum. Divide both by 2. $\frac{0.4}{2} = 0.2$ and $\frac{107.8}{2} = 53.9$. So; 53.9±0.2

-	+	±Tol	
2.15		1.05	
35.00		0.00	
13.00		0.00	
	53.9	0.20	
50.15	53.9	1.25	Totals

$$\begin{array}{r} 53.90 \\ -\ 50.15 \\ \hline 3.75 \end{array} \qquad \begin{array}{r} 3.75 \\ -\ 1.25 \\ \hline 2.5 \end{array} = \text{MIN WALL} \qquad \begin{array}{r} 3.75 \\ +\ 1.25 \\ \hline 5.00 \end{array} = \text{MAX WALL}$$

57.

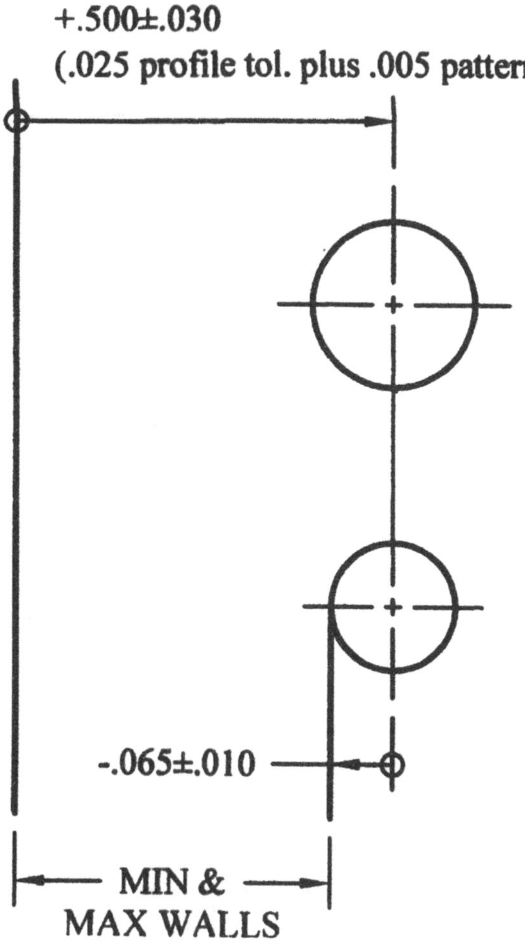

-	+	± Tolerance	
	.500	.030	
.065		.010	
.065	.500	.040	Totals

$$\begin{array}{r} .500 \\ -\ .065 \\ \hline .435 \end{array} \qquad \begin{array}{r} .435 \\ +\ .040 \\ \hline .475 \end{array} = \text{MAX WALL} \qquad \begin{array}{r} .435 \\ -\ .040 \\ \hline .395 \end{array} = \text{MIN WALL}$$

57. (continued)

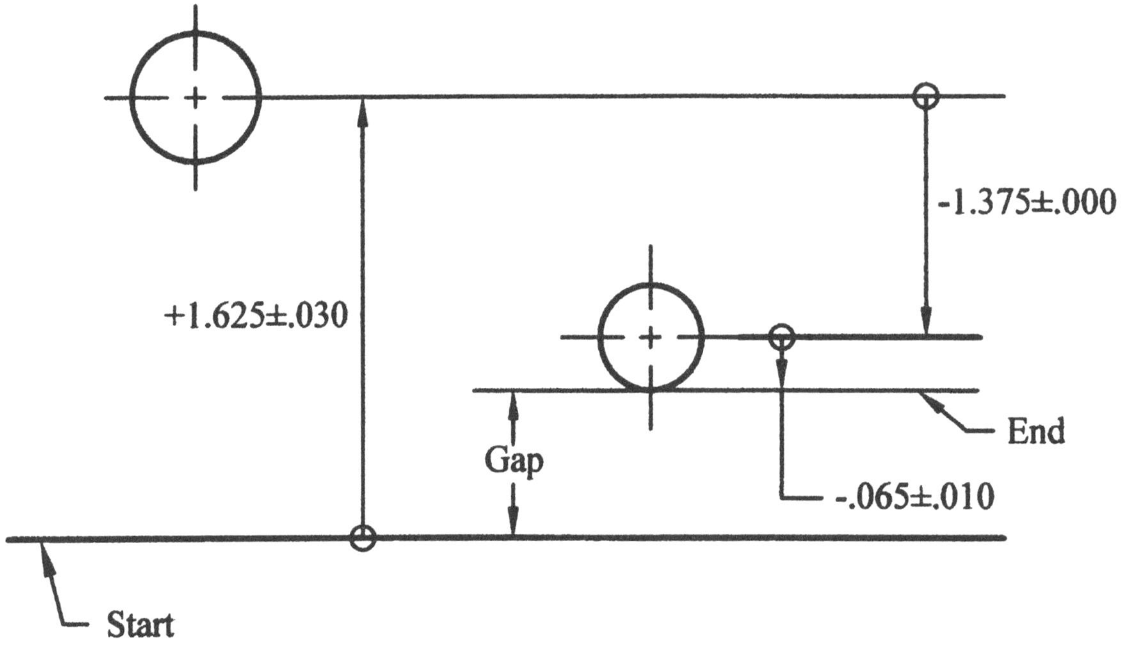

-	+	±Tol	
	1.625	.030	
1.375		.000	
.065		.010	
1.440	1.625	.040	Totals

$$\begin{array}{r} 1.625 \\ -\ 1.440 \\ \hline .185 \end{array} \qquad \begin{array}{r} .185 \\ +\ .040 \\ \hline .225 \end{array} \text{=MAX WALL} \qquad \begin{array}{r} .185 \\ -\ .040 \\ \hline .145 \end{array} \text{=MIN WALL}$$

58.

Step 1: Outside Diameter

2.990 = MMC
\+ .004 = Geo Tol. at MMC
Ø2.994 = Outer Boundary Shaft

2.960 = LMC
\- .034 = Geo Tol. at LMC
2.926 = Inner Boundary Shaft

2.994 = Outer Boundary
\+ 2.926 = Inner Boundary
5.920 = Sum

2.994 = Outer Boundary
\- 2.926 = Inner Boundary
.068 = Difference

$\frac{5.920}{2}$ = 2.960 Mean

$\frac{.068}{2}$ = .034±Tol.

Divide by 2 to convert to radii:

$\frac{2.960}{2}$ = R1.480

$\frac{.034}{2}$ = ±.017

So,
R1.480±.017.

Step 2: Holes

.570 = MMC
\- .020 = Geo Tol. at MMC
Ø.550 = Inner Boundary

.590 = LMC
\+ .040 = Geo Tol. at LMC
Ø.630 = Outer Boundary

.630 = Outer Boundary
\+ .550 = Inner Boundary
1.180 = Sum

.630 = Outer Boundary
\- .550 = Inner Boundary
.080 = Difference

$\frac{1.180}{2}$ = .590 Mean

$\frac{.080}{2}$ = .040±Tol.

Divide by 2 to convert to radii:

$\frac{Ø.590}{2}$ = R.295

$\frac{.040}{2}$ = ±.020

So,
R.295±.020.

58. (continued)

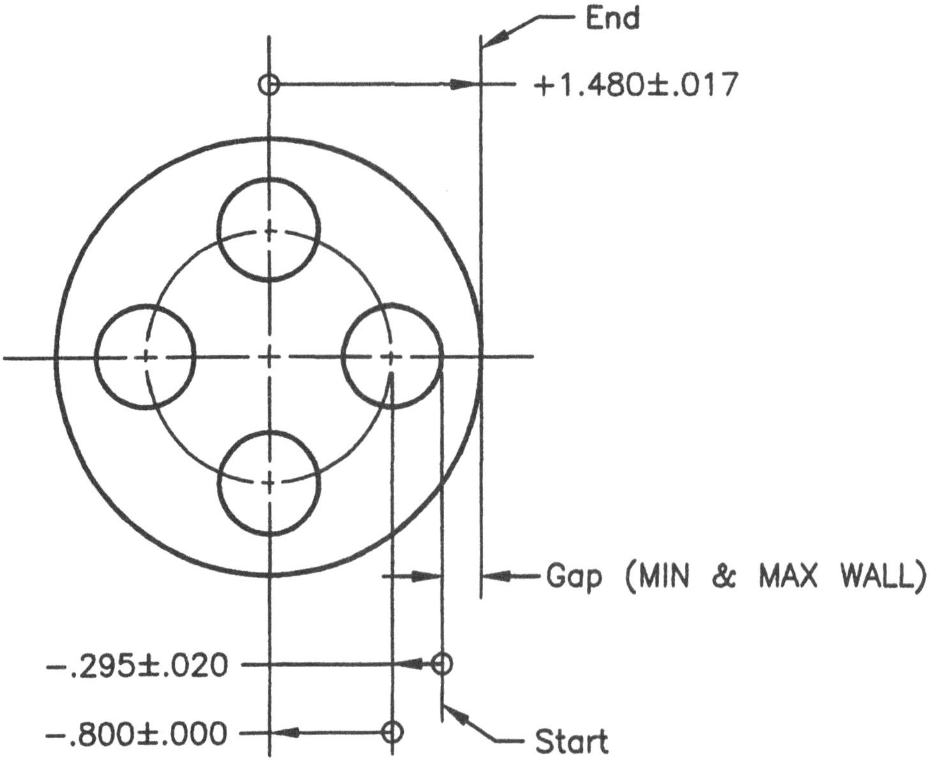

-	+	±Tol	
.295		.020	
.800		.000	
	1.480	.017	
-1.095	+1.480	.037	Totals

$$\begin{array}{r} 1.480 \\ -\ 1.095 \\ \hline .385 \end{array} \qquad \begin{array}{r} .385 \\ -\ .037 \\ \hline .348 \end{array} = \text{MIN WALL} \qquad \begin{array}{r} .385 \\ +\ .037 \\ \hline .422 \end{array} = \text{MAX WALL}$$

59.

Step 1:

.260 = LMC Hole
+ .050 = Geo. Tol. at LMC
Ø.310 = Outer Boundary Hole
+ .130 = ⊕Tol. of D at LMC
.440
+ .130 = Pattern Shift (D Ⓜ)
Ø.570 = Outer Boundary Hole with all Factors

.240 = MMC Hole
- .030 = Geo. Tol. at MMC
Ø.210 = Inner Boundary Hole
- .130 = ⊕Tol. of D at LMC
.080
- .130
Ø -.050 = Inner Boundary Hole with all Factors

+.570
+ -.050
.520

$\frac{.520}{2}$ = .260 = Mean Dimension

and

+.570
- -.050
.620

$\frac{.620}{2}$ = .310 = ± Tol.

So,

Ø.260±.310
R.130±.155

59. (continued)

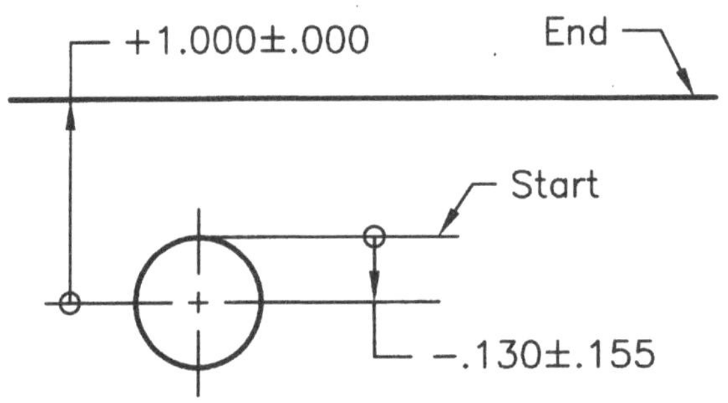

1.000	.870	.870
- .130	+ .155	- .155
.870	1.025 = MAX WALL	.715 MIN WALL to Datum Plane C

1/2 degree = .008 (over this part's maximum thickness)

6.100±.020 = 6.120 - 6.080 = .040 flatness tolerance for datum feature C.

.715 - .048 = .667 MIN WALL to lowest point on the surface of datum feature C.

60.

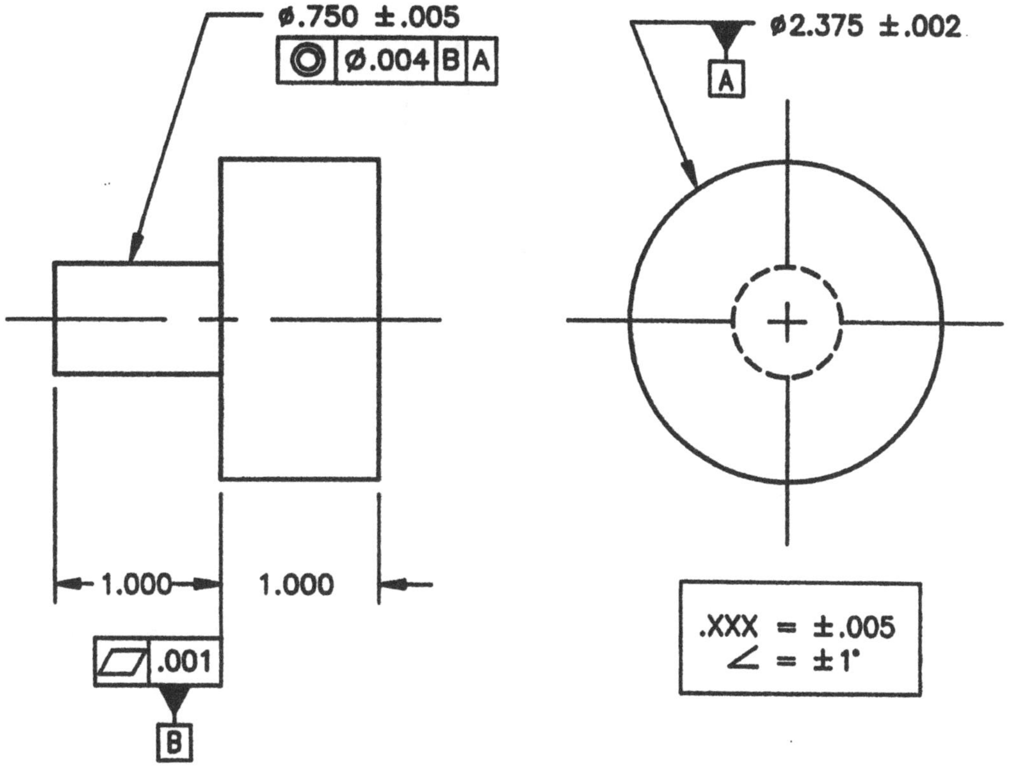

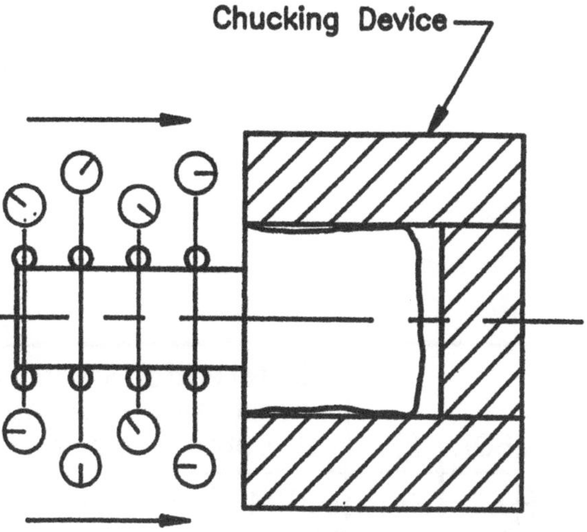

Indicator Readings:

Top
−.001, −.002, −.003, −.002

Bottom
−.001, +.002, −.001, +.005

−.001	−.002	−.003	−.002
− −.001	− +.002	− −.001	− +.005
.000	−.004	−.002	−.007 Bad

61.

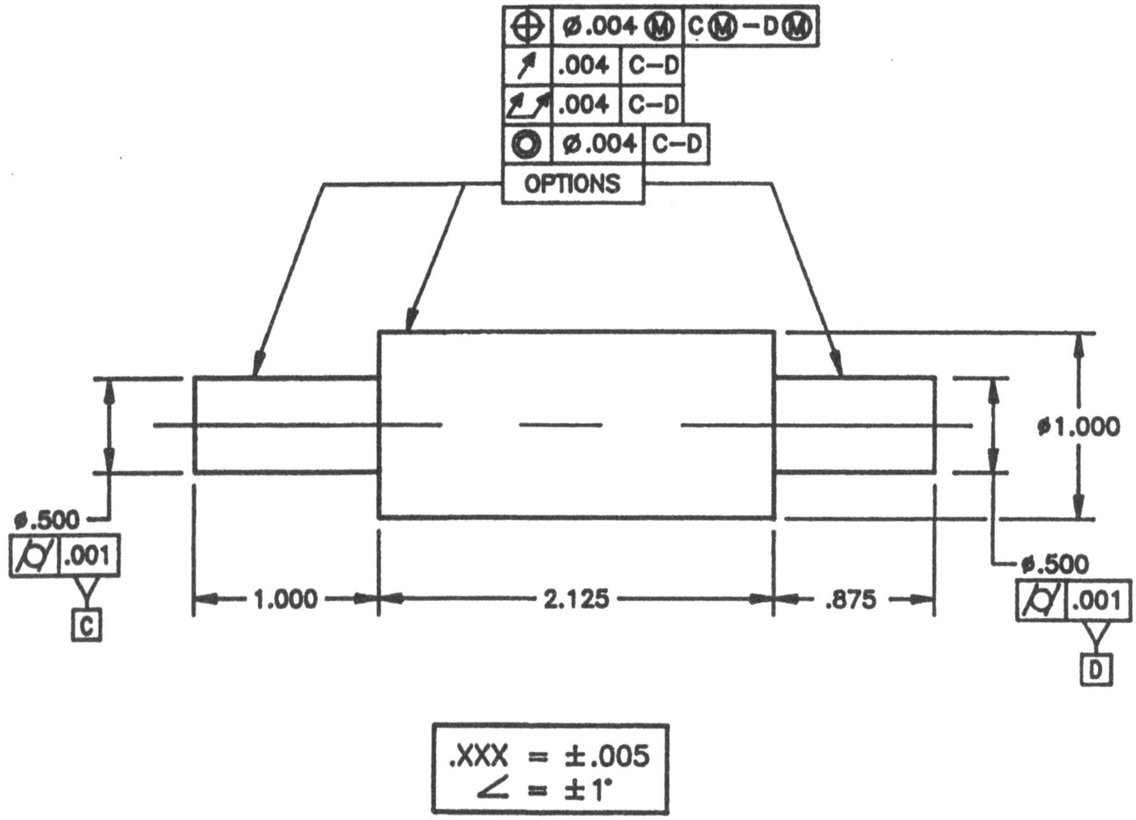

SYMBOL	FIM	Differential Measurements (worst set, part reads zero at some point on surface)	Bonus Tol.	Status
⌖	.004	−.004 (−) −.006 = +.002	.005	Good
↗	.003	N/A	N/A	Good
⌰	.005	N/A	N/A	Bad
◎	.005	−.003 (−) −.006 = +.003	N/A	Good

62. Part 767-23-Stud

1) Okay. Meets size requirement of .499 to .501.
2) Okay.
3) Not okay. Mating size violates MMC envelope. If MIC is .120 and local measurements are no greater than .128, then the axis snakes or bows (is out-of-straightness) by about ∅.008. ↗ to within ,0006 says that isn't the case; therefore, the information is incorrect on one of these three items.
4) Okay.
5) Okay. The axis is out-of-straightness ∅.004, which again conflicts with the FIM not exceeding .0006 (according to the runout measurement).
6) Okay. There is no flatness control on that surface. The dimension origin surface is not constrained by the .004 size tolerance.
7) Okay, but this is in conflict with much of the other measurement data as noted above.
8) The .004 pit is within the profile for conicity limits. Being off center a radius of .003 exceeds the allowed conicity tolerance. It is only allowed to be off center half of what appears in the feature control frame. Since .005 appears in the profile control, the most it may be off center is .0025. Conclusion: not okay.

63. No.

ⓐ 19.0 ÷ 2 = 9.5

ⓑ 130.0 − 9.5 = 120.5

ⓒ 120.50 − 0.03 = 120.47

ⓓ 120.50 + 0.03 = 120.53

ⓔ 120.53 − 120.47 = 0.06; 0.06 ÷ 2 = 0.03 radial dev. from datum centerplane A.

part is within tol. → 0.03 radial dev. from datum centerplane A.

64. No.

ⓐ 19.0 ÷ 2 = 9.5

ⓑ 130.0 + 9.5 = 139.5

ⓒ 139.50 + 0.04 = 139.54 allowed gage block height

(0.04 ← 1/2 tol. zone of 0.08)

ⓓ gage block is 139.54, therefore the part is within ⌖ tol.

65.

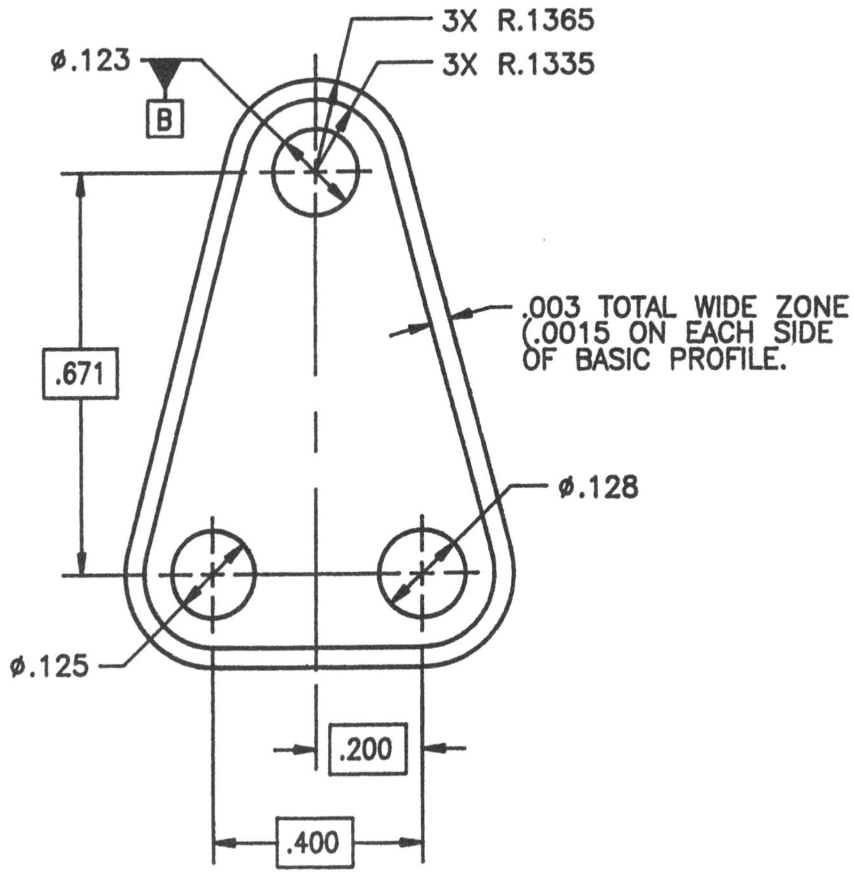

66.

1) .001
2) .002
3) .001
4) A diameter of .010. This is controlled by the size tolerance in the general tolerance note. Circular runout does not control taper.
5) The amount of bonus tolerance for threaded holes allowed by the MMC symbol is not quantifiable.
6) ∅.012 (∅.194 virtual condition – ∅.182 LMC)
7) .010
8) Three sets of two parallel planes .005 apart. These three tolerance zones are oriented at a basic angle of 120° to each other.

67.

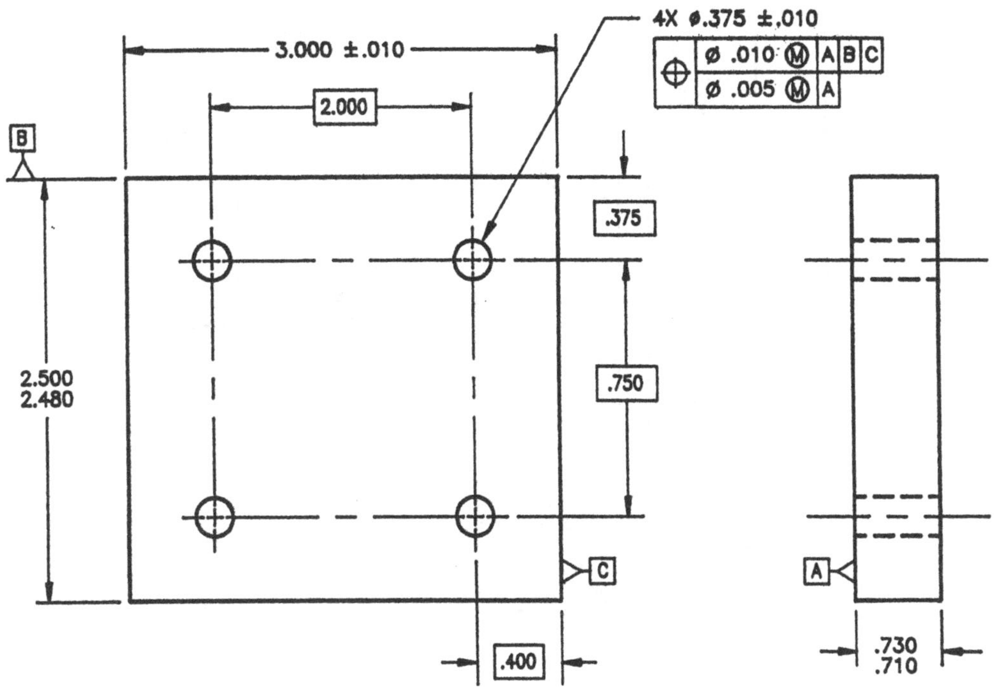

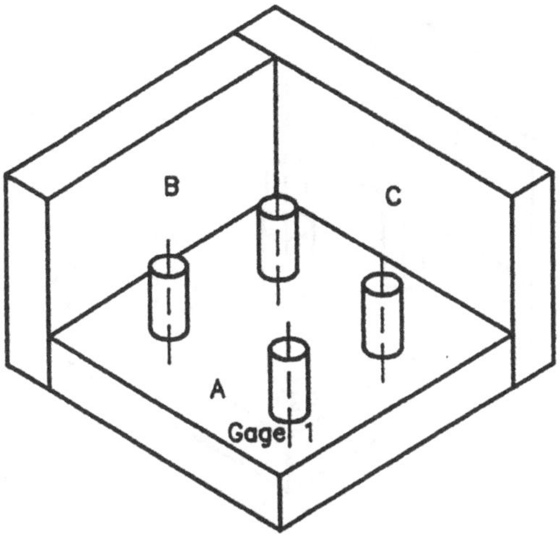

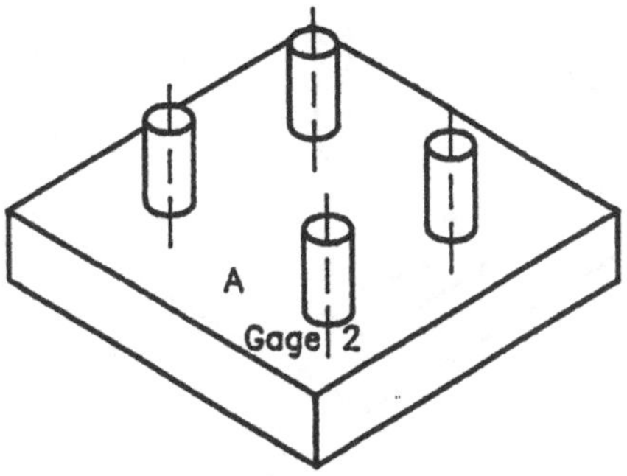

GAGE 1 – Datum to pattern
All four pins are located by the same basic dimensions (from the datum feature simulators) as used on the part drawing.

Pin sizes are ø.355 $^{+.XXX}_{-.000}$

with | ⌖ | Ø.000 Ⓜ | A | B | C |

GAGE 2 – Feature to feature
pin axis to pin axis basic dimensions are the same as gage 1 (and part drawing).

Pin sizes are ø.360 $^{+.XXX}_{-.000}$

with | ⌖ | Ø .xxx Ⓜ | A | | |
| | Ø.000 Ⓜ | A |

68.

GAGE 1	1a) **Yes**	1b) **Ø.870**
GAGE 2	2a) **Yes**	2b) **Ø.900**
GAGE 3	3a) **Ø.900**	3b) **Ø.500**

69.

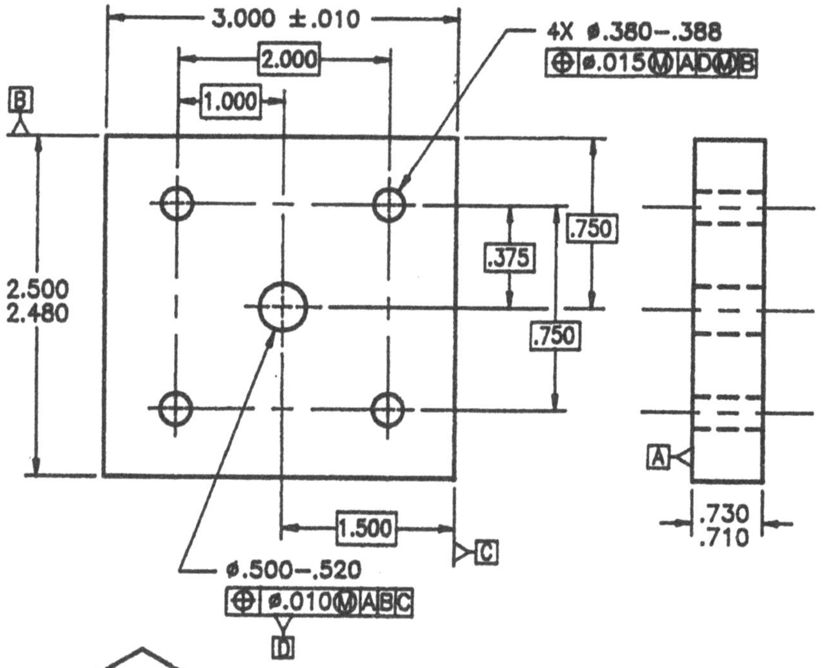

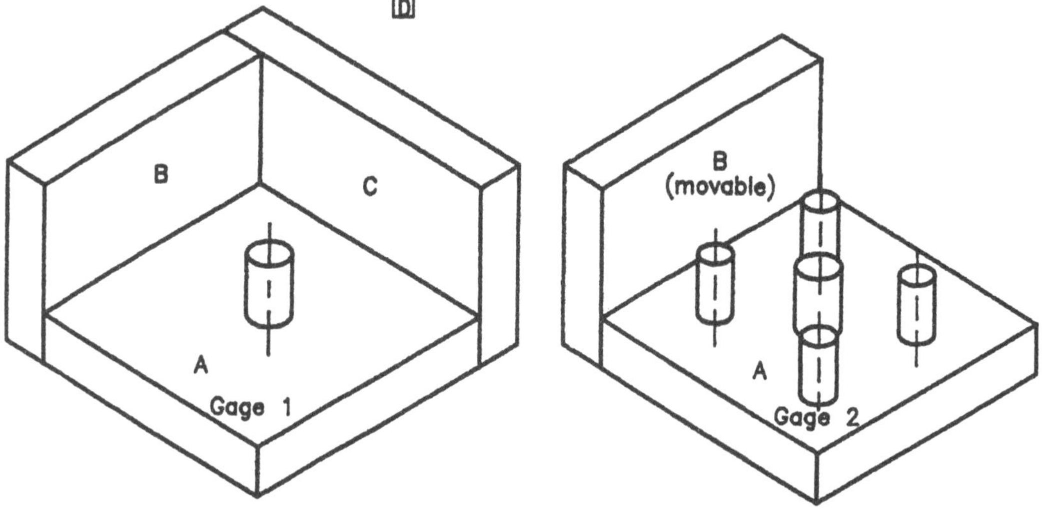

GAGE 1 – Gage inspects location of D. Pin axis is perpendicular to A, .750 basic from B and 1.500 from C.

Pin size is ø.490 +.XXX −.000

with | ⊕ | Ø.000 Ⓜ | A | B | C |

GAGE 2 – Gage inspects four hole pattern. Pin location is from axis of center pin. Each pin is perpendicular to A and .375 basic by 1.000 basic from center pin axis (representing D at its virtual condition of ø.490, see gage 1). B wall is for angular orientation (pattern rotation) only and therefore is movable in a parallel fashion. It requires a 2 high point min. contact from the part.

Pin size is 4X ø.365 +.XXX −.000

with | ⊕ | Ø.000 Ⓜ | A | D | B |

70.

(page 127)

1) .002
2) each other
3) all four – 2 datum holes and 2 Z holes (through SIM REQT)
4) B
5) perpendicularity
6) total of .010
7) X holes = ∅.281; Z holes = ∅.281
8) C
9) ∅.019; ∅.014; ∅.024
10) pattern of 4 holes; harder to find the central axis
11) optical comparator with dedicated comparator chart; ease and accuracy

(page 128)

1) No – it exceeds the envelope of MMC.
2) 1.963 ± .019

1.963	1.963
+ .019	- .019
1.982	1.944
Yes	

3) No
4) 1.123 ± .024

1.123	1.123
+ .024	- .024
1.147	1.099
	Yes

5) No – at ∅.300 they are allowed ∅.019.
6) Yes – it is out of tolerance (profile is all inside) unless datum shift could bring bump into tolerance zone.
7) Yes – they violate Rule #1 (must receive ∅.295).
8) .305 + .295 = .600 — .600 divided by 2 - .300

   .300 + 1.963 = 2.263 ± .019 (half of ∅.038) [2.282 OK]

(page 129)

1) .255 x 2.050 x 3.175
2) 2.010
3) 1.990
4) .500 + .025 (half profile tol.) + .005 (half profile tol.) = .530
5) .500 - .025 - .005 = .470

(page 130)

1) .255 x 2.050 x 3.175
2) 2.010
3) 1.990
4) .500 + .025 (half position tol.) + .005 (half ∅.010 bonus tol. from datum B made at LMC) = .530
5) .500 - .025 - .005 = .470

72.

1) Unless otherwise specified, limits of size are measured in the following manner:
   a) LMC at cross sections (opposing points) and MMC for violations of a full form envelope.

2) When basic dimensions are given to locate a hole,
   c) the location tolerance is found in a position feature control frame.

3) The correct formula for 'bonus tolerance' for a hole controlled with the MMC concept is
   d) actual size minus MMC.

4) Total runout can be inspected the best with
   c) a dial indicator.

5) Position deviation from 'true position' can be calculated using the following formula:
   a) two times the square root of the sum of the squares of the x and y deviations.

6) A coordinate measurement machine can most accurately simulate a primary datum by
   c) probing the surface plate on which the highest points of the datum feature rest.

7) The datum axis generated by an external cylindrical datum feature is the axis of the
   a) minimum circumscribed cylinder.

8) An optical comparator is most often associated with the inspection of
   b) profile.

9) Rocking datum features may
   d) be adjusted to an optimum position.

10) A cylindricity geometric characteristic symbol with a geometric tolerance of .003 allows
   d) size variations for the cylinder's diameter as large as .006.

73.

1) Which of the following would be inappropriate to include in a feature control frame for a straightness control?
   c) a datum reference

2) Runout is what type of control?
   a) surface to datum axis

3) Circular runout of a diameter does not control which of the following characteristics?
   c) straightness of the surface

4) Position is a geometric control that is not capable of controlling
   d) circularity

5) Of the following geometric characteristic symbols, which is the most restrictive in its control of surface characteristics?
   c) total runout

6) Of the following geometric characteristic symbols, which is considered a 2-D line element control?
   a) circularity

7) Of the following geometric characteristic symbols, which is least capable of controlling flatness?
   d) straightness

8) Limits of size on a rectangular part are not a control of which of the following?
   b) perpendicularity

9) Virtual condition (MMC concept) is calculated for a hole by the following formula:
   a) MMC of the hole minus the geometric tolerance applicable at MMC.

10) Positional tolerance for all holes in a floating fastener situation can be calculated in the following manner:
    b) MMC of the hole minus the MMC of the mating fastener.

11) What is the virtual condition of the following shaft?
    ∅8.3 – 8.5

⊕	∅0.3Ⓜ	A	B	C

    c) ∅8.8

74.

1) When a pattern of holes is positioned to a datum feature of size such as a hole and this datum feature is referenced at MMC, as the datum feature (hole) grows,
b) the pattern of holes as a group may shift.

2) In the concept of composite positional tolerancing,
a) the features within a pattern hold a tighter tolerance in relation to one another than to any datums that are listed only in the upper level control.

3) When a pattern of holes is used to establish a datum reference frame from which to locate other part features,
d) an infinite number of holes may be used to form the datum reference frame.

4) When multiple patterns of features are located from the same datum reference frame, they are considered to be
a) a simultaneous requirement.

5) If a pattern of holes is positioned and no datum features are used in their positional feature control frame,
b) the holes are considered positioned to one another.

6) Two hole patterns are positioned to the same primary planar datum feature, and this datum feature, due to imperfections of flatness, tends to rock. If the patterns of holes are considered as separate requirements,
a) the part may be rocked one way to accept one pattern of holes and another way to accept the other pattern of holes during inspection.

7) A four-hole pattern has size limits of ∅10.7 – 11.0 and a position tolerance to a datum reference frame of ∅0.3 at MMC. Its mating pin pattern has size limits of ∅9.7 – 10.0 and is positioned to a representative set of datums. How much position tolerance is available for the pins within the pin pattern at MMC?
d) ∅0.4

8) For the circumstances given in Question 7, the position tolerance that would be allowed for a hole produced at ∅11 is
c) ∅0.6

9) For the circumstances given in Question 7, what is the virtual condition (MMC concept) for the pins within the pin pattern?
d. ∅10.4

75.

Shafts
30.0 = LMC Shaft
\- 1.5 = Geo. Tol. at LMC
28.5 = Inner Boundary of Shaft

Holes
20.00 = LMC Hole
\+ 0.25 = Geo. Tol. at LMC
20.25 = Outer Boundary of Hole

28.50 = Inner Boundary of Shaft
\- 20.25 = Outer Boundary of Hole
8.25 = Material with Hole

8.25 ÷ 2 = 4.125 Minimum Wall Thickness

76. Minimum wall calculation answer:

LMC	4.990
-Geo. Tol. at LMC	- .040
Minimum Inner Shaft Boundary	4.950 = Inner Boundary of the O.D.

Because the 4 holes and the 4.990 - 5.000 O.D. are located to the same datums in the same order of precedence, they are considered one pattern, a Simultaneous Requirement; therefore, as D departs from MMC, they may shift--but all must shift in the same direction. So, if they all shift in the same direction, their relationship to one another does not change. Therefore, the thickness between the two is not further endangered.

4.950 = Inner Boundary of the Shaft

.375 = LMC
+.035 = Tol. at LMC
.410 = Outer Boundary of the (Hole)

1.600 = Bolt Square
x 1.4142 = Square Root of 2
2.26272 = Diagonal of Square (Bolt Circle)

4.950 = Shaft O.D. Inner Boundary
-2.263 = Diagonal of Square (Bolt Circle)

2.687 = Inner Boundary of the Shaft without the Bolt Circle
\- .410 = Outer Boundary of the Hole
2.277 = Total Material

2.277 divided by 2 = 1.1385 Minimum Wall Thickness

77.

4.990 = LMC of O. D.
- .040 = Geo. Tol. at LMC
4.950 = Resultant Condition of O.D.

.375 = LMC of Holes
+.035 = Geo. Tol. at LMC
.410 = Resultant Condition of Holes

1.600 = Bolt Hole Square
x 1.4142 = Square Root of 2
2.263 = Diagonal of Square (Diameter of Bolt Hole Circle)

4.950 = Resultant Condition of O.D.
- 2.263 = Diameter of Bolt Hole Circle
2.687 = Resultant Condition of Shaft w/o Bolt Hole Circle

2.687 = Resultant Condition of Shaft w/o Bolt Hole Circle
- .410 = Resultant Condition of Holes
2.277
- .400 = Movement of D at LMC Allowed to O. D.
1.877

1.877
- .400 = Displacement (Datum feature shift Allowed)
1.477 = Total Material from D at Its As-Produced Location

1.477 divided by 2 = .7385 minimum wall thickness of 4-hole pattern to O. D. of the part (datum feature B)

Logic of adding displacement to 4 holes as a pattern

1. Datum feature D may be out of position a diameter of .400 to the outside diameter of the part if it is produced at its LMC of 1.000.

2. As datum regular feature of size D departs from its MMB of .600 to its LMC of 1.000, the pattern controlled to it may shift with the datum feature axis from the datum axis a like amount (one for one). This additional displacement is allowed due to the MMB symbol next to datum feature D in the positional tolerance for the 4 hole pattern.

78.

PART 1

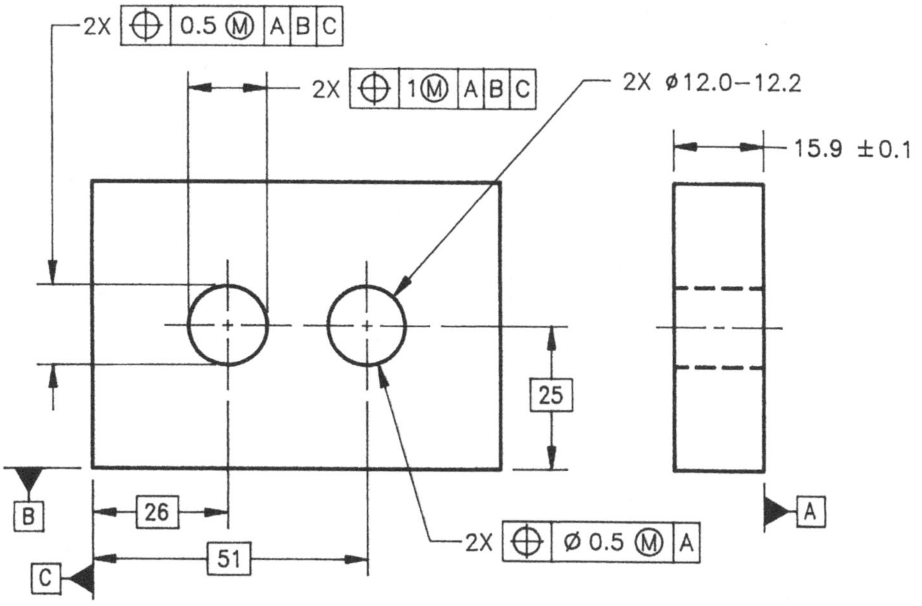

PART 2

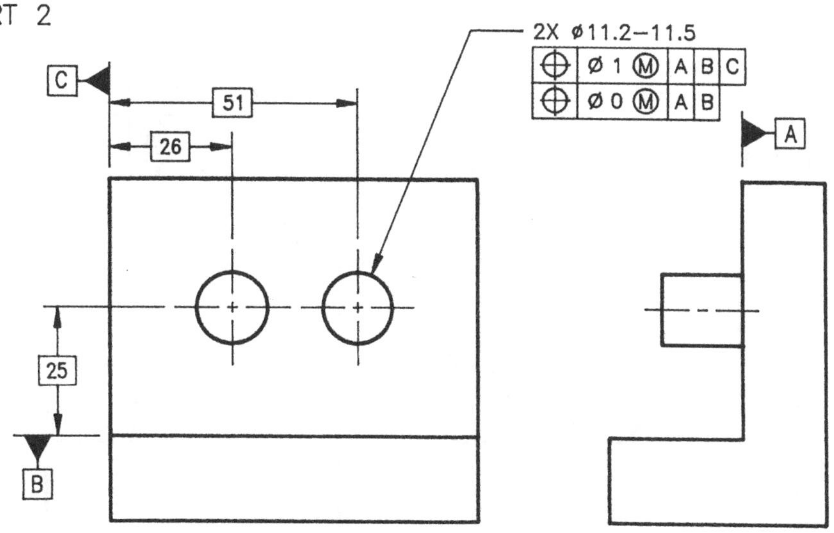

79.

PART 1

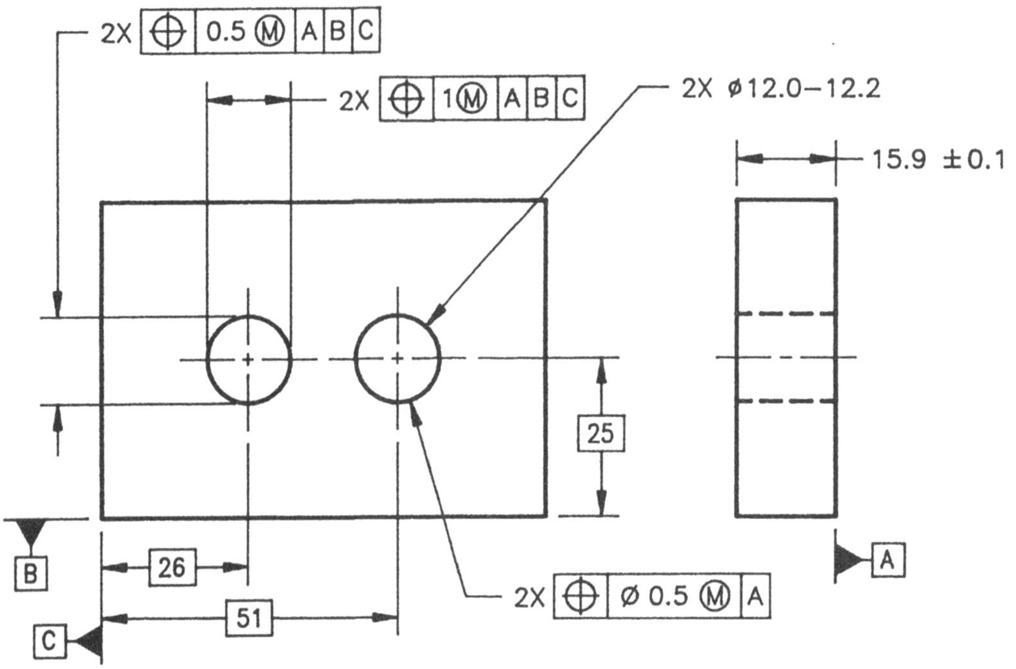

PART 2

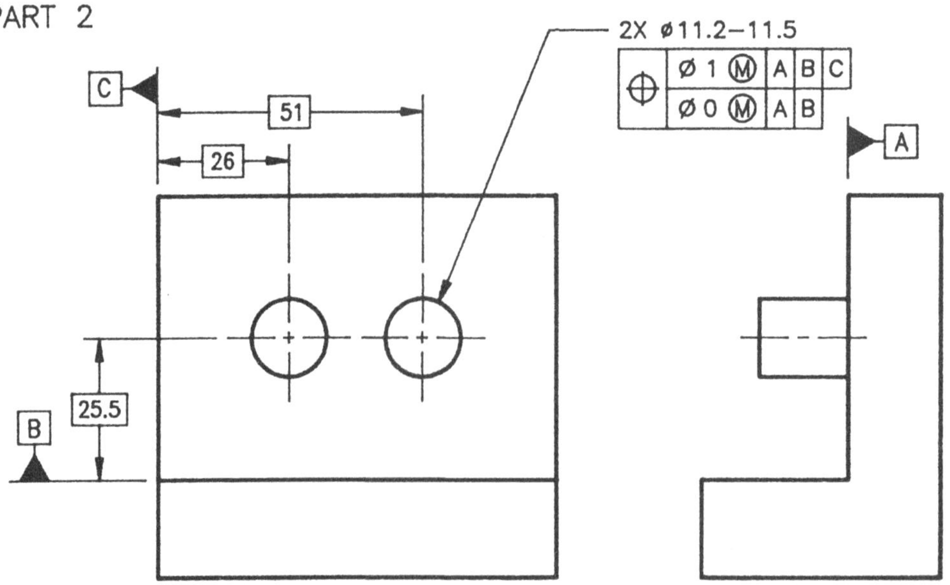

80.

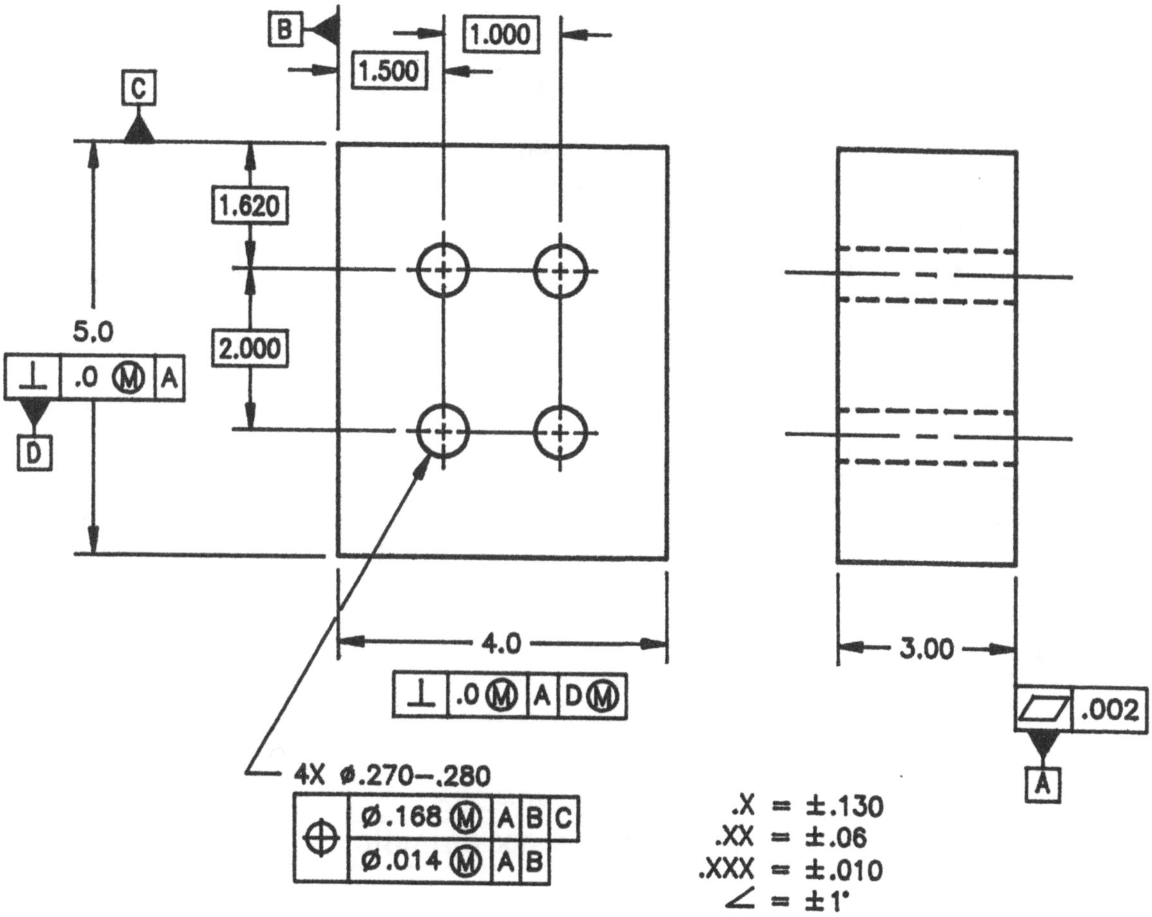

① ±.06 = .120

② .120 × 1.4 = ⌀.168

③ .270 − .256 = ⌀.014

81.

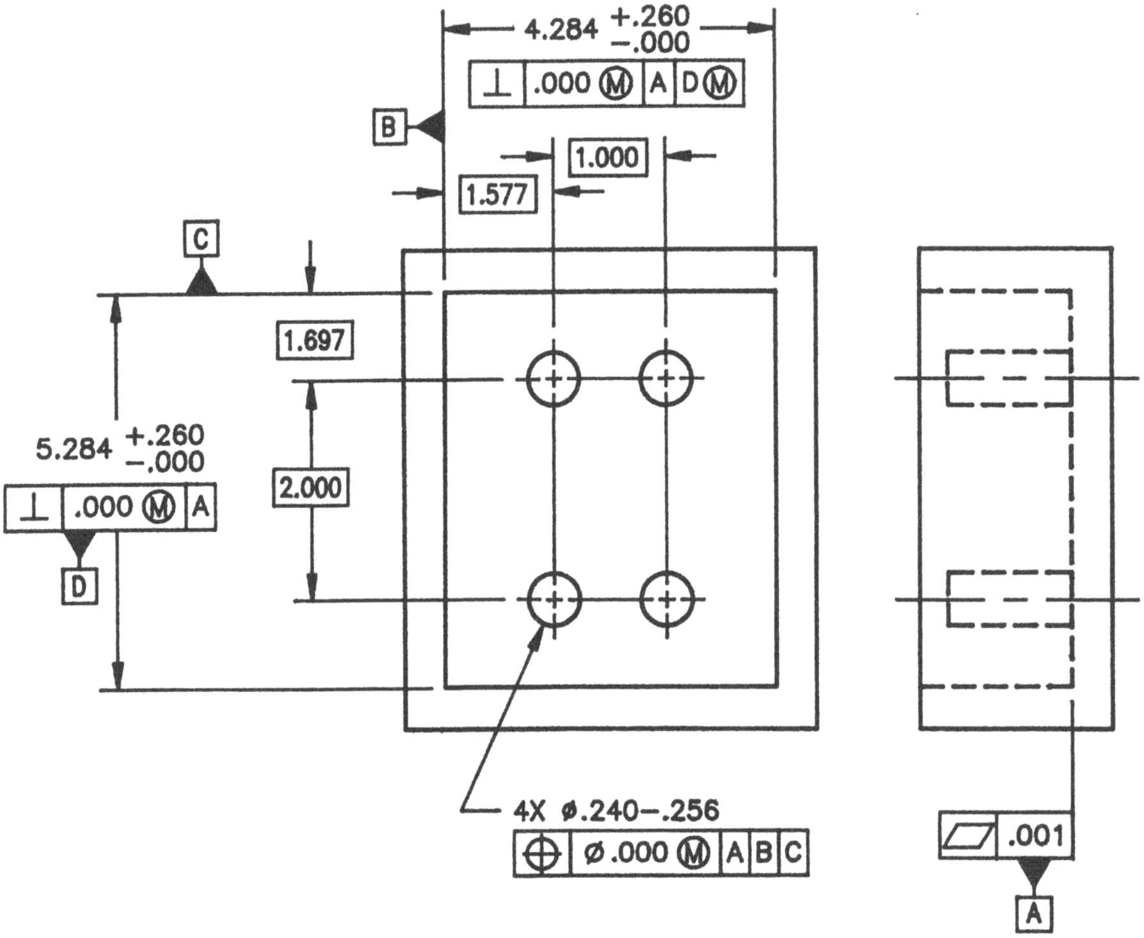

① .168 − .014 = .154

② .154 ÷ 2 = .077

③ 1.500 + .077 = **1.577**

④ 1.620 + .077 = **1.697**

⑤ 5.000 + .130 = 5.130

⑥ 5.130 + .154 = 5.284

⑦ 4.130 + .154 = 4.284

⑧ .270 = MMC
− .014 = Total MMC
.256 = Virtual Cond.

82.

a) .070=LMC of Nail Head
- .007=Position Tol. On Core Board
.063=Diameter of Area for Pogo Pin to Hit

b) .507=LMC of Guide Holes
-.494=LMC of Guide Pins
.013=Diameter of Clearance

c) .050=Position Tol. On Pogo Pins
+.013=Slide of Pogo Pins on Areas to Contact due to Clearance bet. Guide holes & pins
.063=Total Movement of Pogo Pins

d) No. .063
-.063
.000=Possible Contact

83. Terms and Definitions pertaining to Datums and Datum Features

1) Datums are
   b. simulated
   c. perfect
   e. derived from datum features
2) Features are
   a. real
   b. surfaces
   e. sometimes measurable for size
3) Datum features are
   b. real
   d. sometimes size features
   e. surfaces
4) Surfaces are
   a. features
   c. real
   e. flawed
   f. sometimes chosen to serve as datum features
5) Datum features should be chosen based on
   a. function
   c. repeatability
   d. accessibility
   e. representation of mating features
6) List five common physical datum feature simulators.
   surface plates, angle plates, chucking devices, V-Blocks and parallels
7) A dimension with a box around it like [100], is called?
   b. basic dimension
8) What is the [A] symbol called?
   a. datum feature symbol
9) What kind of tolerance is expressed by 50 ± 1?
   c. bilateral plus and minus tolerancing
10) What is the Ⓜ called when used after the geometric tolerance?
   d. maximum material condition symbol
11) What is third datum feature in the feature control frame called?
   b. tertiary
12) What is the Ⓜ called when used after a datum feature of size?
   d. maximum material boundary symbol
13) If a Ø2 appears in a feature control frame, it is
   d. the geometric tolerance
14) List at least two components (other than the perpendicularity symbol) that you feel are needed in a perpendicularity control. **a geometric tolerance and a datum reference**
15) A clearance fit implies the mating features, when produced within size limits,
   c. will never encounter material interference.
16) An interference fit implies the mating features, when produced within size limits,
   a. will always encounter material interference.
17) In a fixed fastener assembly condition, the geometric tolerance to be divided between the two parts containing the mating features is equal to
   b. the difference between the Maximum Material Conditions of mating features.

84. MMC and LMC

Please identify the Ⓜ and Ⓛ of these.

HOLES:

1)	62.5 ± 0.1	Ⓜ	**62.4**
		Ⓛ	**62.6**
2)	59-62	Ⓜ	**59**
		Ⓛ	**62**
3)	83.5 ± 0.05	Ⓜ	**83.45**
		Ⓛ	**83.55**
4)	69.5-70.5	Ⓜ	**69.5**
		Ⓛ	**70.5**
5)	1.245-1.265	Ⓜ	**1.245**
		Ⓛ	**1.265**

SHAFTS:

6)	7.5-8.1	Ⓜ	**8.1**
		Ⓛ	**7.5**
7)	71 ± 0.2	Ⓜ	**71.2**
		Ⓛ	**70.8**
8)	12.45-12.65	Ⓜ	**12.65**
		Ⓛ	**12.45**
9)	21.885-22.875	Ⓜ	**22.875**
		Ⓛ	**21.885**
10)	0.85-0.90	Ⓜ	**0.90**
		Ⓛ	**0.85**

85. Size Controls Form

1) Unless otherwise specified, individual features of size must
c. not violate an envelope of perfect form at MMC.

2) The concept discussed in Question 1 deals with the
b. Taylor Principle.

3) Least material condition on most parts is checked
a. at cross sections.

4) When checking for maximum material condition violations in a hole with size limits of Ø50 - 51, what size and type of gage would you use?
a. a full form plug gage with a diameter of 50

5) What effect on size measurements would adding the note "Perfect form at MMC not required" have? **It would change the envelope/GO gage requirement at MMC to a cross-sectional check requirement.**

6) Ø25 ± 0.1 AVG. This is a hole in metal. It is the I.D. of a thin walled tube.
What is the AVG. MMC? **Ø24.9** What is the AVG. LMC? **Ø25.1**

7) Ø.500 ± .005 This is a flexible rubber shaft.
What is the MMC? **Ø.505** What is the LMC? **Ø.495**

8) .625 - .635 This is a slot milled into a thick rigid plate.
What is the MMC? **.625** What is the LMC? **.635**

9) .010 ± .002 This is a steel tab that is non-rigid.
What is the MMC? **.012** What is the LMC? **.008**

10) Ø100 ± 1
| ⊕ | Ø0 Ⓛ | A | B | C |
This control refers to a glass shaft outside diameter. A hole runs through its center making it look like a drinking straw. It is mounted on a plate.
What is the MMC? **Ø101** What is the LMC? **Ø99**

How does the use of the LMC symbol change the way the size is measured? **An envelope of perfect form at LMC must be verified. No portion of the shaft may be inside of a perfect cylinder with a Ø99. The MMC of Ø101 is verified at cross-sections, instead of a GO gage.**

11) Ø37.5 ± 0.2
| — | Ø0.5 Ⓜ |
This is a metal rod (shaft) that is twenty-five feet long.
What is the MMC? **Ø37.7** What is the LMC? **Ø37.3**

How does the use of straightness of the derived median line change the way size is measured? **All size measurements for both MMC and LMC are checked at cross-sections. The envelope of perfect form at MMC normally required and inspected with a GO gage is eliminated.**

12) 10.0 - 10.1 This forms two sides of a rectangular part. It is considered an external feature of size. It is rigid.
What is the MMC? **10.1** What is the LMC? **10**

86. Rules, Concepts, Characteristics and Untoleranced Dimensions

1) What purpose does referencing a datum in a feature control frame serve? **It stabilizes the part, eliminates various degrees of spatial freedom and controls either orientation or location or both orientation and location.**

2) Please categorize all the geometric characteristics as:
INDIVIDUAL (not related to datums); RELATED (related to datums); INDIVIDUAL or RELATED:
Not related to datums: Straightness, Flatness, Circularity and Cylindricity.
Related to datums: Parallelism, Perpendicularity, Angularity, Circular Runout and Total Runout, Concentricity and Symmetry.
Either related or not related to datums: Profile of a Line, Profile of a Surface and Position.

3) When we check an individual feature for form, what are we checking to?
b. a perfect geometric form of itself

4) Once on a drawing, how can we tell if a feature is individual or related? **If related, a note or feature control frame will reference the relationship between the feature being constrained and the datum or datums to which it is related.**

5) Name three ways to eliminate the requirement for an envelope of perfect form at MMC for a feature of size. **1. With a note such as "PERFECT FORM AT MMC IS NOT REQUIRED." 2. Use of an average dimension such as 1299-1210 AVG. 3. Use of a least material condition symbol inside of the feature control frame for that feature. 4. Use of a straightness of the derived median line or derived median plane may be specified.**

6) If you do one of the three things stated in Question 5, can you exceed MMC at any cross sectional check? **Yes, with the use of AVG dimensioning, the size of the produced feature must only average within the size limits. A minimum of four measurements is taken at each cross section of the feature and then averaged.**

7) If you assign the LMC concept and symbology to the geometric tolerance in the feature control frame, where is the envelope of perfect form size requirement?
c. at LMC

8) Which material condition symbol (modifier) provides the most tolerance to draw from when used? **MMC and LMC symbols**

9) Which parts are considered to be most economical to manufacture?
a. those with features controlled at MMC

10) What material condition symbol provides for the easiest use of functional (receiver-type) gaging? **the MMC symbol**

86. (continued)

11) Is it ever possible to use the Ⓜ symbol both for a feature of size and a datum feature of size in one feature control frame?

    For example: | ⌖ | ⌀0.5Ⓜ | A | BⓂ | C | **Yes, this is quite common.**

12) When features are controlled RFS, they are given an original tolerance of form, orientation, profile, runout or location. Where can we draw extra geometric tolerance from?

    c. we can't

13) As a minimum, the feature control frame must contain

    b. a geometric characteristic symbol and a geometric tolerance.

14) Draw the symbols for the following geometric characteristics and categorize them as form, orientation, profile, runout or position.

    a. Flatness ▱
    b. Straightness —
    c. Roundness ○
    d. Cylindricity ⌭
    e. Profile of a Line ⌒
    f. Profile of a Surface ⌓
    g. Parallelism //
    h. Perpendicularity ⊥
    i. Angularity ∠
    J. Feature Control Frame | ⌖ | ⌀0.5Ⓜ | A | BⓂ | C |
    k. Position ⌖
    l. Runout (Circular) ↗
    m. Concentricity ◎
    n. Runout (Total) ⌰

15) Identify the following symbols:

    a. Ⓜ = **maximum material condition**

    b. ↧ = **depth**

    c. ⌀ = **diameter**

    d. Ⓟ = **projected tolerance zone**

    e. ⟨ST⟩ = **statistical tolerance**

    f. Ⓛ = **least material condition**

    g. ⌴ = **counterbore or spotface**

87. Datums

1) What geometrically perfect elements are representative of datums? **points, axis and planes**
2) How are datum features simulated in manufacturing or inspection? **with datum feature simulators, such as surface plates, angle plates and chucks**
3) What criteria are used to choose a datum feature when tolerancing? **Functionality, features that are representative of seating, mating and alignment features, their accessibility and repeatability.**
4) What is a datum reference frame and what is it used for? **Three mutually perpendicular planes. A coordinate system is used to orient to and from which to measure angle and distance/location.**
5) What six degrees of spatial freedom does a part have before it is restrained? **A part must have three linear degrees of freedom (x, y and z) and three rotational degrees of freedom (u, v and w).**
6) **a minimum of three; a minimum of two; a minimum of one**
7) Are three datums always necessary for part orientation? If yes, why? If no, why not? **No. Some relationships require less that full part stabilization. Coaxiality and symmetry are two simple examples of this.**
8) Is a datum feature perfect? **No. A datum feature is real and therefore flawed.**
9) Is a datum feature ever given form control of its own? If not, why is it not necessary? If so, why it is necessary? **Yes, it is necessary to make it more repeatable and better shaped for part stabilization.**
10) How is the order of precedence of datum features shown in a feature control frame? **Read from left to right with the primary first, etc. It has nothing to do with the order of the letters as they appear in the alphabet. It has only to do with the order in which they are referenced.**
11) Why does the order of precedence of datum features make such a difference in feature orientation? **The primary dictates the angle of the secondary. The primary and secondary dictate the angle of the tertiary.**

12a) **with a minimum circumscribed cylinder**
12b) **with a maximum inscribed cylinder**

13) Of what use is an angular orientation datum in a datum reference frame that includes a cylindrical datum feature? **It orients the datums that have preceded it. It can be used to stop pattern rotation and keep the desired angular relationship between the pattern and the datums.**
14) How many datum planes are derived from a cylindrical datum feature? **two**
15) Why are datum targets necessary? **They are necessary to stabilize parts that are rough or warped or irregularly shaped or to duplicate partial areas of mating part contact.**
16) How are datum features controlled RMB simulated as datums? **by contacting the surface to establish an actual mating envelope axis or centerplane**

87. (continued)

17) At what size or condition are secondary datum features of size that have their own tolerance of orientation or position simulated as datums? **At their applicable virtual condition, if they are referenced at MMC: at their actual properly oriented mating size, if referenced at RFS.**

18) At what size is a primary datum feature of size represented for the features controlled to it? **At its MMC-if referenced at MMC (unless a straightness control generates a virtual condition boundary-MMC concept); at actual mating size if referenced RFS.**

19) What is a compound datum feature and what is it used for? **Compound datum features (or multiple datum features) is when more than one diameter or surface is used to establish a single axis or plane.**

20) When a pattern of holes is controlled to a datum feature hole, as the datum feature hole departs from MMB within its limits, what happens to the controlled hole pattern and how is it calculated? **The pattern of holes is allowed an apparent shift from the datum feature axis. Usually, this shift is one for one. Actually, the datum feature axis (actual feature axis) is allowed to shift away from the datum axis (imaginary datum axis). This appears as if the pattern of holes has shifted away from the datum feature axis.**

21) Do screw threads make good datum features? **Generally, they should be avoided because they generate a datum axis from the pitch diameter (also know as a pitch cylinder) that is hard to simulate in inspection.**

22) If more than one datum reference frame is specified for a part, what effect will this have on manufacturing and inspection? **It implies a different set-up for each datum reference frame.**

23) Why would a designer choose to use only a portion of a surface as a datum feature instead of the entire surface? **Perhaps only a portion of the surface is functional or contacts the other parts in the assembly.**

24) In what forms are datum targets specified? **points, lines and areas**

25) If a designer wants to show a datum target as originating from the opposite side of a part than the view it has been shown in, how may it be done without another view? **with a dashed radial leader line that terminates at the datum target**

26) What kind of contact is required from the physical datum feature simulation equipment, in a datum target area, on a non-cylindrical surface? **Full flat contact with the entire area is desired. High point contact within the area is attempted.**

27) How is the shape and location of a datum target area defined? **either by basic or toleranced dimensions and sometimes through the use of chain lines and hatched lines**

28) What does the dimension origin symbol look like and of what use is it? **It is a circle used in place of an arrowhead on one end of a dimension line, and it designates that surface as generating an origin plane for measurement.**

88. Virtual Condition and Resultant Condition Boundaries

1) When a feature has errors of orientation, location and certain tolerances of form, its effective mating boundary is virtually altered at that orientation or at that location or by its form.
True

2) In virtual condition (MMC concept), the effective mating boundary of a hole is virtually increased and the shaft size virtually decreased, even though they still measure within their limits of size.
False. The hole is virtually decreased and the shaft is virtually increased.

3) A virtual condition boundary (MMC concept) of the feature represents the worst condition of assembly of the mating components.
True

4) If a feature is allowed a virtual condition, it may exceed its limits of size as an individual feature when measured.
False

5) How does calculating the virtual condition of an external feature of size differ from calculating the virtual condition (MMC concept) of an internal feature of size?
For holes: MMC minus the geometric tolerance applicable equals the virtual condition.
For shafts: MMC plus the geometric tolerance applicable equals the virtual condition.

6) Functional gages are made to the virtual condition (MMC concept) of features they inspect.
True

7) Functional gages are usually only made to check features controlled with MMC. Otherwise, for checking features controlled RFS, a functional gage would have to be made to take into account every size the feature could be produced at.
True

89. Form Controls

FLATNESS

1) Describe the configuration of a flatness tolerance zone? **two parallel planes separated by the geometric tolerance**
2) Is flatness considered individual or related to datums? **individual**
3) Is a datum feature reference allowable in the feature control frame? **No**
4) May a diameter sign be included in the feature control frame? **No**
5) Is flatness a surface or an axis control? **a surface control**
6) Is flatness a two-dimensional or a three-dimensional control? **three-dimensional**
7) To obtain a high level of certainty (for example: 90% certainty) of how flat a surface is, when measuring flatness will a few points from the surface suffice, or does this require hundreds of points per square inch? **Approximately 191 points per square inch (between 80 and 357 points) for a 90% certainty of flatness according to a study done by Electronic Measuring Devices, Inc. They also said that for a 75% certainty an average of 102 points (between 48 and 227 points) per square inch was required. For their study (done in 1988) they scanned in 2500 evenly spaced points on 6 one-inch blocks to represent 100% certainty of flatness. Each block had a different deformity (convex, concave, gouged, etc.)**
8) If a surface is controlled with flatness, may the controlled surface in combination with its opposing side on the part violate an envelope of perfect form at MMC? **No**
9) Must the flatness tolerance refine the size tolerance or may it be larger? **It must refine (be smaller than) the size tolerance.** Why? **If the size tolerance was tighter than the flatness tolerance, then the flatness of the surface would be controlled better by the size tolerance (and the flatness tolerance would be meaningless).**
10) May a MMC symbol be included in the feature control frame if flatness is applied to a single surface? **No**
11) How is the flatness control directed to the controlled surface? **by a leader line to the surface or an extension line off of the surface**
12) May flatness be applied on a unit or a rate basis? **Yes**
13) May flatness be applied to control a derived median plane? **Yes**

STRAIGHTNESS

General:

1) Are datum feature references allowable/required in the feature control frame? **No, they are not allowable.**
2) What view should the straightness tolerance be applied in? **For a surface control, it is shown in a view that the tolerance zones run parallel to. Straightness of a surface applies only in the plane of the view in which it is shown.**

89. (continued)

Surface Straightness:

3) In straightness of a surface, what is the shape of the tolerance zone, and what must lie within it if the feature is to be accepted? **Two parallel lines that are separated by the geometric tolerance. A separate tolerance zone applies to each line element. Each tolerance zone may be on a different height or on a different angle to the other tolerance zones. Each point on the line under test must reside within its tolerance zone.**

4) What material condition symbols may be included in the feature control frame? **none for surface straightness**

5) May a diameter sign appear in the feature control frame? **not for surface straightness**

6) Without a straightness feature control frame, what controls the straightness of a part surface? **The limits of size. The part must have perfect form if produced at MMC (if the feature is rigid and no other notes or controls have eliminated the requirement that size tolerance controls form).**

7) Is perfect form of the controlled surface and its opposing surface required at MMC? **Yes**

8) Is bonus tolerance ever possible? **not for surface straightness** If not, why not? **No surface control is allowed to use a maximum material condition symbol after the geometric tolerance.**

Straightness of the Derived Median Line:

9) In straightness of a derived median line, what is the tolerance zone shape and what must be contained within it for the feature to be acceptable? **The tolerance zone is cylindrical in shape. The derived median line must reside within this tolerance zone. The derived median line is an imperfect line (abstract) that passes through the center points of all cross sections of the feature. These cross sections are normal to the axis of the actual mating envelope.**

10) Is perfect form of the controlled feature required at MMC? **No. Perfect form at MMC is no longer required when this control is used.**

11) Is this control considered individual or related to datums? **individual**

12) How might straightness of a derived median line that is controlled with a MMC symbol be gaged? **with a functional gage made at the virtual condition of the feature under test**

13) How might straightness of a derived median line that is controlled with the RFS concept be gaged? **with differential measurements to determine the derived median line as it is described in Question 9**

14) Is a virtual condition allowed for both controls described in Questions 12 and 13? **No. A virtual condition is defined as a constant boundary. A constant boundary is created that can be gaged only for features controlled at MMC, not at RFS.**

15) Must the straightness tolerance be smaller than the feature's size tolerance? **No. For straightness of the derived median line, the geometric tolerance may be smaller or larger than the size tolerance.**

89. (continued)

CIRCULARITY

1) Is circularity a two- or a three-dimensional control? **two-dimensional**
2) What is the tolerance zone shape? **The tolerance zone shape is two concentric circles that are the geometric tolerance apart radially. Each circular cross-section has its own tolerance zone. Each tolerance zone may have a different optimal center offset from the others. Although the zones are the same distance apart (the size of the circularity tolerance), the diameters of the zones are determined optimally by the circles at each cross-section.**
3) What material condition symbols are allowed in the feature control frame? **none**
4) Are datum feature references allowable? **No**
5) Is a feature controlled with circularity allowed to exceed its size limits? **No**
6) Is any extra tolerance derived from the controlled feature's size limits as the feature departs from MMC? **No**
7) How is circularity checked? **The full indicator movement in a 360° revolution for each circular cross-section tested may not exceed the circularity tolerance in the feature control frame. The inspector is responsible for every circular cross-section of the surface being controlled.**
8) Does the tolerance in the feature control frame have to be smaller than the feature's size limits? **Yes**
9) Is circularity considered individual or related to datums? **individual**
10) Can circularity controls be assigned to cones? **Yes**
11) May a diameter sign be included in the circularity feature control frame? **No**
12) How is the feature control frame connected to the controlled surface? **with a leader line to the surface**
13) Circularity controls the surface **b. at every cross section.**
14) Without a circularity feature control frame, what would normally control a feature's roundness? **the size tolerance**
15) A circularity control is assumed to apply **b. RFS.**

CYLINDRICITY

1) Name at least two differences between circularity and cylindricity. **Cylindricity controls straightness and taper. Circularity does not.**
2) List three different ways cylindricity can be checked. Order them with the best way first and the least desirable method last.
   1. **Mounted in a roundness checking machine while points are collected and graphed while the part rotates and the electronic probe moves across (or up, depending on the orientation of the machine) the surface.**
   2. **Mounted between centers and a full indicator movement taken while the part rotates quickly and the dial indicator probe moves slowly down the surface**
   3. **With a coordinate measuring machine with a constant contact or non-contact scanning probe.**

89. (continued)

3) What is the tolerance zone configuration a cylindricity control generates? **two concentric cylinders separated radially by the amount specified in the feature control frame**

4) Is cylindricity a surface or an axis control? **a three-dimensional surface control**

5) May a least material condition symbol be used in the feature control frame? **No**

6) Can cylindricity and circularity be applied to the same surface? If so, which tolerance must be smaller and why? **Yes they can, if the circularity tolerance is smaller because cylindricity controls circularity, straightness and taper.**

7) Does a cylindricity tolerance ever reference datum features inside the feature control frame? **No**

8) Must the controlled feature, if produced at MMC, have perfect form? **Yes**

9) Is the cylindricity tolerance additive to the controlled feature's MMC size limit? **No**

10) Is bonus tolerance ever allowed? **No**

11) A cylindricity tolerance is considered **a. form.**

12) A produced feature's cylindricity is compared to a perfect geometric counterpart of
c. itself.

13) List all items allowed in the cylindricity feature control frame. **a geometric characteristic symbol (cylindricity) and a geometric tolerance**

14) A cylindricity tolerance controls roundness and
c. straightness and taper.

15) The tolerance applies simultaneously to circular and ____ elements of the controlled surface.
a. longitudinal

90. Orientation Controls

1) What is the configuration of the tolerance zone defined by a parallelism of a planar surface control? **The configuration is two parallel planes that are parallel to the datum. The two parallel planes are separated by the geometric (parallelism) tolerance.**

2) What does that tolerance zone take its orientation from? **the datum**

3) If we wanted an orientation tolerance zone | ∠ | 0.5 | for a planar surface to be exactly 35° from datum plane R as well as perfectly perpendicular to datum plane S, what could we do to the given feature control frame to reflect that? **| ∠ | 0.5 | R | S | and make the 35° a basic angle.**

4) If a part has a maximum height of 1.000. Is a .002 parallelism control additive to that height? **No**

5) Is at least one datum feature reference always required in the feature control frame for angularity? **Yes**

6) How does the tolerance configuration for the angularity control in Question 3 differ from the tolerance zone that would be generated if the 35° was ± 2° ? **The zone for Question 3 consists of two parallel planes that are 0.5 apart and oriented 35° from datum R and 90° from datum S. This tolerance zone of ± 2° consists of a wedge-shaped zone originating at a vertex allowing the surface to vary a total of 4° .**

7) What material condition symbols would be appropriate in feature control frames for orientation controls if they were used on a single planar surface? **None would be appropriate after the orientation geometric tolerance, but either Ⓜ or Ⓛ would be after any datum features of size that are referenced.**

8) What is usual configuration and orientation of a controlled feature's perpendicularity tolerance zone if it is used on a planar surface? **Two parallel planes separated by the perpendicularity tolerance. Both planes are 90° to the datum or datums referenced.**

9) What does the use of a datum feature reference add to the orientation geometric control? **It shows to what the desired relationship is to be held.**

10) ∅.250±.005 Hole
| ∠ | ∅.005Ⓜ | A |

If the controlled hole was produced at a diameter of .250, what angularity tolerance would apply? **∅.010**

11) Does the controlled hole in Question 10 have a virtual condition? If so, what is it? **The controlled hole in Question 10 has a virtual condition of ∅.240.**

12) Would it be appropriate to include a maximum material boundary symbol in the feature control frame for datums that are planar surfaces? Please explain. **No. Planar surfaces don't have a maximum material condition. Only features of size have a maximum material condition.**

13. How many contact points are required to establish datum A from a planar surface if A is primary? **3 high points of contact are required.**

90. (continued)

14) Is an orientation geometric control related to datums or individual? **always related to datums**

15) For parallelism used on a planar surface to a datum plane generated by an opposing planar surface, must the tolerance in the feature control frame be smaller than the size tolerance of the part's opposing sides that are included in the control? **Yes. If the size tolerance was smaller, then it would control parallelism better than the parallelism control would and the parallelism control would be meaningless.**

16) ∅25±0.5 Hole

⊥	∅0.5Ⓜ	AⓂ	B

If the controlled hole departs from MMC to LMC, how much **extra** perpendicularity tolerance can be drawn from the size? **1 millimeter**

17) List the MMC and LMC of the datum feature in Question 16. **MMC=24.5 and LMC=25.5**

18) If the datum feature A referenced in question 16 is a hole and has a size tolerance of ∅50±0.1, as it departs from MMB to LMC, how much additional shift of the controlled hole's tolerance zone is gained? **∅0.2**

19) If both the controlled hole and datum feature A were made at LMC, how much could their axes be out of perpendicularity to each other? **1 + 0.5 + 0.2 = ∅1.7**

20) What is the shape of the tolerance zone generated by this geometric control? **cylindrical tolerance zone that is oriented to datum axis A and datum B that must contain the axis of the hole**

91. Profile

1) In a profile of a curved surface control, the profile must be dimensioned with **d. basic dimensions.**

2) What is the difference between profile of a line and profile of a surface? **Profile of a line is a two-dimensional line element control. Profile of a surface is a three-dimensional surface control.**

3) For profile controls, datum feature references are **a. allowable but not required.**

4) If a datum feature reference was used for a profile control, what purpose would be served? **to orient or locate (or both locate and orient) the tolerance zone**

5) Are material condition symbols allowed for the feature being controlled? Why or why not? **No. Profile is not allowed to use MMC or LMC symbols after the geometric tolerances. However, MMC or LMC symbols may be used after any datum features of size that are referenced in the control.**

6) Is profile of a surface more often individual or related to datums? **related**

7) Is profile of a line more often individual or related to datums? **individual**

8) If both profile of a surface and profile of a line are used on the same feature, which must have the smaller profile tolerance? **Profile of a line must be smaller.**

9) How is profile checked? **Profile is checked in many different ways. Coordinate measuring machines, optical comparators, video comparators, templates with dial indicators, etc.**

10) What is the tolerance zone shape of profile of a line? **two uniform lines separated by the geometric tolerance**

11) What is the tolerance zone shape of profile of a surface? **two uniform boundaries separated by the geometric tolerance**

12) How is profile tolerance designated as bilateral or unilateral? **If phantom lines are shown, they designate a direction of tolerance. ASME Y14.5-2009 and ASME Y14.5-2018 prefer that a circled U symbol for unilateral or unequal tolerance zones be added to the profile feature control frame.**

13) How is the extent of where the profile control applies on a surface designated? **with notes, between symbols, all-around symbols or the ALL OVER note**

14) Suppose the entire outer periphery of the part profile must be controlled. How may it be done? **It may be done wither with an all-around symbol or with an "ALL OVER" note. All-around means that the control is view dependant. ALL OVER means that the control applies in all views.**

92.

1) a
2) b
3) d
4) a
5) For cylindrical surfaces: roundness, straightness, taper, coaxiality, angularity and surface profile.
   For flat surfaces: flatness and perpendicularity to the datum axis.
6) Roundness and coaxiality for circular features.
7) When balance is important but 3-D form is not as important.
8) d
9) Related
10) Required
11) No
12) No
13) Yes
14) No
15) Circular runout must be smaller

93.

1) c
2) a
3) No
4) Yes, a diameter sign.
5) A diameter. Generally, it is a cylindrical tolerance zone that must contain the median points.
6) Related
7) c
8) a
9) When balance for a rotating feature is important, surface form is not important and the designer is willing to allow inspection to spend the time necessary to take differential measurements.
10) No
11) No
12) b
13) a
14) No
15) No
16) Two parallel planes
17) Yes
18) e
19) b

93. (continued)

20) Yes
21) Datum features, basic dimensions and a feature control frame
22) Three mutually perpendicular planes
23) Datum features
24) The form of a datum feature being irregular
25) Because basic dimensions are used, there is no tolerance accumulation.
26) The datum reference frame serves as the origin of measurement and orientation and stabilizes the part's six spacial degrees of freedom.
27) A position symbol, perhaps a diameter sign, a geometric tolerance, specified or implied material condition symbols and perhaps datum features
28) Size tolerance
29) A floating fastener situation is two parts with clearance holes in each. A fastener floats through holes in both parts, and a nut binds the parts together.
30)

MMC Hole	Geo. Tol. at MMC	MMC Hole
-MMC Fastener	+MMC Fastener	-Geo. Tol. at MMC
Geo. Tol. for All Holes on Both Parts	MMC Hole	MMC Fastener

31) a) .265 - .250 = .015
    b) .520 - .500 = .020
32) a) .775 - .025 = .750
    b) 1.030 - .030 = 1.000
33) a) .3125 + .010 = .3225
    b) .625 + .020 = .645
34. The maximum height of the screw body or stud that protrudes from the threaded hole
35) Because of variations in orientation of the threaded holes pitch diameter and how this is magnified over the height of the screw when the screw is threaded in
36) Threaded holes and tight fitting holes, such as dowel holes, which dictate the angle of the screw or dowel
37) MMC Hole
    <u>- MMC Screw</u>
    Geo. tol. to be divided between parts

'NEW RULES' Answerbook Section

1. Please complete the following feature control frame to indicate that the datum feature simulator for datum feature C translates (moves).

   21X Ø 4.0-4.3

⊕	Ø 0 Ⓜ	A	B Ⓜ	C Ⓜ ▷

2. Complete the following feature control frame to indicate a unilateral profile of a surface control that is 0.8 all inside of the basic profile.

⌓	0.8 Ⓤ 0	A	B Ⓜ	C Ⓜ

3. Eliminate the envelope of perfect form at maximum material condition by using the appropriate symbol after the size limits of the following hole.

   Ø 10.5-10.7 Ⓘ

4. Use the correct symbology to make the following two diameters treated as though they are one diameter.

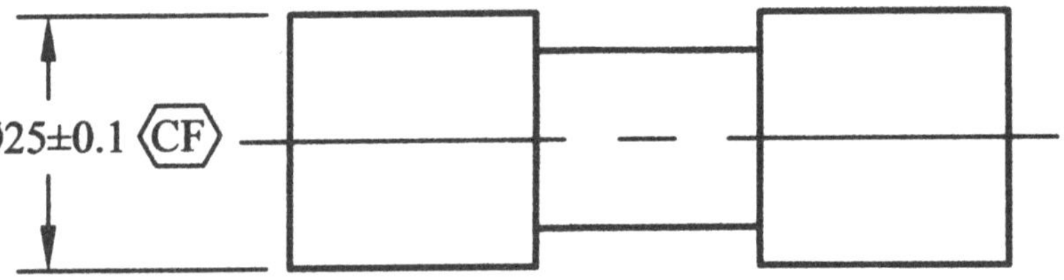

5. When the symbol Ⓜ is used after a datum feature of size referenced in a feature control frame, it stands for: (Circle correct answer)
   a. Minimum Material Condition
   b. Maximum Material Condition
   c. Maximum Material Boundary
   d. Minimum Material Boundary

6. When the symbol Ⓛ is used after a geometric tolerance in a feature control frame, it stands for: (Circle correct answer)
   a. Least Material Boundary
   b. Least Material Condition
   c. Least Squares Center
   d. Least Boundary Condition

7. Put a symbol in the blank area on the elbow of the leader line from the feature control frame that means ALL OVER, APPLIES IN ALL VIEWS SIMULTANEOUSLY.

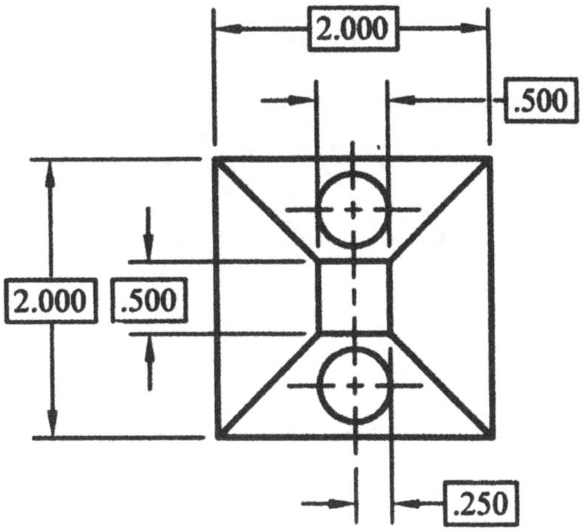

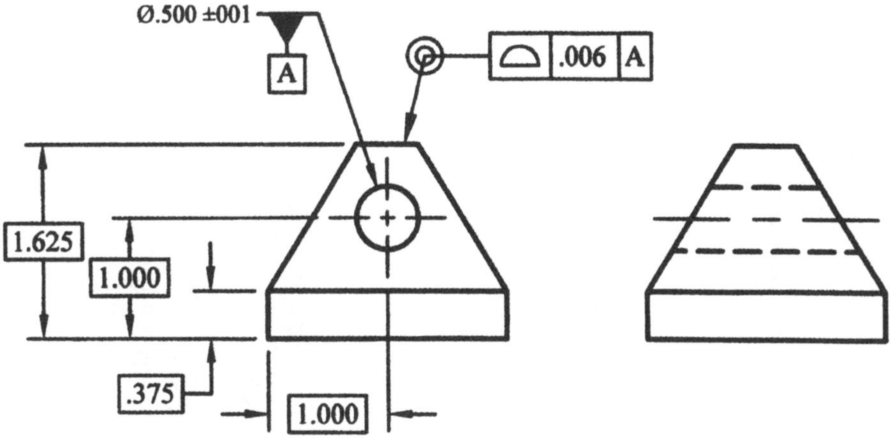

8. When no material condition symbol is used inside a feature control frame after any datum features of size referenced, what is the abbreviation for the symbol that is implied? (Circle correct answer)
   a. MMC
   b. RMB
   c. LMC
   d. MMB
   e. RFS
   f. LMB

9. What does each of the abbreviations in question 8 stand for?
   a. MMC stands for **Maximum Material Condition.**
   b. RMB stands for **Regardless of Material Boundary.**
   c. LMC stands for **Least Material Condition.**
   d. MMB stands for **Maximum Material Boundary.**
   e. RFS stands for **Regardless of Feature Size.**
   f. LMB stands for **Least Material Boundary.**

10. What does the symbol that includes the SF letters in the following drawing stand for?
Spotface

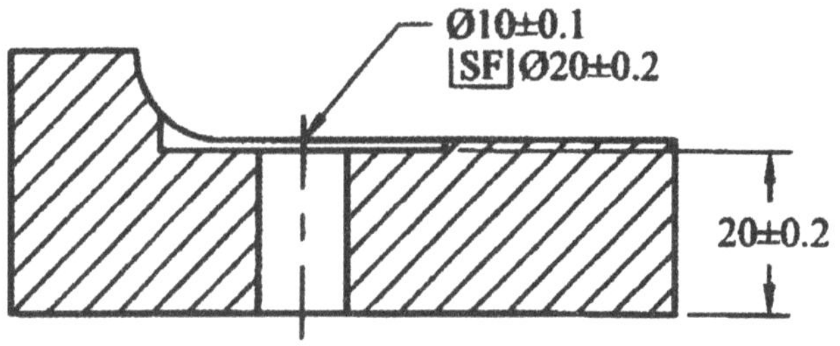

11. According to the rules in ASME Y14.5-2009, may the Ⓢ symbol be used in a feature control frame? **No. It is implied, so it may not be specified.**

12. Label each of the envelopes to which arrows point in the following illustrations as either the "actual mating envelope" or the "actual minimum material envelope".

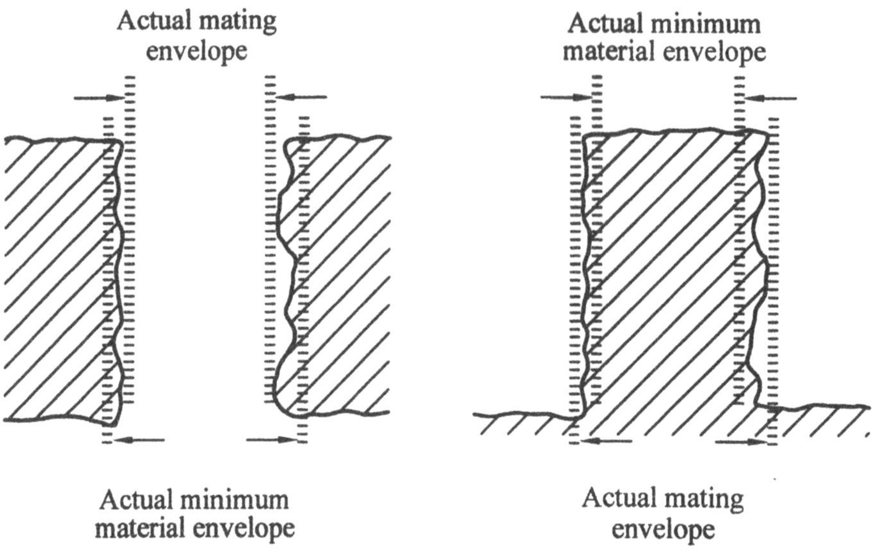

INTERNAL FEATURE

EXTERNAL FEATURE

13. What geometric characteristic symbol (if placed below the 25±0.5) would be capable of creating the following tolerance zone for the part shown? (Circle the correct answer below the illustrations)

Part Specification

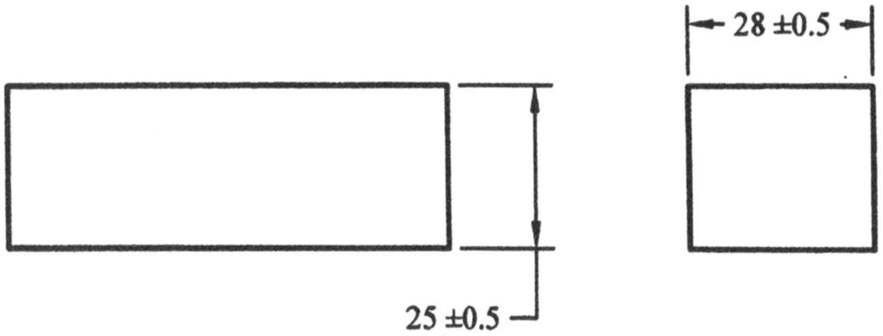

Tolerance Zone

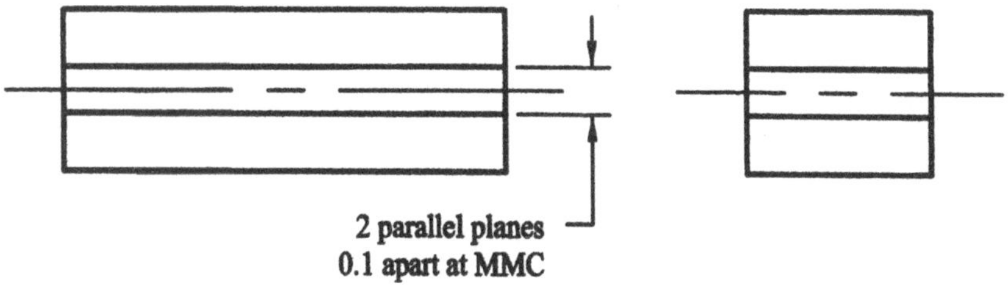

a. | ⏤ | 0.1Ⓜ |
b. | ○ | 0.1 |
c. | ⌒ | 0.1 |
d. | ⏥ | 0.1Ⓜ |

14. An "Actual Mating Envelope" is a similar perfect feature(s) counterpart of smallest size that can be contracted about an external feature(s) or of largest size that can be expanded within an internal feature(s) so that it coincides with the surface(s) at the highest points. What is the difference between an "Unrelated Actual Mating Envelope" and a "Related Actual Mating Envelope"? (Circle one)
   a. Unrelated is not tied to the feature control frame, but related is.
   b. Related is oriented or located to a datum(s), while unrelated is not.
   c. Unrelated is not a "Simultaneous Requirement" with other feature patterns, but Related is related to other feature patterns.
   d. Related is a combined control using one feature control frame stacked on another.

15. Which geometric characteristic symbol that is categorized as "Orientation" may be used to tolerance any angle no matter how many degrees it is specified as?
   a. Angularity
   b. Perpendicularity
   c. Parallelism
   d. Profile of a Surface

16. Where axes, center planes or surfaces are shown coincident on a drawing and geometric tolerances establish the relationship between the features, what applies?
    a. a plus and minus tolerance
    b. a zero basic dimension
    c. a general tolerance note
    d. a reference dimension

17. Where a coordinate system is shown on the drawing, unless otherwise specified, it shall be:
    a. left handed
    b. labeled in the negative direction
    c. neither labeled in the positive or negative directions
    d. right handed

18. Per ASME Y14.5-2009, unless otherwise specified, a datum feature simulator is:
    a. a true geometric counterpart
    b. physical
    c. theoretical
    d. rails, machine table, surface planes, collet chucks and parallels used to fixture parts.

19. Match the following symbols with what they represent by listing the correct letter of the definition after each symbol.
    a. Continuous Feature of Size (treats multiple features of size as one)
    b. Translation symbol (datum feature simulator translates/moves)
    c. All Over (control applies in all views-commonly used with profile of a surface)
    d. Unequally Disposed or Unilateral Tolerance
    e. Independency (size tolerance does not control form)

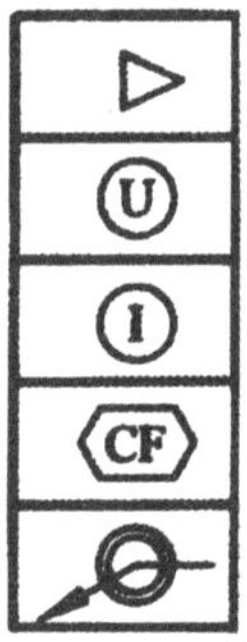

= b
= d
= e
= a
= c

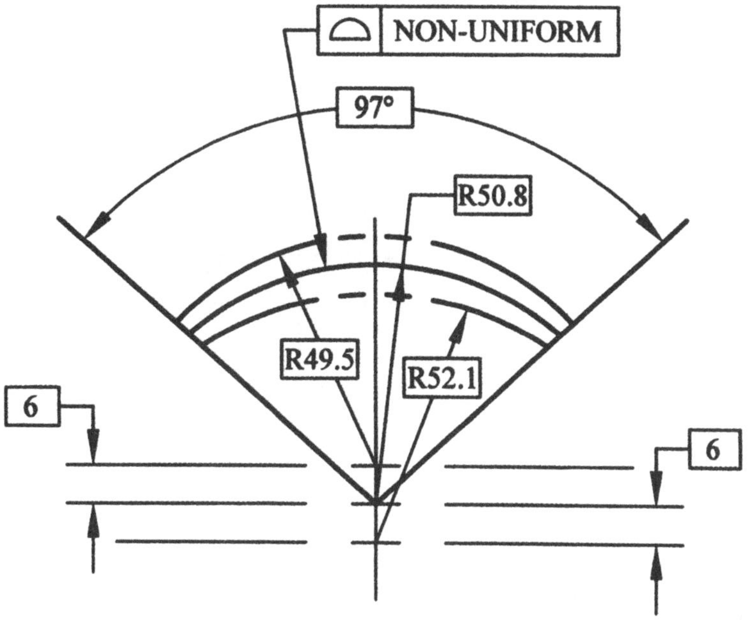

A non-uniform profile tolerance zone can be used to smooth the trans-action between profile segments. Instead of abrupt surface variations being allowed at the end of one profile tolerance zone and the beginning of another, the zones can now be shown to narrow at each end. The difference in the radii, taken in conjunction with the offset of the radii (as shown in the figure above) are considered when calculating the profile tolerance. The extent of each profile tolerance zone must be made clear using the usual methods, such as between symbols.

20. The illustration above is of a NON-UNIFORM profile of a surface control. The surface being toleranced by this control must reside within the tolerance zone depicted by the phantom lines. The purpose of such a tolerance zone can be to: (Circle all correct answers)

  a. smooth the transition between profile segments

  b. allow greater variations in the surface form at the ends of the zone

  c. allow abrupt surface variations at the ends of the profile zone

  d. allow smaller surface variations at the ends of the profile zone

  e. allow larger surface variations in form in the middle of the zone

21. The following illustration compares Composite Profile Tolerancing (one profile symbol) with Two Single Segment Profile Tolerancing (two profile symbols).

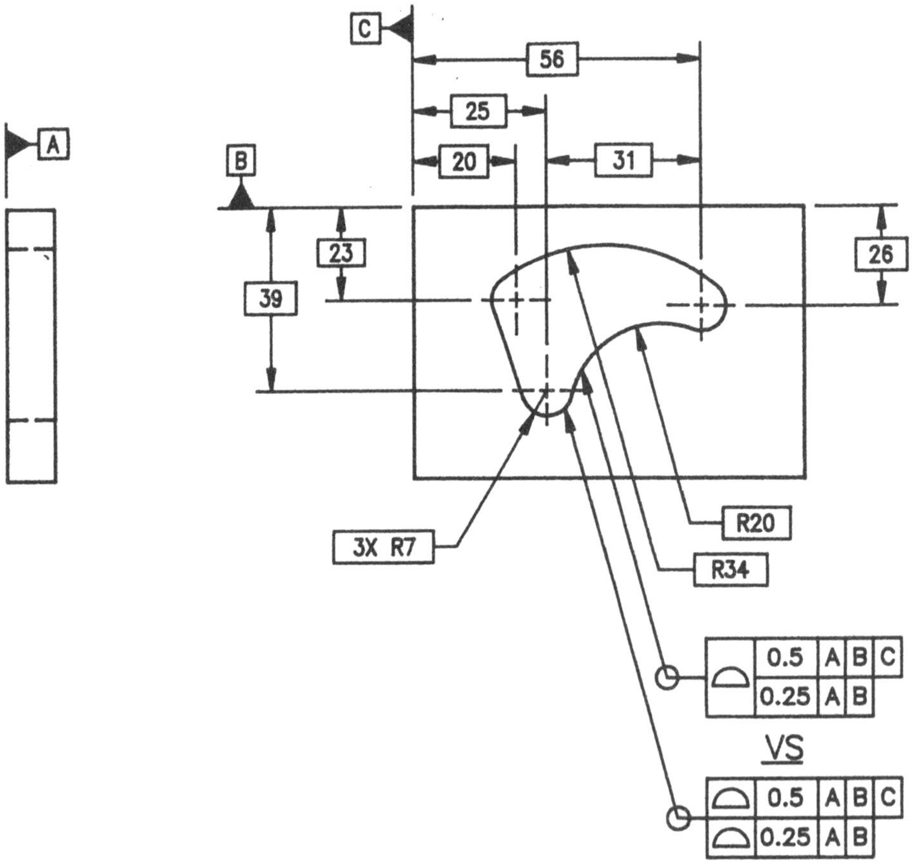

In the composite profile control:

a. What is the tolerance on the size and form of the hole? **0.25**
b. What is the tolerance on the perpendicularity of the hole to datum plane A? **0.25**
c. What is the tolerance on the angular orientation of the hole to datum plane B? **0.25**
d. What is the tolerance on the location of the hole to datum plane B? **0.5**

In the two single segment profile control:

e. What is the tolerance on the size and form of the hole? **0.25**
f. What is the tolerance on the perpendicularity of the hole to datum plane A? **0.25**
g. What is the tolerance on the angular orientation of the hole to datum plane B? **0.25**
h. What is the tolerance on the location of the hole to datum plane B? **0.25**

22. The following illustration a composite profile control with another profile control that applies INDIVIDUALLY to each hole.

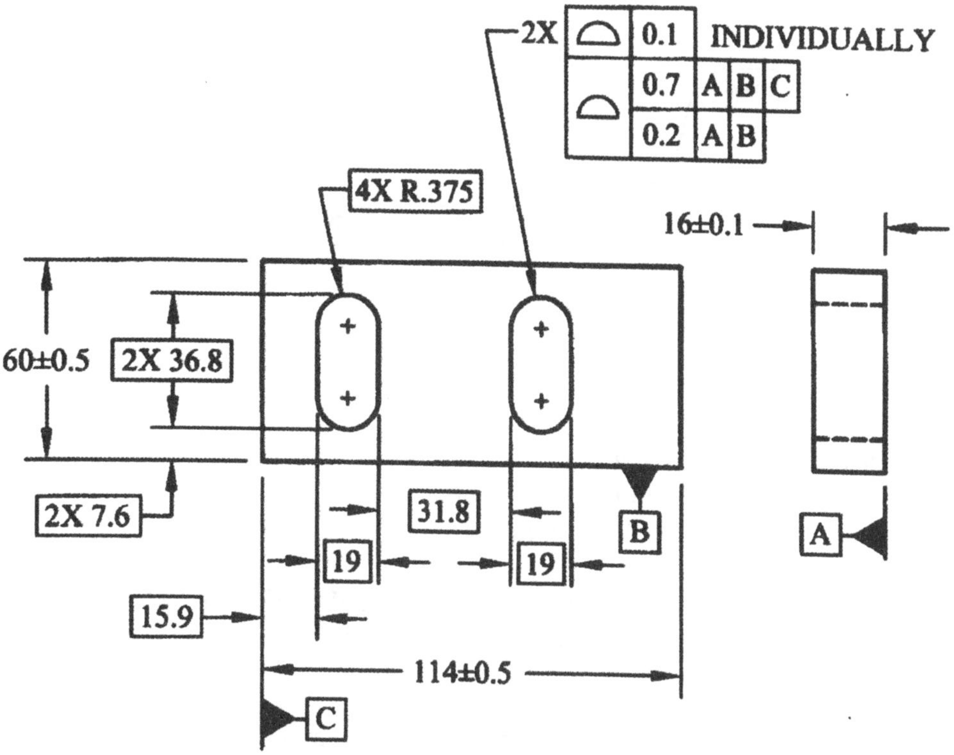

a. What is the tolerance on the size and form of the holes? **0.1**
b. What is the tolerance on the location of the holes to datum B? **0.7**
c. What is the tolerance on the location of the holes to datum C? **0.7**
d. What is the tolerance on the location of the holes to each other? **0.2**
e. What is the tolerance on the perpendicularity of the holes to datum plane A? **0.2**
f. What is the tolerance on the angular orientation (rotation) of the 2-hole pattern to secondary datum plane B? **0.2**
g. What is the tolerance on the angular orientation (rotation) of the 2-hole pattern to tertiary datum plane C? **0.2 (If you cannot rotate more than 0.2 to B, you cannot rotate more than 0.2 to C, since datum planes B and C are perpendicular to each other.)**

23. The following illustration shows a conical feature used as a datum feature.

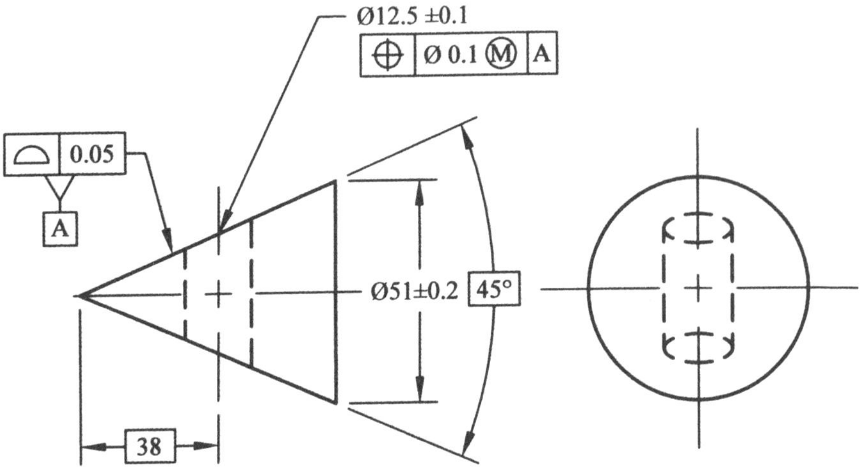

a. How many spatial degrees of freedom are controlled by referencing datum feature A in the position control on the hole? **5**
b. Does datum feature A generate:
   1. an axis only
   2. **an axis and a plane (from the point at the apex of the cone)**
   3. an axis and a plane from the right edge in the view on the left
   4. a plane only

24. The following illustration shows a conical datum feature. In the position feature control frame the spatial degrees of freedom that each datum reference controls are listed in brackets after the datum reference.

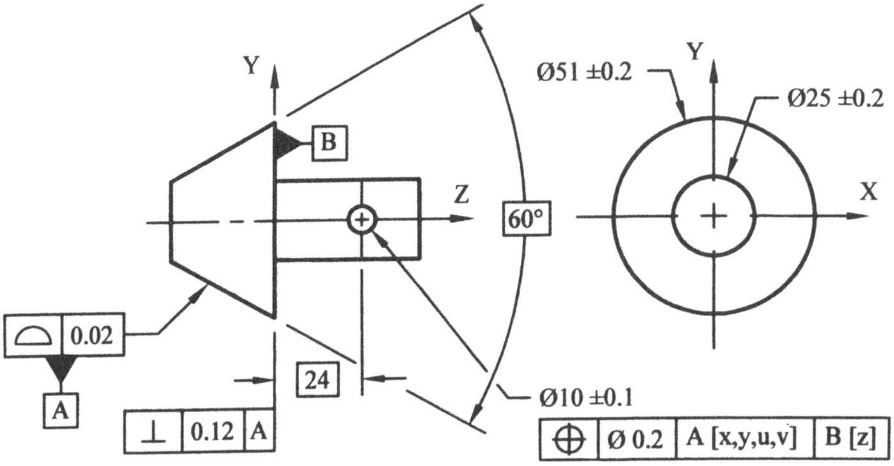

a. What controls the rotational spatial degree of freedom known as w? **Nothing**
b. If datum feature B was referenced as primary instead of secondary, would it be necessary to specify the bracketed z after B in the feature control frame? **No**
c. If datum feature B was not referenced in the position control, how would the dimensions on the part drawing change? **The 24mm basic dimension would go away and be replaced by a basic dimension from the apex of the cone.**

25. **Detail Drawing**: Datum Feature Pattern Referenced at Regardless of Material Boundary (RMB) **Fixture:** The fixture shown below is to be used to mount the part on for inspection of the profile and position controls. The leader line points at one of the two fixturing pins. Describe how the pins would be sized to represent the two-hole pattern at RMB.

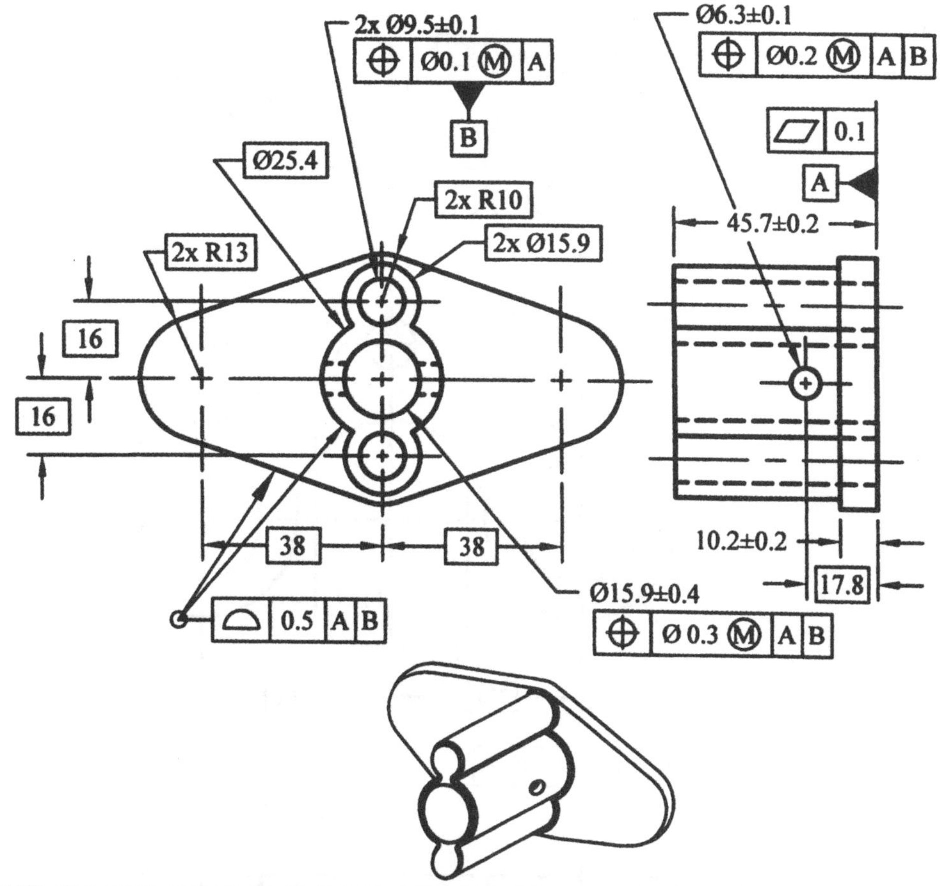

Fixture to simulate datum feature A and datum pattern B regardless of material boundary:

2 expanding fixturing pins to simulate datum pattern B. These pins would expand simultaneously to engage both holes. Datum axis B is perpendicular to datum plane A and halfway between the two fixturing pins.

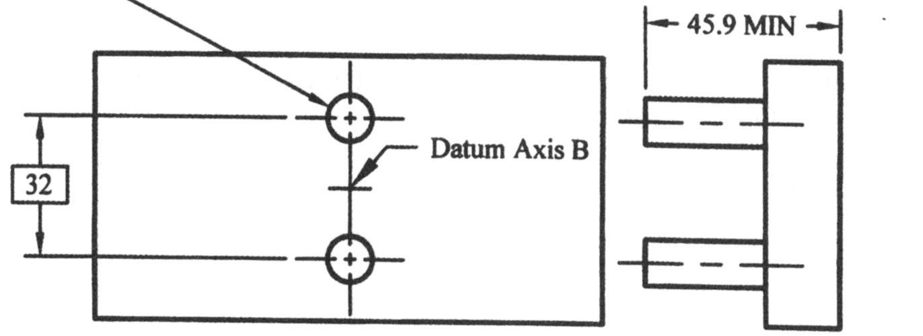

26. Detail Drawing of Datum Feature Pattern Referenced at Maximum Material Boundary (MMB).

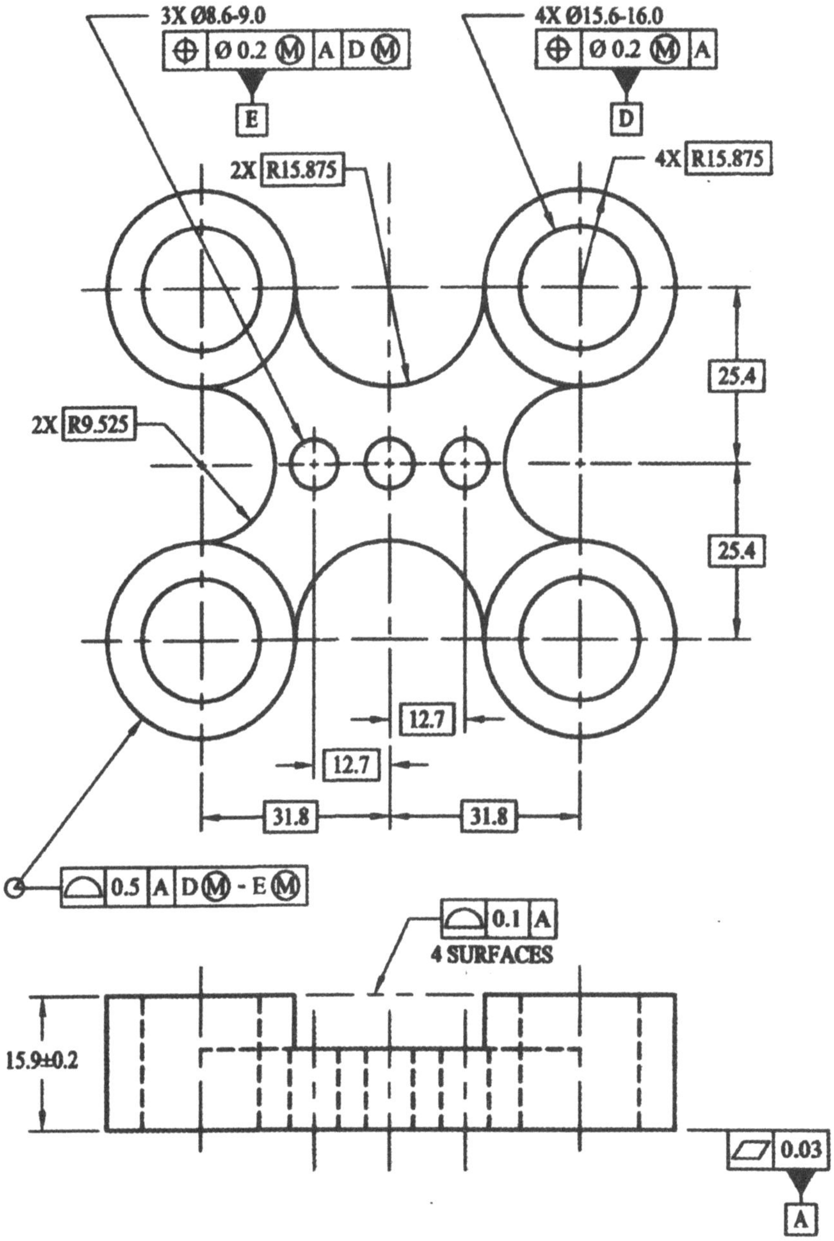

Describe or draw the fixture that could be used to mount the part shown above to inspect its profile of a surface control.

A plate with 7 pins:
4 pins are a ∅15.4
3 pins are a ∅8.4

27. Detail Drawing with Degrees of Spatial Freedom Indicated by Showing X, Y and Z Axes on the Drawing.

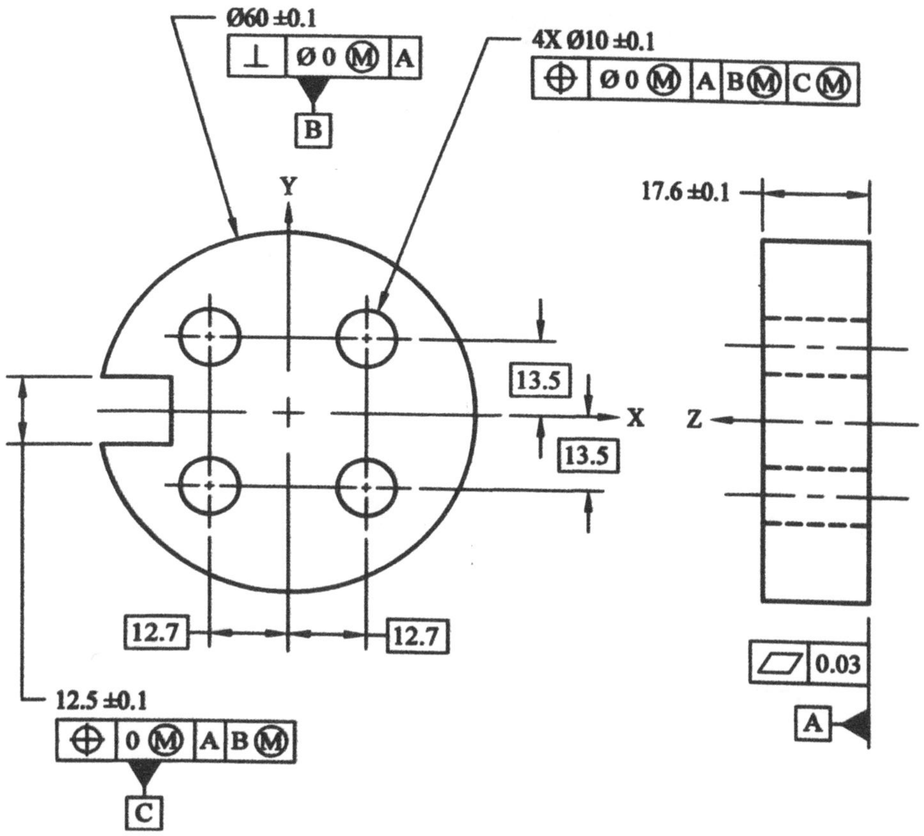

There are 6 spatial degrees of freedom. 3 are linear (x, y and z) and 3 are rotational (u, v and w). Which datum features control each of these? Specify the datum feature that controls each degree of spatial freedom listed below.

a. Movement in the x direction is controlled by datum **B.**
b. Movement in the y direction is controlled by datum **B.**
c. Movement in the z direction is controlled by datum **A.**
d. Rotation in the u direction is controlled by datum **A.**
e. Rotation in the v direction is controlled by datum **A.**
f. Rotation in the w direction is controlled by datum **C.**

28. Overriding the Degrees of Spatial Freedom Controlled by Datums.

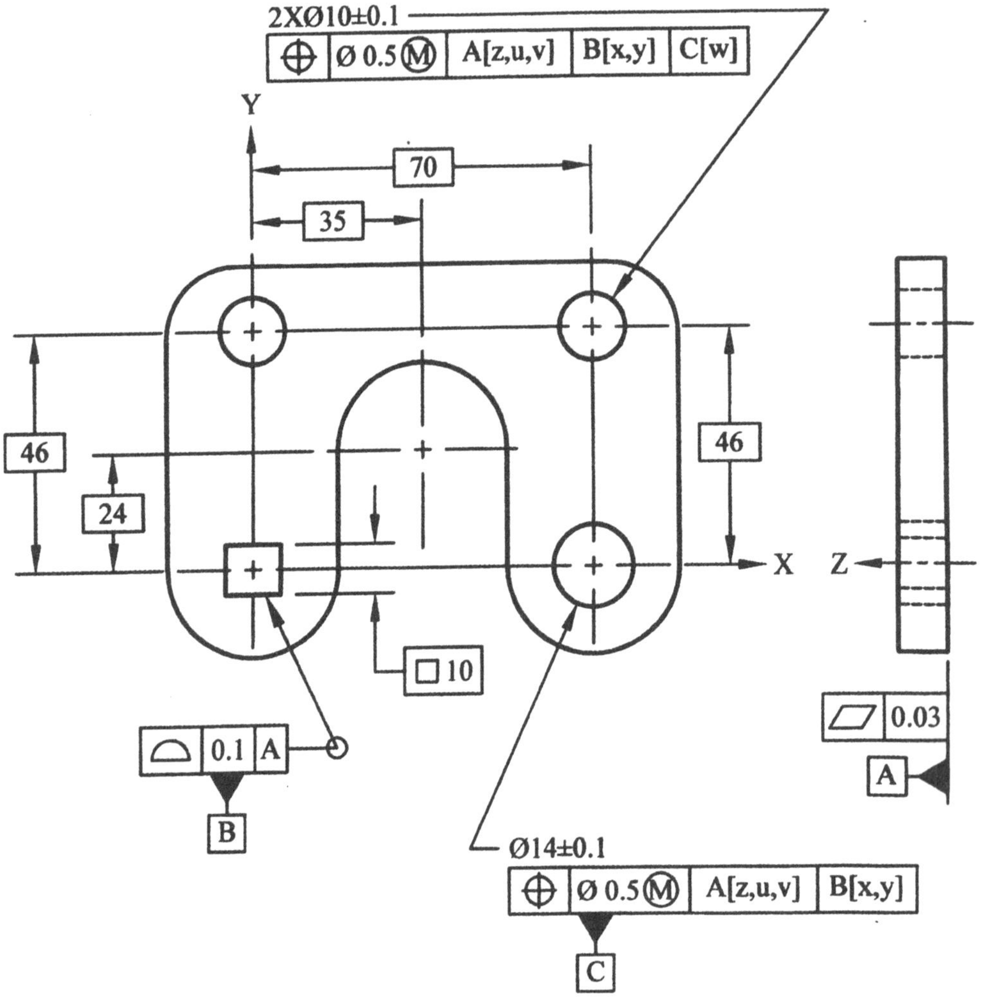

The positional controls shown in the illustration above designate which spatial degrees of freedom are controlled by the datum references. In the 4-hole positional control, what degrees of spatial freedom would each datum reference have controlled if not overridden by the bracketed designations?

a. Datum A would have controlled the following degrees of spatial freedom: **z, u, v**
b. Datum B would have controlled the following degrees of spatial freedom: **x, y, w**
c. Datum C would have controlled the following degrees of spatial freedom: **None**

29. Multiple Datum Reference Frame Identification. Sometimes it is desirable to use more than one datum reference frame on a single part. When this is done, the spatial degrees of freedom may be delineated on the field of the drawing. This can be done by showing the X, Y and Z axes created for each datum reference frame and stating the datum reference frame each applies to in brackets after each X, Y and Z shown.

For Example:

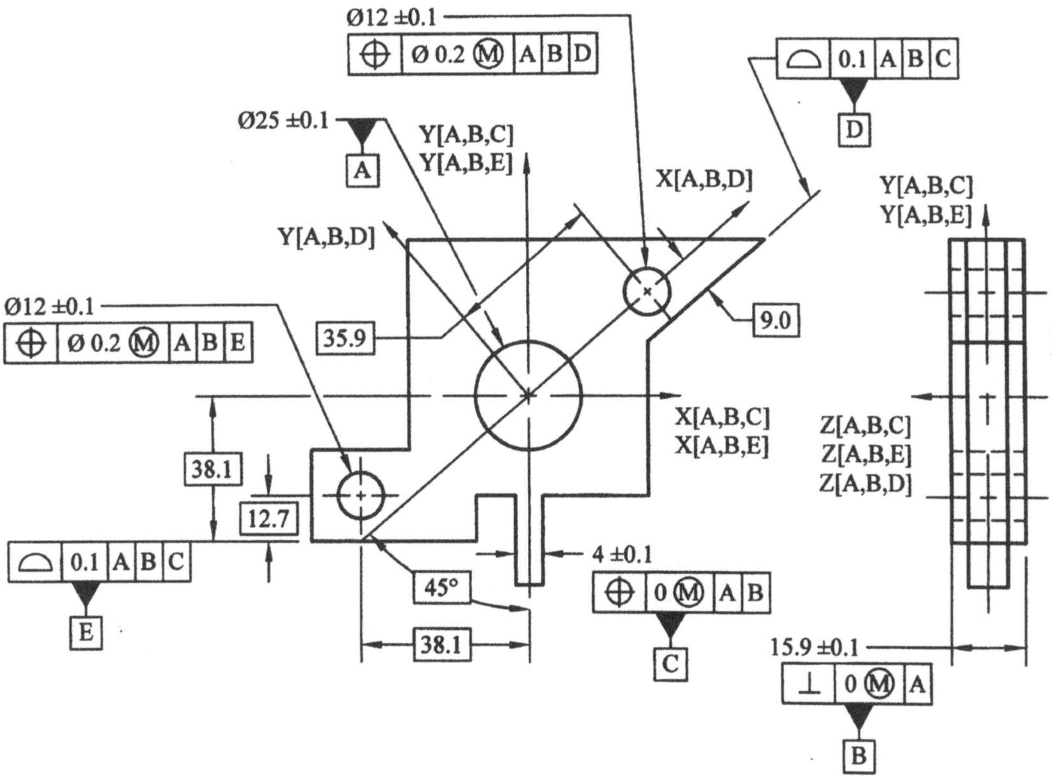

NOTE 1: UNLESS OTHERWISE SPECIFIED
ALL SURFACES [profile of a surface] 0.5 | A | B | C

NOTE 2: FOR A COMPLETE DEFINITION
OF THIS PART SEE THE CAD MODEL

In this example:

a. Could an angularity geometric characteristic symbol be used in place of the perpendicularity symbol in the control on datum feature B? **Yes**

b. Which datum planes do the datum reference frames use to control angular orientation?
   1. Datum reference frame A,B,C uses **C.**
   2. Datum reference frame A,B,D uses **D.**
   3. Datum reference frame A,B,E uses **E.**

c. What is the virtual condition of the ∅12±0.1 holes? **∅11.7**

d. What is the flatness tolerance on datum feature E? **0.1**

e. What is the location tolerance on datum feature D? **0.1**

30. Specifying the correct Maximum Material Boundary.

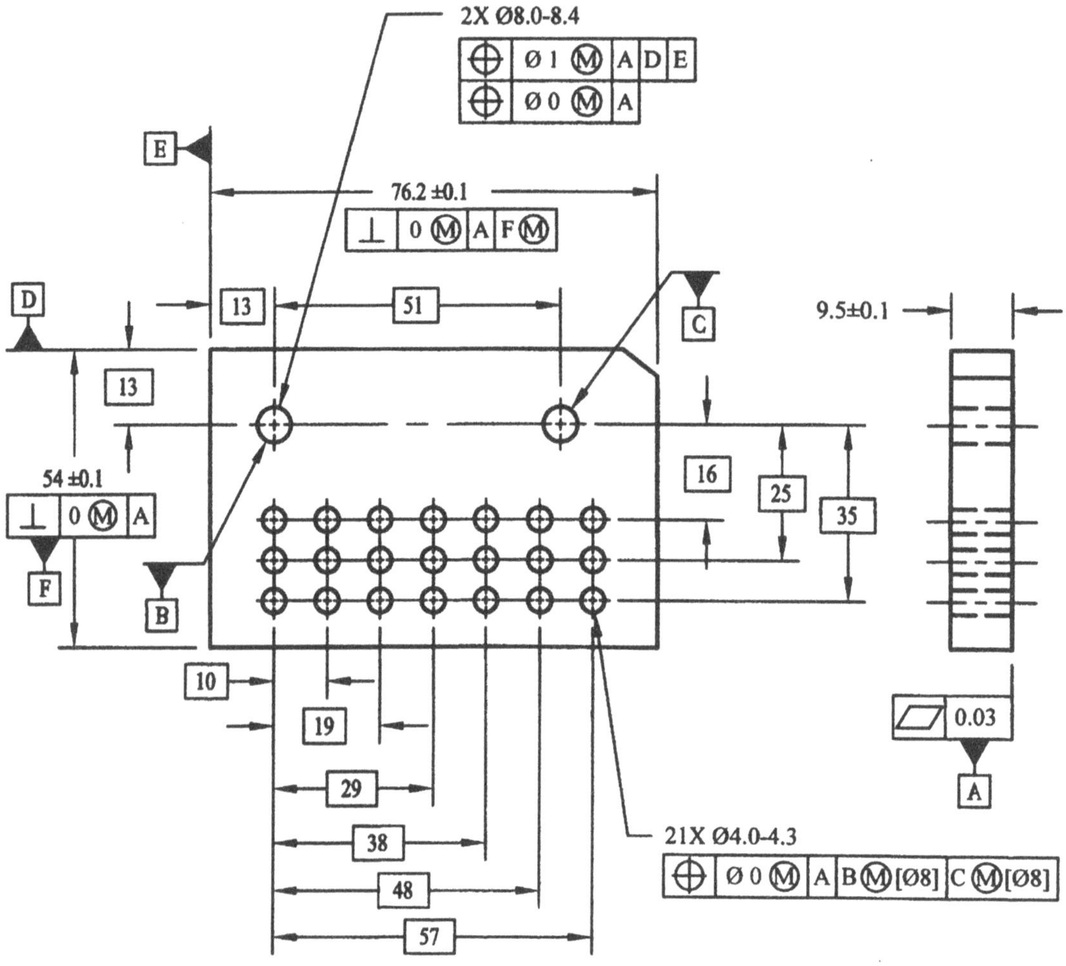

Sometimes it may desirable to specify the applicable boundary generated by a datum feature. This done by putting the correct MMB or LMB in brackets after the datum feature and its material boundary symbol as shown in the 21-hole pattern's feature control frame above.

a. The two hole pattern has two virtual conditions. What are they? **∅7 and ∅8**
b. If the ∅8 was not specified in brackets after datum features B and C, how would one know what the applicable Maximum Material Boundary for each is?
The applicable MMB is the one that applies between B and A (that precedes B). The MMB for C is the one dictated by the A and B (that precedes C).
c. What are the inner boundaries for the 54 and 76.2 widths? **53.7 and 75.9**
d. What size would the gage pins be for a gage that inspects the position of the 2-hole pattern to datums A, D and E? **∅7**
e. If a gage was designed for the 21-hole pattern to inspect its position tolerance, how many gage pins on a plate would be needed and what would their sizes be?
23 pins: 21 at ∅4 and 2 at ∅8
f. What would the size and shape be of a gage meant to inspect the perpendicularity of the 76.2 millimeter width? **A rectangular cavity that is 54.1 by 76.3 with a minimum depth of 9.6.**

31. Please fill in the correct MMB or LMB in the blank spaces on the following illustrations.

Correct Maximum Material Boundary Calculations

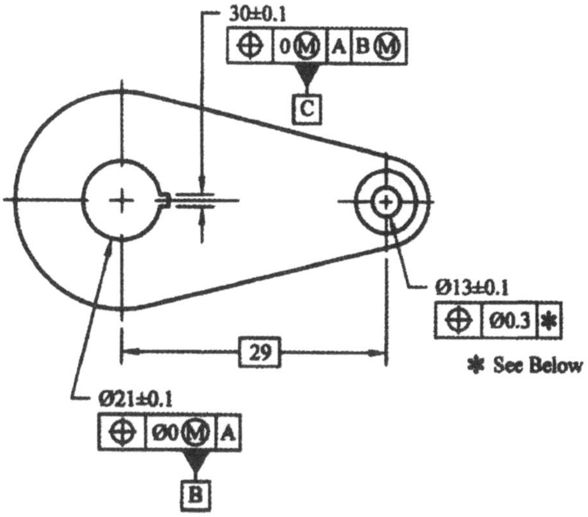

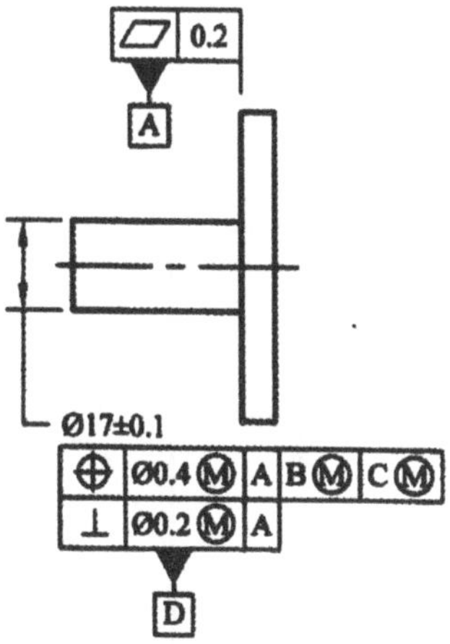

Option	Geometric Control	Maximum Material Boundary of datum feature D
(a)	⊕ | Ø 0.3 | DⓂ	17.1
(b)	⊕ | Ø 0.3 | A | DⓂ	17.1+0.2 = 17.3
(c)	⊕ | Ø 0.3 | A | B | DⓂ	17.1+0.4 = 17.5

Correct Least Material Boundary Calculations

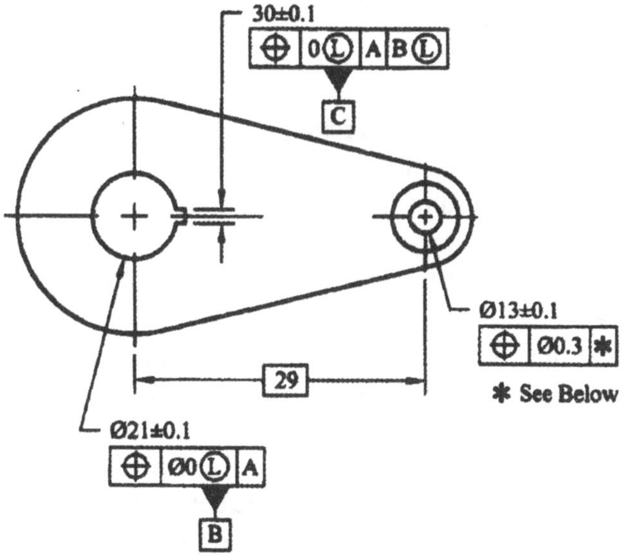

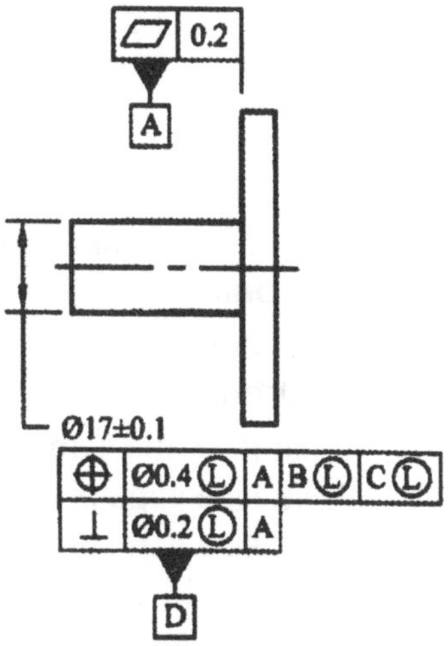

Option	Geometric Control	Least Material Boundary of datum feature D
(a)	⊕ | Ø 0.3 | DⓁ	16.9
(b)	⊕ | Ø 0.3 | A | DⓁ	16.9-0.2 = 16.7
(c)	⊕ | Ø 0.3 | A | B | DⓁ	16.9-0.4 = 16.5

32. Answer the questions for the following illustration.

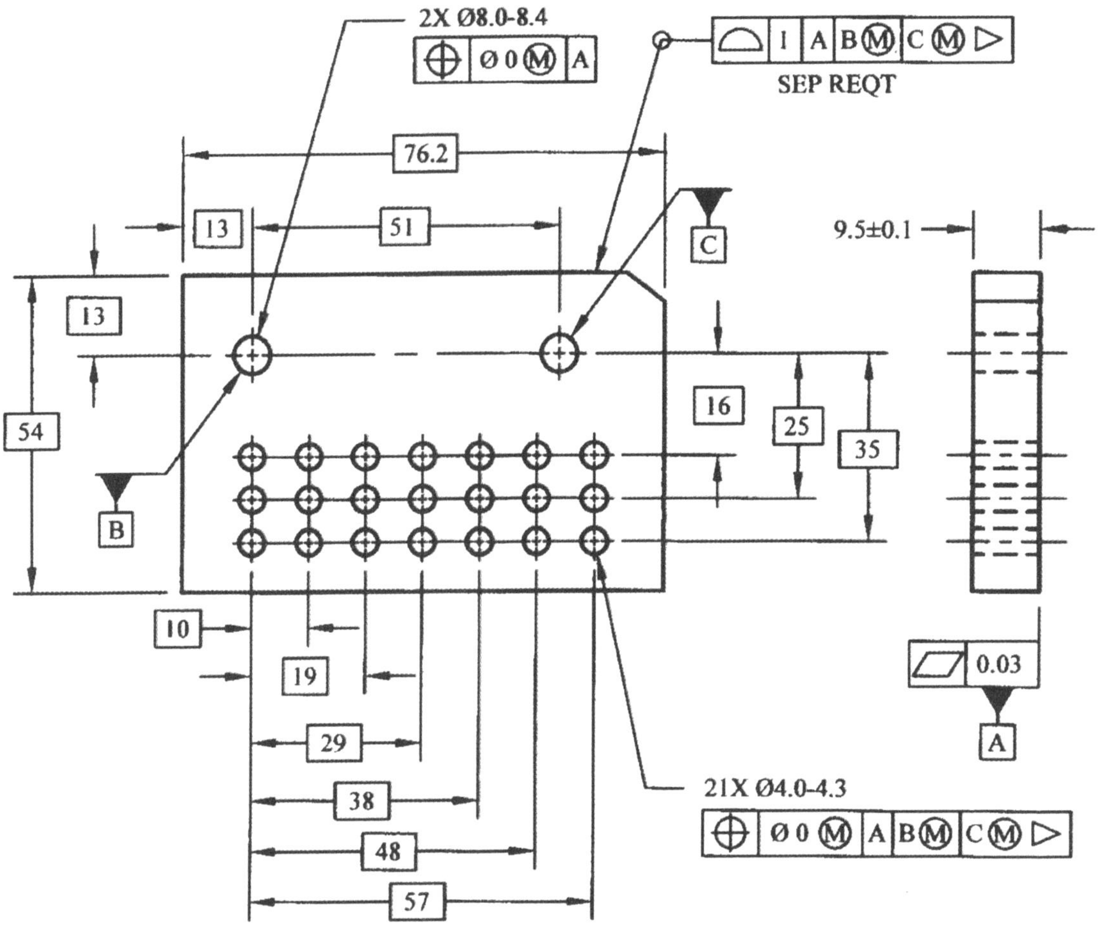

For the 21-hole pattern:

a. What does the symbol Ⓜ used after the zero tolerance in the feature control frame (FCF) stand for? **Maximum Material Condition**

b. Would the datum feature simulator for datum feature B be stationary or moveable? **Stationary**

c. What does the symbol Ⓜ used after datum feature B in the FCF stand for? **Maximum Material Boundary**

d. What does the symbol used after the Ⓜafter datum feature C stand for? **Translation**

e. What size gage pin would be used to represent datum feature B? **Ø8**

f.. What size gage pin would be used to represent datum feature C? **Ø8**

g. How would the datum feature simulator used to represent datum feature C differ from the one used for datum feature B? **The datum feature simulator for C would translate (be moveable) toward or away from the stationary datum feature B simulator.**

h. What does the SEP REQT note stand for? **SEPARATE REQUIREMENT**

i. How would datum feature shift due to datum feature B's departure from its virtual condition differ if the SEP REQT note was not on the drawing? **With SEP REQT, the 21 holes (as a pattern) could shift in a different direction than the profile. Without SEP REQT, the 21 holes and the profile would constitute a SIMULTANEOUS REQUIREMENT and have to shift together.**

33. Using BASIC or BSC after a datum feature reference in a feature control frame.

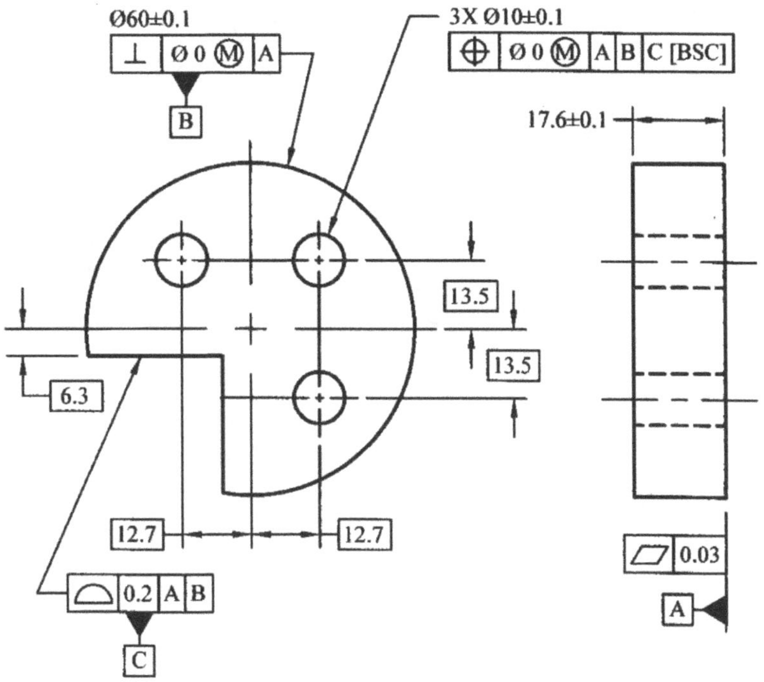

1. The datum feature simulator for datum feature C:
   a. Is represented at 6.4.
   b. Is represented at 6.3.
   c. Progresses normal from the MMB until it makes maximum contract with datum feature C.
   d. Is represented at 6.2.
2. What is the inner boundary of datum feature B? **59.7**
3. What is the outer boundary of datum feature B? **60.1**
4. What is the inner boundary of each of the 3 holes? **9.9**
5. What is the outer boundary of each of the 3 holes? **10.3**
6. What is the minimum wall thickness between one of the three holes and the outside diameter of the part?

59.700000 = I.B. of O.D.
- 37.069664= Bolt Circle
22.630336
- 10.300000 = O.B. Holes
12.330336 = Total Material

12.330336 /2 = 6.165168 = MIN WALL

34. Using MMB after a planar angular orientation datum feature reference in a feature control frame.

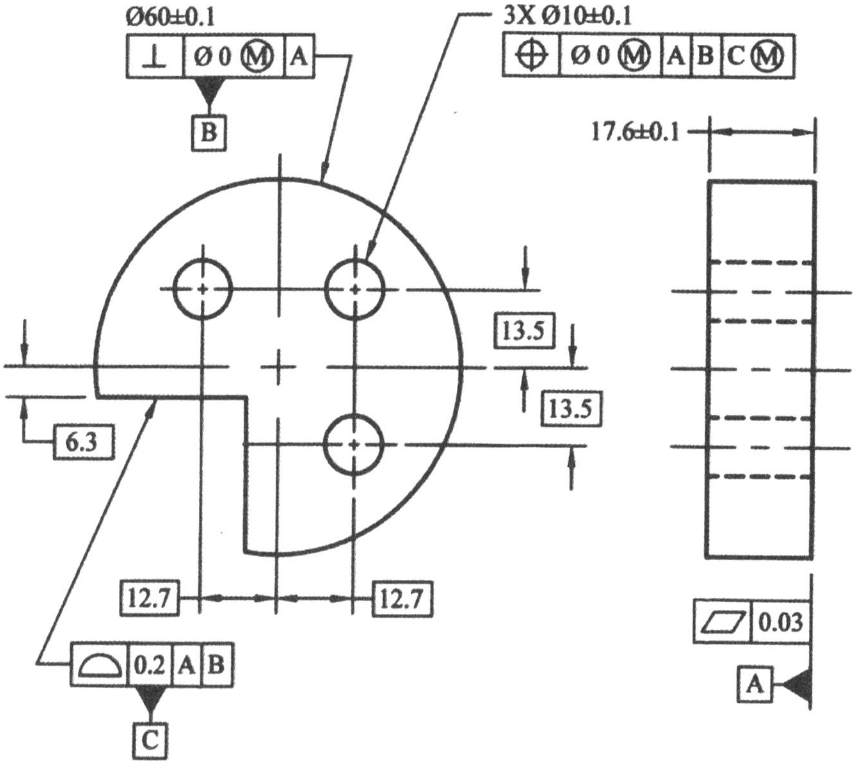

1. The datum feature simulator for datum feature C:
   a. Is represented at 6.4.
   b. Is represented at 6.3.
   c. Progresses normal from the MMB until it makes maximum contract with datum feature C.
   d. Is represented at 6.2.

35. Implying RMB after a planar angular orientation datum feature reference in a feature control frame.

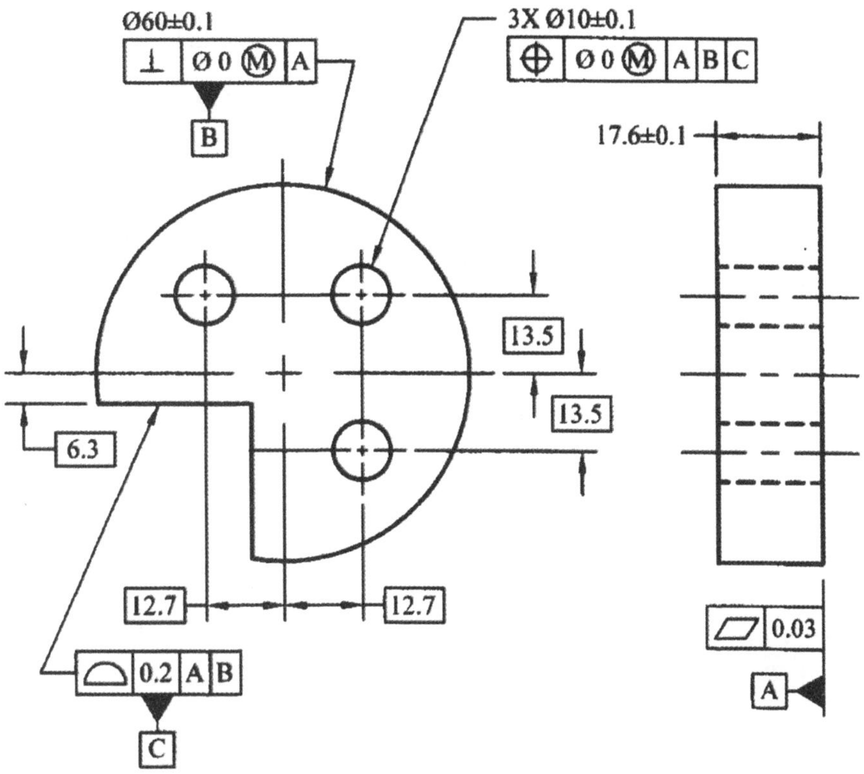

1. The datum feature simulator for datum feature C:
   a. Is represented at 6.4.
   b. Is represented at 6.3.
   c. Progresses normal from the MMB until it makes maximum contract with datum feature C.
   d. Is represented at 6.2.

36. Using a Translation Modifier after an offset angular orientation datum feature width.

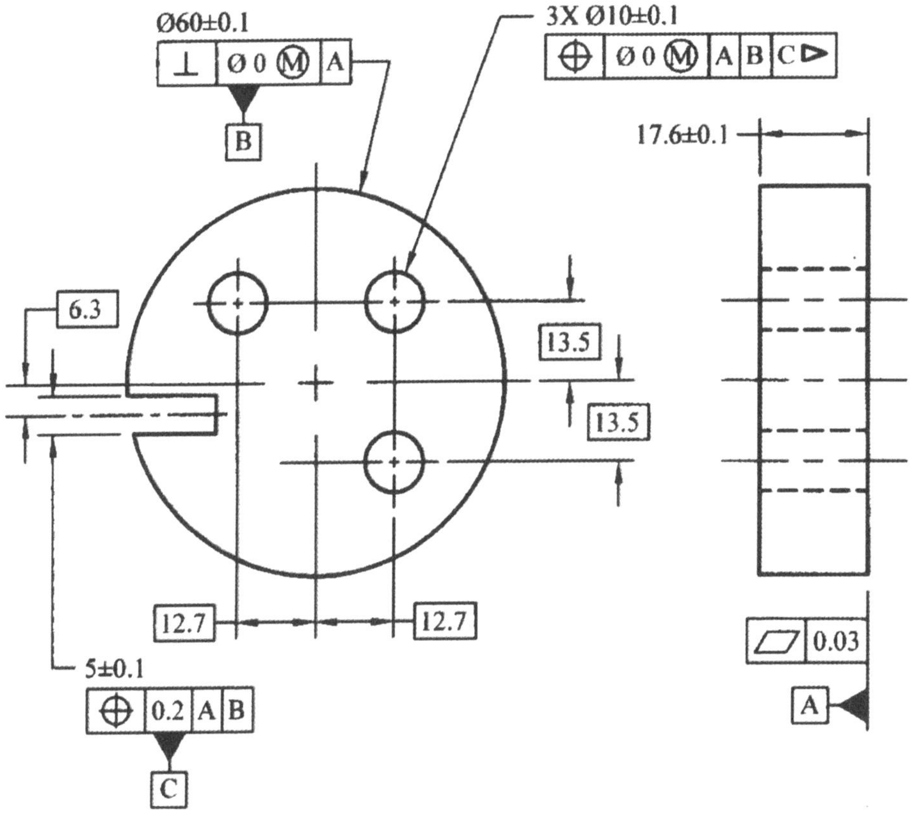

1. The datum feature simulator for datum feature C:
   a. Is represented at its maximum separation, perpendicular to datum plane A and its center plane is located 6.3 from datum axis B.
   b. Is represented at its maximum separation, perpendicular to datum plane A and its center plane is free to move within the position tolerance zone from datum axis B.
   c. Is represented at its maximum material boundary of 4.7, perpendicular to datum plane A and its center plane located 6.3 from datum axis B.
   d. Is represented at its least material boundary of 5.3, perpendicular to datum plane A and it center plane located 6.3 from datum axis B.

37. Using RMB after an offset angular orientation datum feature width.

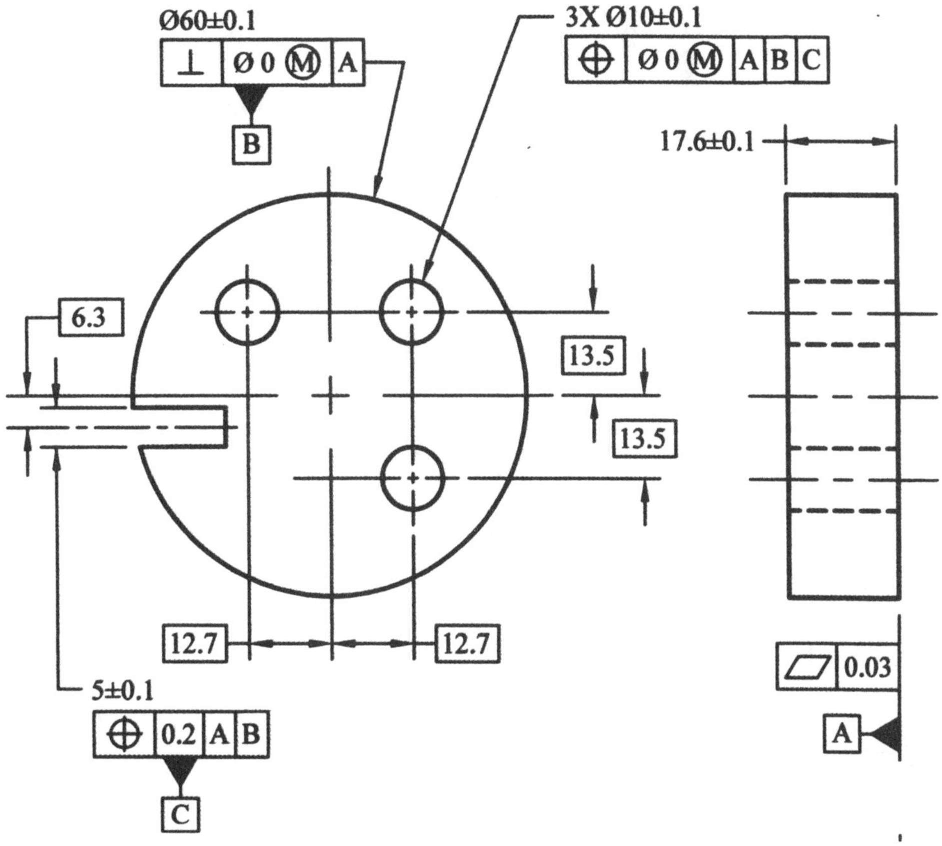

1. The datum feature simulator for datum feature C:
   a. Is represented at its maximum separation, perpendicular to datum plane A and its center plane is located 6.3 from datum axis B.
   b. Is represented at its maximum separation, perpendicular to datum plane A and its center plane is free to move within the position tolerance zone from datum axis B.
   c. Is represented at its maximum material boundary of 4.7, perpendicular to datum plane A and its center plane located 6.3 from datum axis B.
   d. Is represented at its least material boundary of 5.3, perpendicular to datum plane A and it center plane located 6.3 from datum axis B.

38. Using MMB after a centered semi-cylindrical angular orientation datum feature.

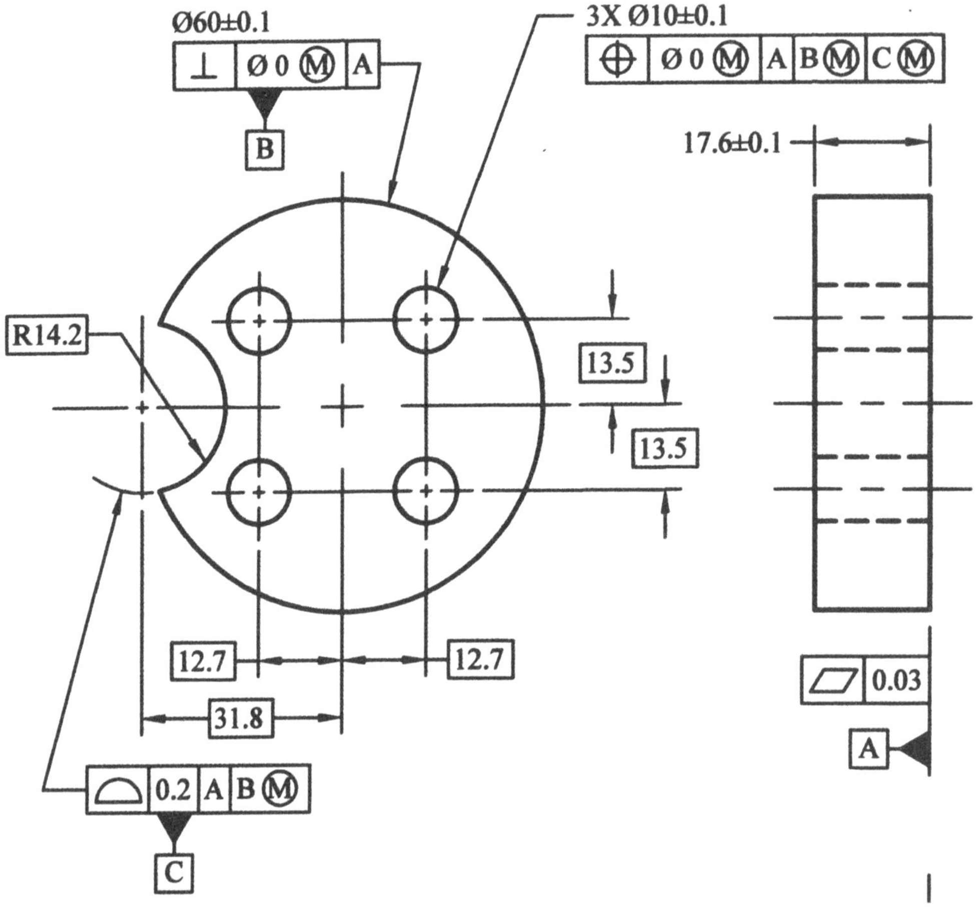

1. The datum feature simulator for datum feature C:
   a. **Is represented at a radius of 14.1, perpendicular to datum plane A and its center plane is located 31.8 from datum axis B.**
   b. Is represented at a diameter of 14.1, perpendicular to datum plane A and its center is free to move within the profile tolerance zone from datum axis B.
   c. Is represented at its maximum inscribed cylinder size, perpendicular to datum plane A and its center plane located 31.8 from datum axis B.
   d. Is represented at its least material boundary of 14.3, perpendicular to datum plane A and it center plane located 31.8 from datum axis B.

39. Using RMB after a centered semi-cylindrical angular orientation datum feature.

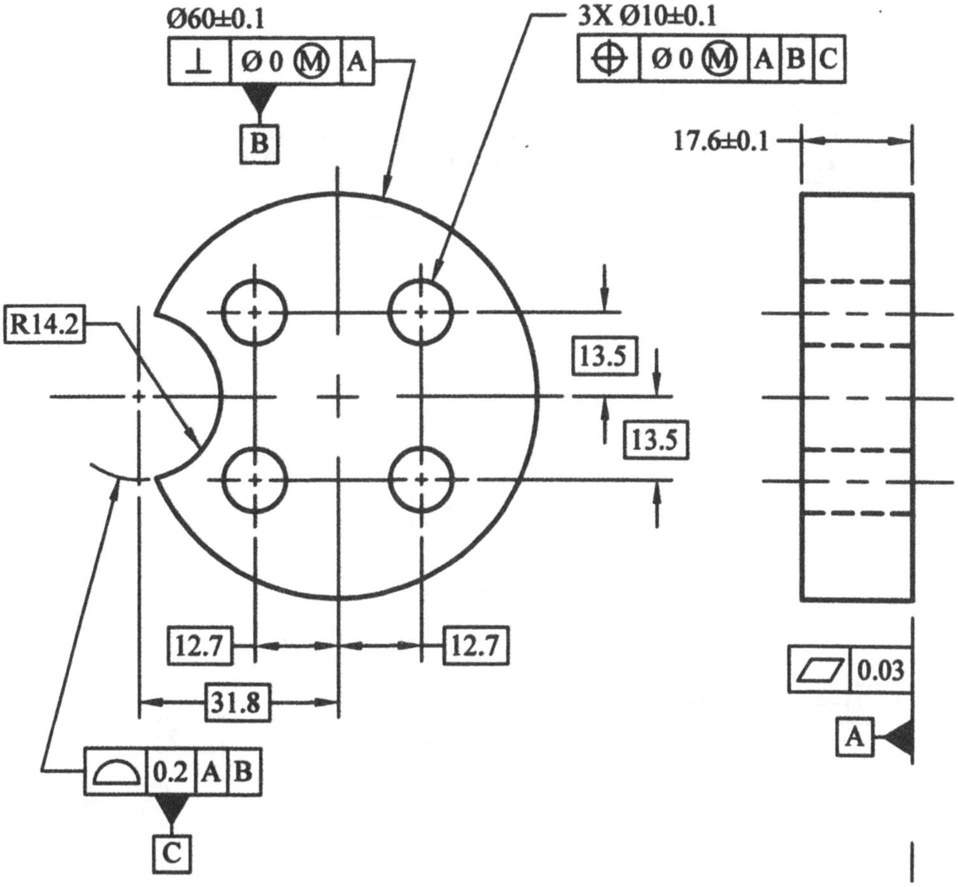

1. The datum feature simulator for datum feature C:

   a. Is represented at a radius of 14.1, perpendicular to datum plane A and its center plane is located 31.8 from datum axis B.
   b. Is represented at a diameter of 14.1, perpendicular to datum plane A and its center is free to move within the profile tolerance zone from datum axis B.
   c. Is represented at its maximum inscribed cylinder size, perpendicular to datum plane A and its center plane located 31.8 from datum axis B.
   d. Is represented at its least material boundary of 14.3, perpendicular to datum plane A and it center plane located 31.8 from datum axis B.

40. Using the illustration of the part detail shown at the top of the page, please dimension (but don't tolerance) the diameter of the gage pins, the gage inside diameter and the distance from the center of the gage (datum feature simulator B) to datum feature simulator C.

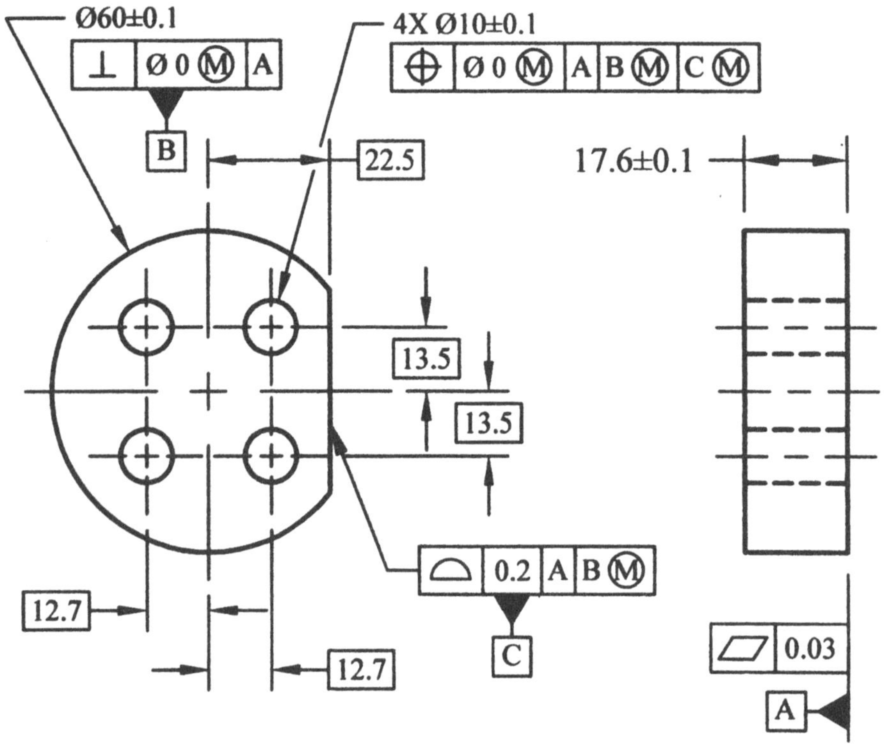

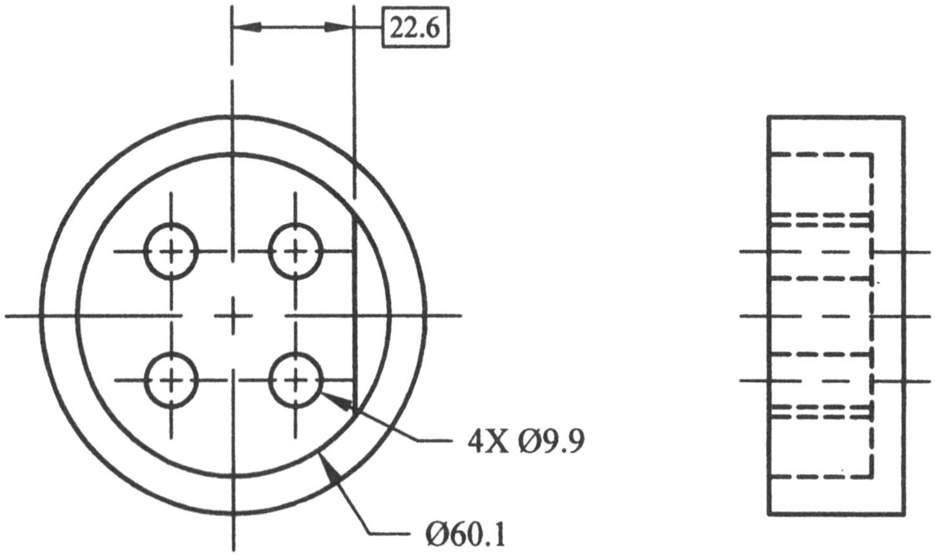

Functional Gage

41. For the detail drawing shown in item (a) please dimension (not tolerance) the partial gage shown in item (b) to represent the datum feature simulator and also the gage element to represent the outside diameter of the part.

(a) Outer Boundary of a Pattern of Pins Referenced at MMB.

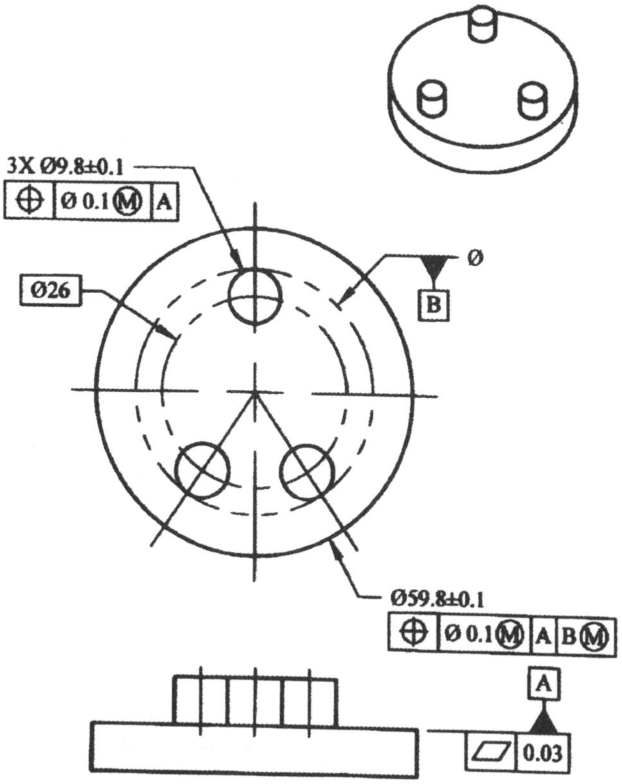

(b) A Partial Gage Design for the part in Illustration a.

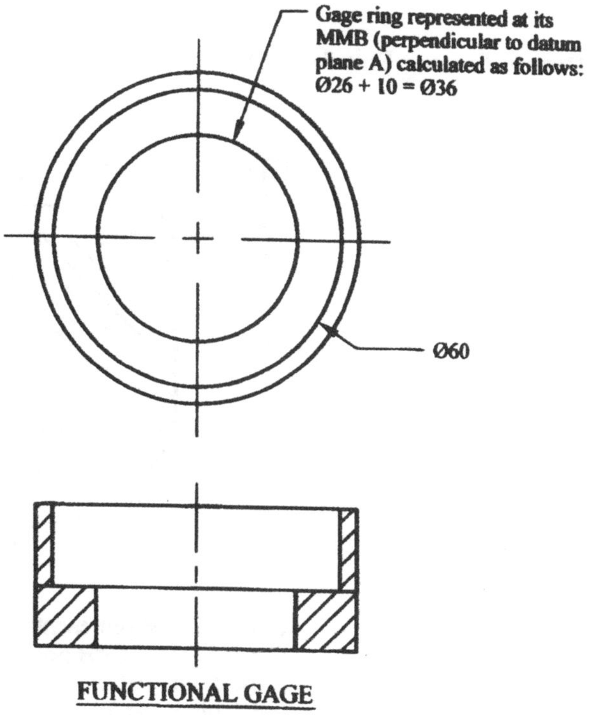

42. For the detail drawing shown in item (a) please dimension (not tolerance) the gage element shown in item (b) to represent datum feature simulator B.

(a) Inner Boundary of a Pattern of Pins Referenced at MMB.

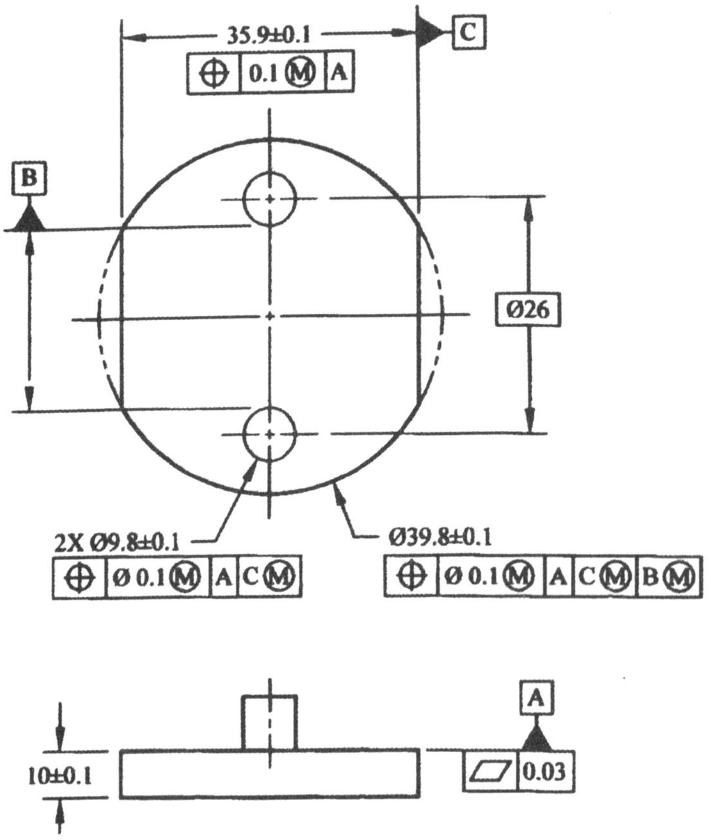

(b) Datum Feature Simulator B.

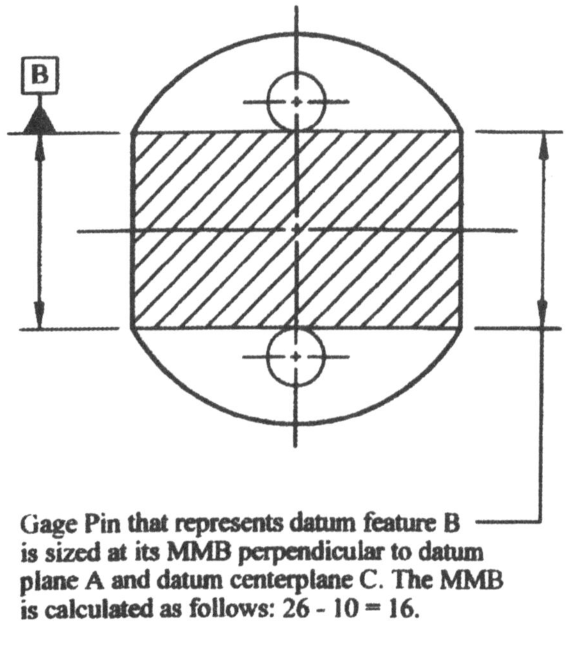

43. Using the correct datum feature designations, feature control frames and local notes:
    a. Designate each of the 12 millimeter holes as a datum feature using one datum feature symbol (the letter D).
    b. Locate the smaller 6 millimeter holes to the left and right of each large hole to that 12 millimeter hole only, using datum plane A for perpendicularity, datum axis D for location and datum plane B for angular orientation.
    c. Use the local note 4X INDIVIUALLY twice.

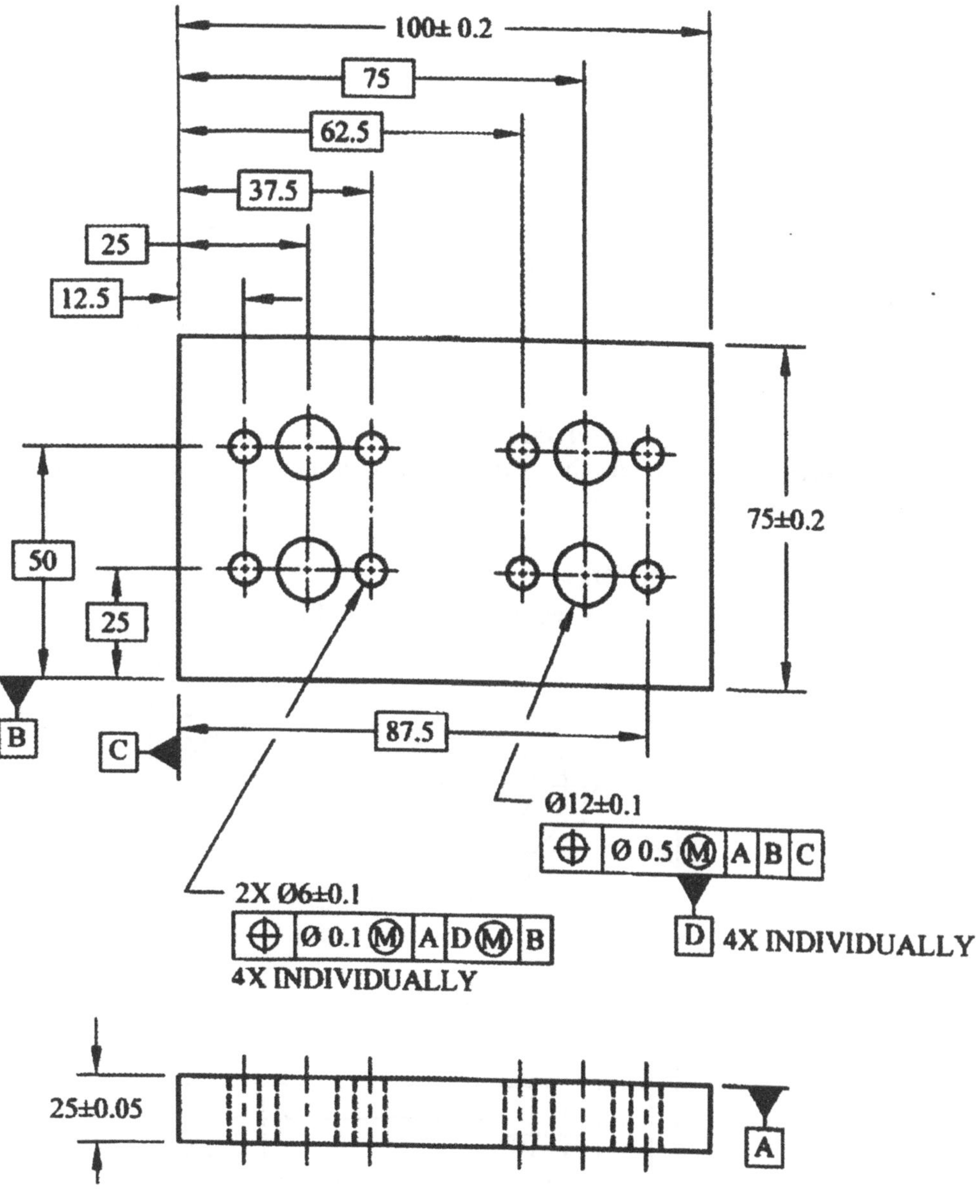

44. Moveable Datum Target Symbol.

Workpiece Mounted in Fixture

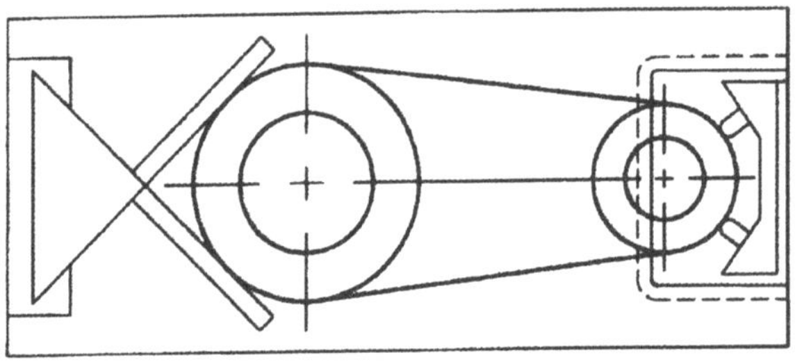

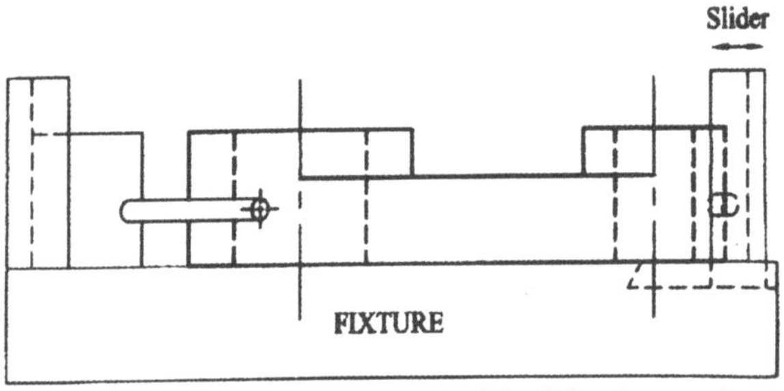

Moveable datum target simulators for C1 and C2 slide through tight fitting holes on the fixture and contact the part surface normal to the desired geometric configuration.

44. Spherical Moveable Datum Targets.

Workpiece Mounted in Fixture

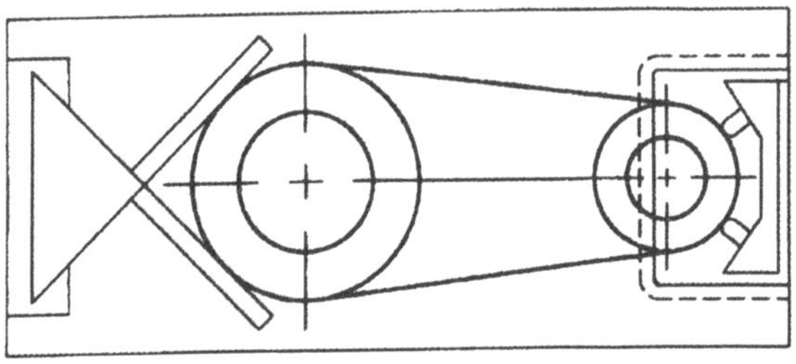

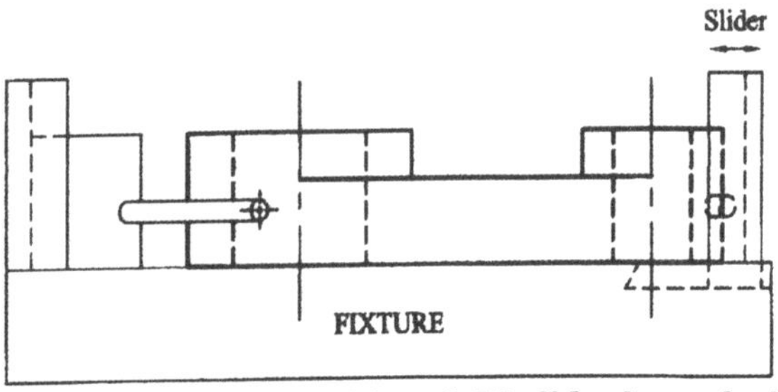

Moveable datum target simulators for C1 and C2 slide through tight fitting holes on the fixture and contact the part surface normal to the desired geometric configuration.

45. Spherical Moveable Datum Targets.

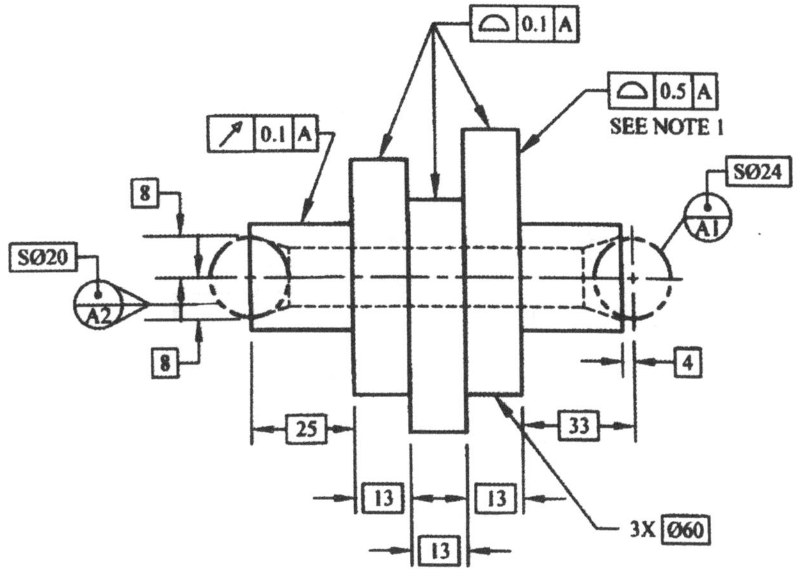

NOTE 1:
THIS CONTROL APPLIES TO ALL PLANER SURFACES LOCATED BY BASIC DIMENSIONS IN THIS VIEW.

46. Please dimension the size and location of the gage pins for the functional gage shown below the part drawing.

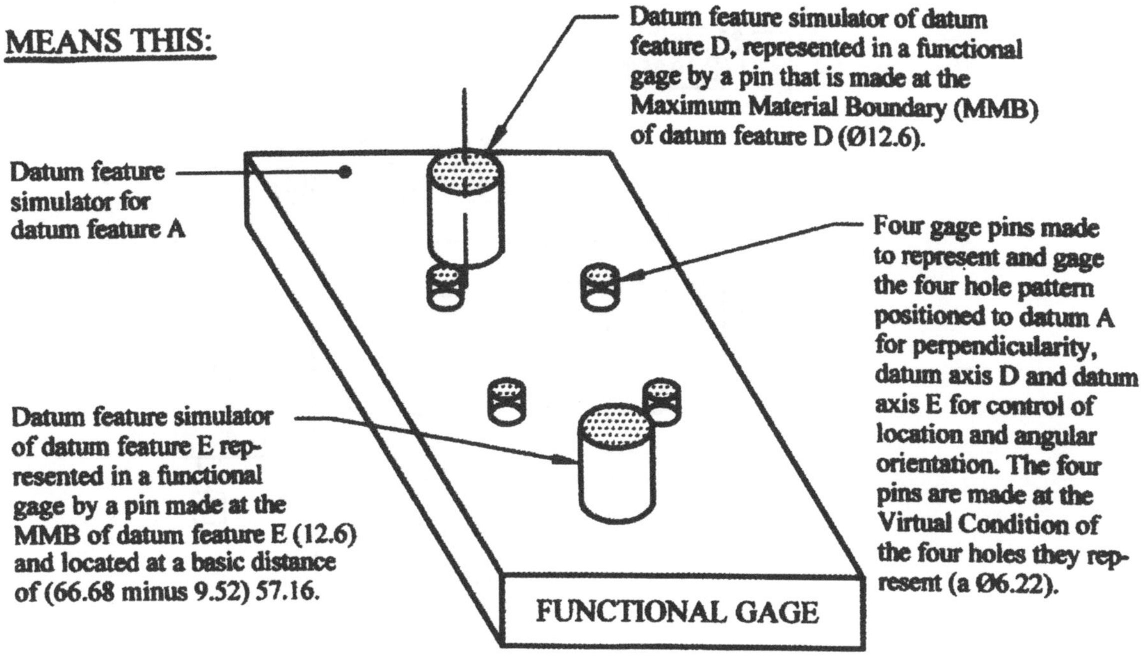

47. Please dimension and tolerance the functional gage shown beneath the part drawing.

Part Detail Drawing

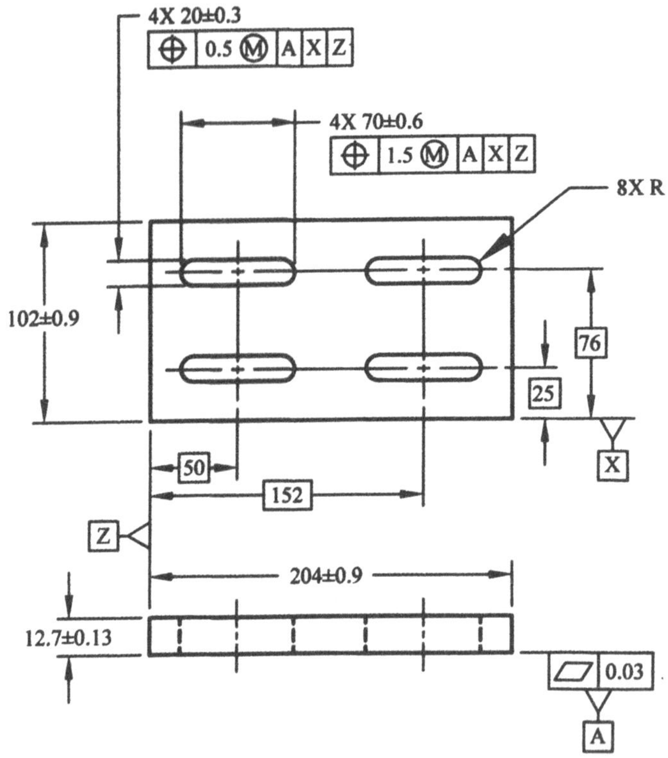

Functional Gage Drawing

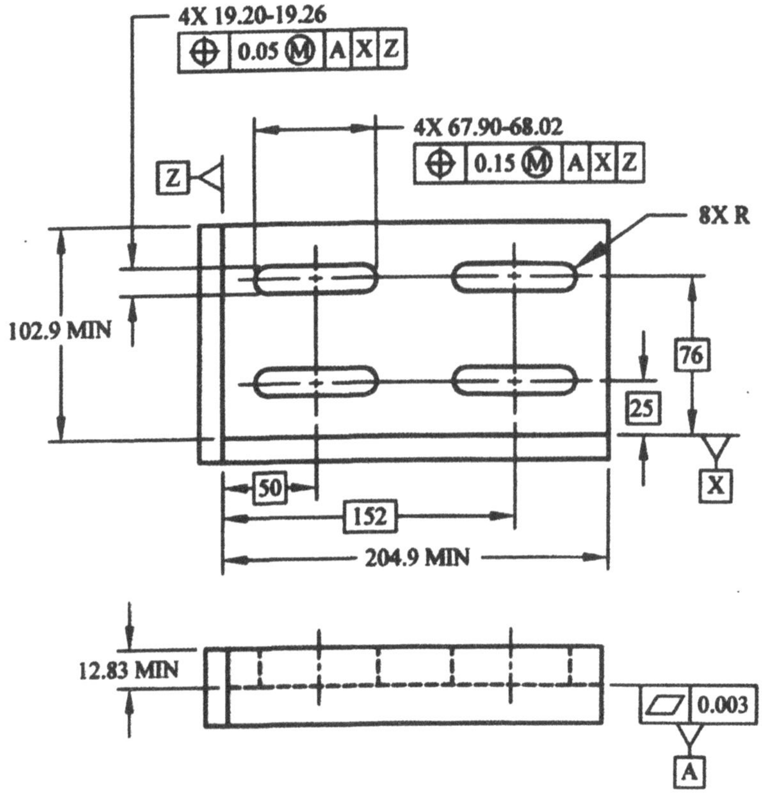

48. Please apply a three-level composite positional tolerancing control to the three-hole pattern to mean the following:

1. The top level control locates the three holes to datums B and C. Regardless of the material boundary of datum feature B, the three holes must be centered to datum centerplane B and hold the 19 millimeter basic dimension from datum plane C to within a diameter of 3 millimeters if the holes are produced at their maximum material condition.
2. The middle level control is ultimately only holding the orientation of the three holes (as a pattern) to datum centerplane B and datum plane C to within a diameter of 1.5 millimeters if the holes are produced at their maximum material condition.
3. The third level control is aligning the three holes' axes to one another for coaxiality to within a diameter of 0.3 millimeters if the holes are produced at their maximum material condition. It is worth noting that no datum feature references are used here, in that, none are necessary to position the three holes to each other. This tolerance is the one that allows the 8 millimeter screw to pass through the three holes. It generates a virtual condition (inner boundary) for the three holes that is calculated by subtracting the 0.3 millimeter diameter position tolerance from the 8.3 millimeter maximum material condition of the holes. 8.3 minus 0.3 equals an 8 millimeter diameter virtual condition. This is the worst mating condition and is compatible with the 8 millimeter screw that must pass through. Since 8 millimeter screws are actually a little smaller than 8 millimeters, there is even a little clearance built into this assembly condition.

Detail Drawing

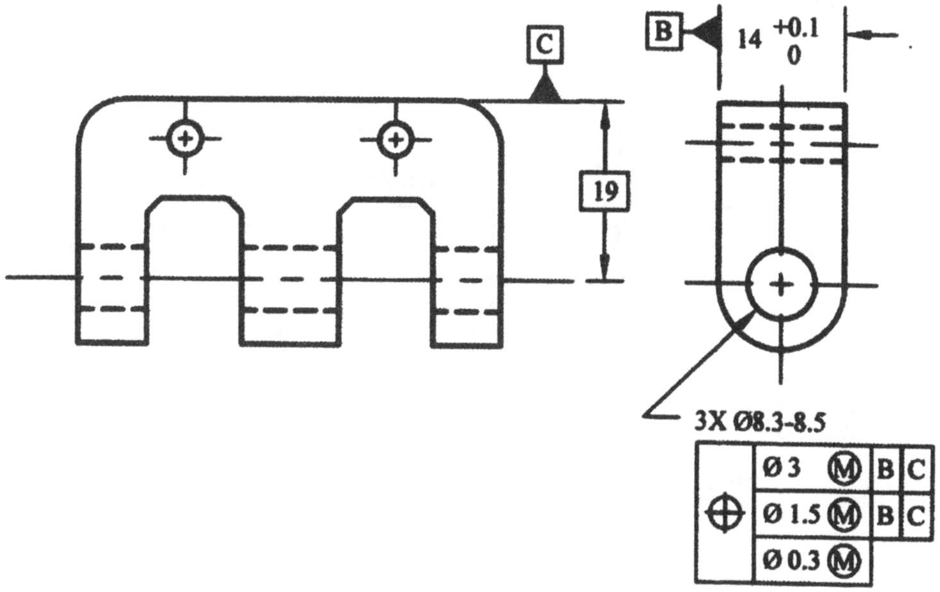

49.

Detail Drawing

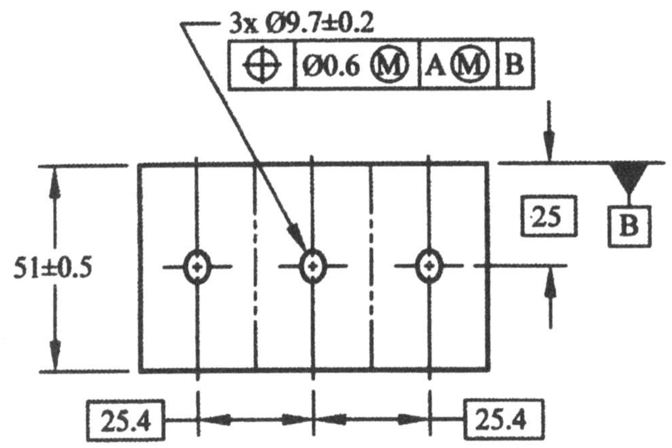

NOTES:
1. MATERIAL THICKNESS 2.5 MILLIMETERS
2. UNLESS OTHERWISE SPECIFIED ALL ANGLES ARE ±1 DEGREE
3. ALL RADII REFER TO DATUM FEATURE A

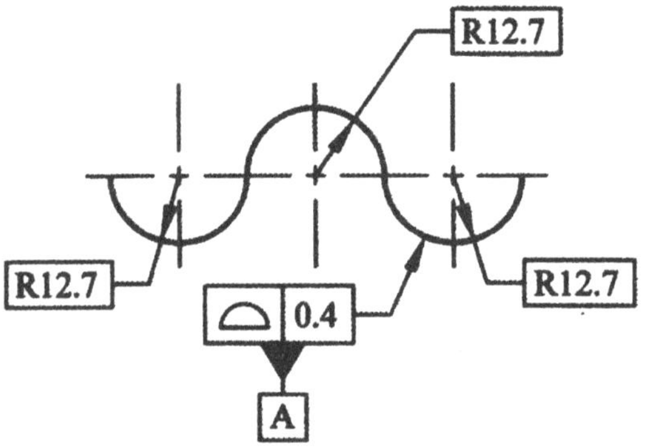

Fixture

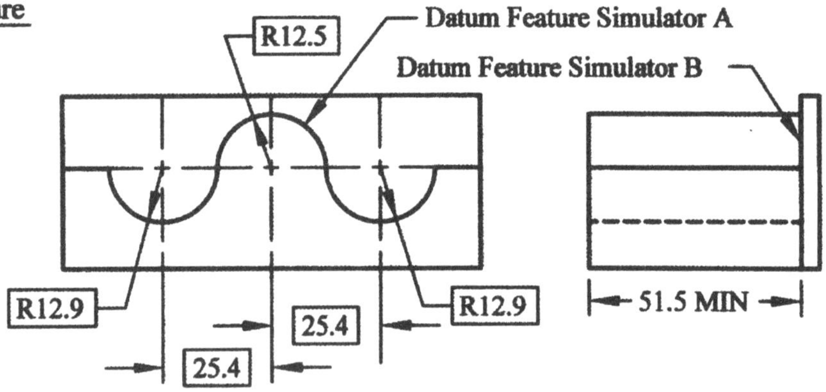

VI. Copyrighted GD&T Books and Course Materials written by James D. Meadows

Copyrighted books and course materials written by James D. Meadows:

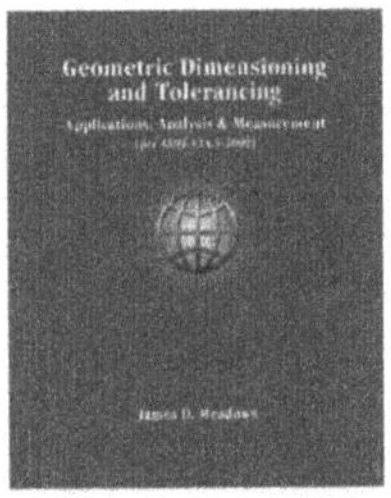

GEOMETRIC DIMENSIONING AND TOLERANCING
Applications, Analysis & Measurement
(per ASME Y14.5-2009)
March 2009
574 pgs, hardcover, illus.
ISBN: 0-9714401-6-6

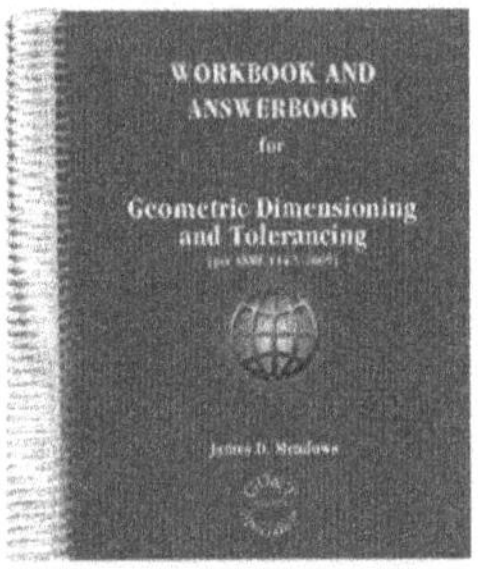

WORKBOOK and ANSWERBOOK for GEOMETRIC DIMENSIONING and TOLERANCING
(per ASME Y14.5-2009)
March 2009
367 pgs, softcover, illus.
ISBN: 0-9714401-7-4

'NEW RULES' in GD&T
(per ASME Y14.5-2009)
March 2009
160 pgs, softcover, illus.
ISBN: 0-9714401-8-2

'NEW RULES' in GD&T
(per ASME Y14.5-2009)
DVD Training Series
March 2009
Includes DVD set & 'NEW RULES' t
ISBN: 0-9714401-9-0

GEOMETRIC DIMENSIONING AND TOLERANCING IN 2007
(per ASME Y14.5M-1994)
February 2007
560 pgs, illus.
ISBN: 0-9714401-2-3

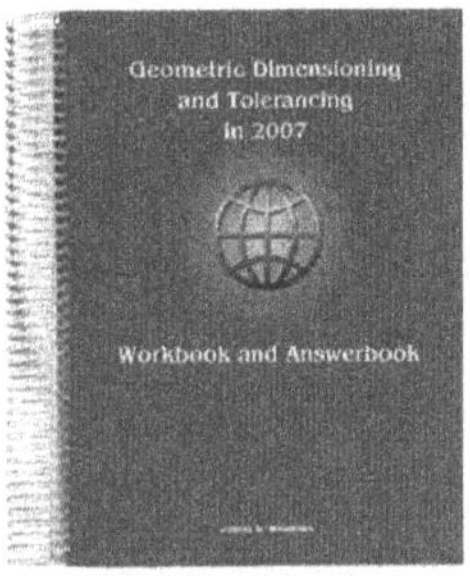

GEOMETRIC DIMENSIONING AND TOLERANCING IN 2007
Workbook and Answerbook
(per ASME Y14.5M-1994)
292 pgs, illus.
ISBN: 0-9714401-5-8

TOLERANCE STACK-UP ANALYSIS
[for Plus and Minus and Geometric Tolerancing]
Second Edition, Published 2010
300 pages, illus. [per the ASME Y14.5-2009 and 1994 standards]
ISBN: 978 0971 440142

DIFFERENCES AND SIMILARITIES BETWEEN ASME AND ISO DIMENSIONING AND TOLERANCING STANDARDS
December 2006 English Edition
80 pgs, illus.
ISBN: 0-9714401-3-1

GEOMETRIC DIMENSIONING AND TOLERANCING
Workbook and Answerbook
(per ASME Y14.5M-1994)
340 pgs, illus.
ISBN: 0-8247-0076-7

GEOMETRIC DIMENSIONING AND TOLERANCING
Applications and Techniques for Use in Design, Manufacturing and Inspection
(per ASME Y14.5M-1994)
624 pgs, illus.
ISBN: 0-8247-9309-9

MEASUREMENT OF GEOMETRIC TOLERANCES IN MANUFACTURING
(per ASME Y14.5M-1994)
496 pgs; illus.
ISBN 0-8247-0163-1

APPLICATIONS-BASED GD&T DVD TRAINING SERIES II
(per ASME Y14.5M-1994)
"BASICS" Set: 4 DVD GD&T Lesson Series (4 hours)
plus Book ***and*** Workbook/Answerbook

APPLICATIONS-BASED GD&T DVD TRAINING SERIES I
(per ASME Y14.5M-1994)
Complete Set: 12 DVD GD&T Lesson Series (14 hours)
plus Book ***and*** Workbook/Answerbook